ICT Acceptance, Investment and Organization:
Cultural Practices and Values in the Arab World

Salam Abdallah
Abu Dhabi University, UAE

Fayez Albadri
ADMO–OPCO, UAE

INFORMATION SCIENCE REFERENCE

Hershey · New York

Director of Editorial Content:	Kristin Klinger
Director of Book Publications:	Julia Mosemann
Acquisitions Editor:	Lindsay Johnston
Development Editor:	Mike Killian
Publishing Assistant:	Natalie Pronio
Typesetter:	Natalie Pronio
Production Editor:	Jamie Snavely
Cover Design:	Lisa Tosheff

Published in the United States of America by
Information Science Reference (an imprint of IGI Global)
701 E. Chocolate Avenue
Hershey PA 17033
Tel: 717-533-8845
Fax: 717-533-8661
E-mail: cust@igi-global.com
Web site: http://www.igi-global.com

Library of Congress Cataloging-in-Publication Data

ICT acceptance, investment and organization : cultural practices and values in the Arab world / Salam Abdallah and Fayez Albadri, editors.
       p. cm.
  Includes bibliographical references and index.
  Summary: "This book is a unique source of information outlining the importance of Information Communication Technology (ICT) adoption and diffusion, covering the Arab world's strong need for access to information systems, while still paying close attention to their culture and localization of practices"--Provided by publisher.
  ISBN 978-1-60960-048-8 (hbk.) -- ISBN 978-1-60960-050-1 (ebook) 1. Information technology--Arab countries. 2. Information technology--Social aspects--Arab countries. I. Abdallah, Salam, 1958- II. Albadri, Fayez.
  HC498.9.I55I38 2011
  303.48'3309174927--dc22
                          2010019860

British Cataloguing in Publication Data
A Cataloguing in Publication record for this book is available from the British Library.

All work contributed to this book is new, previously-unpublished material. The views expressed in this book are those of the authors, but not necessarily of the publisher.

Editorial Advisory Board

Edward A. Stohr, *Stevens Institute of Technology, USA*
Khalid Soliman, *Hofstra University, USA*
Ernest Jordan, *Macquarie University, Australia*
Gerald Grant, *Carleton University, Canada*
Mohamed Baka, *Abu Dhabi Water and Electricity Authority, UAE*
Nabeel Al-Qirim, *United Arab Emirates University, UAE*
Rami Gharaibeh, *Jordan University of Science and Technology, Jordan*
Sami Akabawy, *The American University in Cairo, Egyp*t
John Venable, *Curtin University of Technology, Australia*
Ezendu Ariwa, *London Metropolitan University, UK*

List of Reviewers

Ashraf Khalil, *Abu Dhabi University, UAE*
Sami Akabawi, *The American University of Cairo, Egypt*
Tarek Hatem,*The American University of Cairo, Egypt*
Elham Metwally, *The American University of Cairo, Egypt*
Atef Abuhmaid, *Middle East University, Jordan*
Norita Ahmad, *American University of Sharjah, UAE*
Alanoud Alhaj, *Student at American University of Sharjah, UAE*
Bassam Hammo, *University of Jordan, Jordan*
Basem Saraireh, *University of Jordan, Jordan*
Hamed Al Hinai, *Student at University of Sunderland, UK*
Helen Edwards, *University of Sunderland, UK*
Ali Al Kinani, *King Saud bin Abdulaziz University for Health Sciences, KSA*
Sonia Kawas, *British Council, Jordan*
Abdelnasser Abdelaal, *Sur College of Applied Sciences, Oman*
Sonda Fakhfakh, *Univeristy of Tunis,Tunisia*
Nasim Matar, *Anglia Ruskin University, UK*
Ziad Hunaiti, *Anglia Ruskin University, UK*
Shahid Halling, *Talal Abu Gazala Group, Jordan*

Table of Contents

Detailed Table of Contents

Chapter 1
A Perspective on ICT Diffusion in the Arab Region... 1
Salam Abdallah, Abu Dhabi University, UAE
Fayez Ahmad Albadri, ADMO-OPCO, UAE

This chapter examines the extant literature plays on ICT in the Arab world in order to glean and understand the level of ICT investment, acceptance and the interaction of cultural practices and values and to set the direction for ongoing research in this developing region.

Chapter 2
Deciding to Use ICT in the Arab Culture: The Influence of Obedience to Authority
and Collectivism 16
Sonda Bouattour Fakhfakh, University of Tunis, Tunisia

This chapter reports on two dominant values which are assumed to be descriptive features of the Arab world: obedience to authority and collectivism. It examines the hypothesized influence of the two variables on ICT use within an Arab context (Tunisia) and at an individual level.

Chapter 3
Embracing ICT by the Jordanian Education System... 34
Atef Abuhmaid, Middle East University, Jordan

This chapter discusses the experience of adopting information and communication technologies (ICT), by the Jordanian Education System. The discussion tackles the Jordanian Ministry of Education's endeavor in reforming the education system for the Knowledge Economy.

This chapter presents the results and experiences of a study carried out at the University of Jordan to evaluate an accredited one-year high diploma program called "ICTE" which aims at preparing Jordanian teachers to use information and communication technology in education.

This chapter discusses the ICT in Education Project jointly sponsored by Jordan's Ministry of Education and the British Council, with the participation of four other Middle Eastern countries (Egypt, Syria, Lebanon and the Palestine). It shows how ICT in education brought positive impact on the delivery of teaching in class, and also how online forums can be utilized as opportunities to exchange invaluable information and knowledge in this sector.

This chapter focuses on SISP approaches and their potential impacts on IS&T investment with emphasis on the importance of SISP to UAE organizations' ability to achieve their business goals and objectives.

This chapter explores the barriers of adoption and adaptation of Enterprise Resources Planning (ERP) systems at the post implementation phase within Developing Countries (DCs) business culture. This research is meant to highlight a new perspective to the understanding of the success model of ERP use and adoption.

This chapter demonstrates, and suggests solutions for the challenges that face the preparation of organizations' stakeholders in managing and supporting the implementation and using enterprise systems in an Arabian Gulf context.

This chapter reports the results of a case study that covers a successful project of IT implementation in International Commercial Bank from the Egyptian banking industry, highlighting leadership actions, and other related factors that are linked to strategic competitiveness and value creation.

This chapter examines the importance of understanding the components of IS and how they affect one another and on stressing the importance of IT selection. Through a case study, the consequences of failing to understand IS and the users' needs are discussed.

This chapter explores information related to the influence of Saudi culture on the adoption of ICT systems related to business & communication language, culture and ICT R&D, government support, Internet consistency with local culture, openness of the country's culture to foreign influence, the impact of employees' culture on their work, ICT and the protection of the Saudi culture.

This chapter investigates issues related to E-Health applications and its implementations in developing countries, in particular Saudi Arabia, and evaluates the country's readiness of ICT in the health organizations.

This chapter discusses the status and quality of e-learning in Arab Universities located in the Middle East and draws into different solutions and recommendations in order to make a successful match that will result in a better adoption of e-learning technology.

This chapter addresses the adoption of Internet applications by an Arabic Student Association in North America (ASANA). ASANA uses the Internet to integrate its members, promote the Arabic culture, bridge with the American society and transfer knowledge to its native country.

This chapter identifies specific areas in which Social software (Web 2.0) applications are being used innovatively in Western governments. Later it turns its focus to the Arab World and puts forward thoughts on the potential, opportunities and challenges for both citizens and governments for embracing this new wave of web-based services.

Foreword

It is rather hard to find trusted publications addressing the raw information required by ICT researchers and policy makers in governmental, educational and large corporate institutions that would support them in contriving facts-based policies and decisions. The deficiency is far more acute when it comes to the Arab world. Hence, the exceptional value of this tome.

Although Arab talent played a substantive role in shaping the building-bricks of the information society, it is rather sad to see that the fruits of this endeavour scarcely realised in this part of the world. It is therefore very inspiring to see researchers and practitioners coming together to produce an excellent reference material on promoting and integrating ICT culture in the region, with strong emphasis on applicable knowledge and real-life, locally relevant, case studies.

Apart from the first chapter, which discusses the current status and obstacles facing the adoption of ICT in the Arab world, the reader may wish to fast-forward to select the studies of particular relevance to their industry, areas of interest, or region of choice. Nonetheless, it is rather likely that all topics covered would ultimately come to the forefront of attention at some point in time.

In short, this is a most welcome volume, providing a significant contribution to required knowledge infrastructure imperative to the enablement of ICT for a thriving Arab region.

Mohamed Baka
Visiting Scholar
Waterloo University, Canada

Mohamed Baka *is the ICT Advisor to the Abu-Dhabi Water and Electricity Authority (ADWEA). He is also the Chairman of the Emirates Center for Innovation and Entrepreneurship at the UAE University. He earned his PhD in Computing from Bristol University in the UK and received executive education at MIT Sloan and Harvard Business School. Dr. Baka has more than 20 years experience in the field of Information Technology Management. He has led several global ICT Programs for blue-chip firms as part of the European Strategic Project in Information Technology (ESPRIT). His academic experience includes a research tenure at Imperial College – London and is currently a visiting Scholar at Waterloo University in Canada. Dr Baka is on the editorial board of the "Springer" book series on Technology and Knowledge Management.*

Preface

Transforming into an information society is essential for the economic and social development of countries or regions. The transition to a digital economy can provide the foundation for growth, competitiveness and job creation. This social transformation is driven predominantly by Information and communication technology (ICT). Information Society may be defined as "A society characterized by a high level of information intensity in the everyday life of most citizens, organizations and workplaces; by the use of common or compatible technology for a wide range of personal, social, educational and business activities, and by the ability to transmit, receive and exchange digital data rapidly between places irrespective of distance" (IBM Community Development Foundation in a 1997 report, "The Net Result - Report of the National Working Party for Social Inclusion)

The 'Arab Countries' region is one of the regions partaking in this form of society by taking proactive measures to adopt policies, frameworks and skilful application of ICT. Countries of the Arab region are striving, each at its own pace, to develop and implement ICT initiatives at both local and regional levels, with the aim of economic growth, motivated by a view of ICT acceptance as a key determinant of productivity and human capital advancement. However, the success of such ICT measures and initiatives has been countered by important challenges when it comes to its adoption and diffusion.

Research has indicated that the adoption and diffusion of ICT in general is strongly influenced by cultural practices and values at both, societal and organizational levels, and that it is not a mere replication of a global standardized template. The Arab world has its own array of cultural values and beliefs, and it has its own distinct business, financial and socio-technical issues and challenges that are reflected on its practices of ICT development, implementation and management. An insight of these challenges is essential considering their far-reaching impact on the success or failure of ICT systems adoption and diffusion at both public and the private sectors.

This book combines a collection of ICT investigations and reflections of ICT implementations encountered at Arab countries, viewed from a cultural perspective. The book covers through its different chapters a wide range of ICT domains including; ERP systems, E-readiness, Internet applications, ICT Strategic Planning, ICT Infrastructure, ICT Education, E-learning and Industry applications. The investigation of cultural practices and values in the Arab World is specifically projected to ICT investment, acceptance and organization.

In addition to the diversified ICT domains that are covered, the book is enriched further by the fact that important ICT applications and experiences that are captured in the book are related to different businesses and industry sectors in the Arab region including; Energy, Utility, Education, Health, Real Estate and Banking. Although it is beyond the scope of this book to examine the whole Arab region and all the related factors of adoption and diffusion, it does however cover a wide range of ICT issues related to Arab countries that can be fitted to the categories of developed, emerging, or developing countries.

This book enriches our understanding of technology adoption and the challenges facing many of the Arab countries in their efforts to transform into information society. The main contribution of the book is the range of ICT applications discussed in different contexts focusing at three levels: Arab region as a whole, country level and organization level. The book should be a valuable source boosting the ICT literature in the developing countries and specifically in the Arab world. There are many valuable lessons to be learned from the implementation experiences discussed throughout the various chapters, and specifically on how Arab countries attempt to adopt and benefit from ICT applications. It aims primarily at understanding the nature, contributions and the potential of ICT development in the Arab world and does highlight the most prevalent success factors for ICT acceptance in the region.

Although the main focus of the book is on the Arab countries, the work should stimulate researchers, practitioners and consultants from other regions to benefit from the captured experiences and findings when considering ICT transfer, adoption, or adaptation in developing countries either by the local governments, private sector or international organizations who may have interest in investing in the developing countries.

The book is organized in 15 chapters, discussing various aspects of ICT applications adoption and diffusion in different contexts. The first Chapter (1) introduces broad ICT status and trends in the Arab world and highlights some of the important challenges and problems hindering ICT adoption. The chapter is an examination of the extant literature on ICT adoption to glean the level of ICT investment, acceptance and to understand the interplay of cultural practices and values on the successful implementations of ICT initiatives. It is suggested to read this chapter first since it sets the background for the rest of the chapters in the book.

The second chapter (2) takes us to Tunisia, the northernmost country in Africa, where the author investigates two denominate cultural values that may have impact on the use of ICT. Data collection was carried out on 378 Tunisian students in three quasi-experimental settings measuring the degree of obedience, and by a pilot questionnaire determining the level of collectivism. The findings are interesting and have implications on Arab institutions to take into account the cultural dimensions in order to redirect their action and improve their social and cultural awareness about technology.

Arab countries have recognized that need to invest heavily in ICT education and training to build their capacity to meet the needs of the emerging information society. Three chapters were selected on ICT in education focusing on the Near East and North Africa region, with a focus on Jordan as having a high maturity level in this area.

Chapter 3 discusses the Jordanian education system's approach to reforming itself through the adoption of ICT. Jordan is envisioned by its leaders to become an IT hub for the region. However, the frequent conflicts and wars in the Middle East have disturbed development and reform. Reform is a challenging task to any education system, and it is even more challenging to systems with limited resources as they are pressured to seek and accept external assistance. The chapter argues that Jordan like other developing countries faces serious challenges while it is trying to update its education system to prepare itself for the information society. The author highlights that ICT initiatives at national level must rely heavily on external aid, but this carries with it a different type of challenge leading sometimes to disputes due to conflicting agendas. In the progress of the chapter the author illustrates how the Ministry of Education established local and international partnership to support technological development, however, the motivation behind such programs has been debated. The chapter reflects on two major large ICT projects at national level to expound the various challenges. The chapter highlights many essential factors that should be considered for a successful ICT integration including; maintenance, technical support, training, upgrade, new software, and replacement of old technologies and building the capacity of teachers.

Capacity building of stakeholders is essential when adopting and implementing reform in ICT projects. Substantial efforts and resources should be directed to building capacity at all levels of the education system. Chapter 4 focuses on capacity building at the Ministry of Education in Jordan. The chapter discusses a successful program that has been in operation for the past four years in partnership with Ohio University. The program had positive impact on teachers' lives, and had altered their pedagogical beliefs and practices. The chapter discusses in details the specifics of the program and the challenges faced by students and instructors. The chapter concludes with a list of recommendation that should act as a useful guide for researchers and agencies who may have interest in investing in education in the developing countries and specifically in the Arab countries.

Chapter 5 reflects on the experience of a knowledge and information practitioner engaged in building an online community of practice in the education sector in the Near East and North Africa (NENA) region. Countries in this region share important common similarities, such as language, culture, the traditional values of their education systems, low internet connectivity, a shortage of computers, educational resources and skilled teachers. The author reflects on the three year project that was filled with successes and pitfalls. The chapter looks at both the early stages of the project, and how it evolved to include strategic leadership in ICT, together with the technicalities and infrastructures, culture, outcomes and challenges. The project discussed in the chapter shows how ICT in education has brought impact on the delivery of teaching in class and also how online forums creates an opportunity to exchange information and knowledge across different countries. The author spells out clearly the challenges be it with technicalities, infrastructures or skills, and calls on countries that have these limitations to consider carefully changing their procedures, as technology is advancing at a speedy pace.

The next several chapters shift the focus to various business and industry specific sectors to examine the issues, problems and encountered challenges. Chapter 6 reports on the findings of an exploratory research to investigate Strategic Information Systems Planning (SISP) approaches in UAE organizations, and their potential impacts on IS&T investment. The study uses a survey and simple model to characterize, classify and examine SISP approaches in 17 UAE organizations. Typical cases representative of the main SISP categories were put to the test to determine specific strengths and weaknesses, and to appraise the suitability and effectiveness of each category considering the organization culture and business environment. On the positive side, the study confirms that some of the UAE organizations do have SISP approaches and which can be improved in both formalization and effectiveness. However, it is a major concern to know that the majority of ICT investment was not based on proper strategic planning, and therefore carrying risks of problems and failures. The study highlights the importance of SISP to UAE organizations' ability to achieve their business goals and objectives. It also suggests that the SISP classification model can be benefited from by UAE and other Arab organizations to pin-point inadequacies in their existing approaches and to understand the cultural shift needed to adjust these approaches, so they can guide effective ICT investment, ensuring alignment of technology solutions to business strategy, and focusing on the realization of business goals and objectives. The study also anticipates that major ICT investment savings could be realized should a suitable SISP approach is implemented by UAE and Arab organizations.

Chapter 7 discusses a topic of importance to many Arab organisations who have implemented ERP systems or planning to do so. The chapter reflects the findings of a study to explore the challenges and barriers pertinent to ERP adoption and adaptation in post the implementation phase within Developing

Countries organization culture and business environment. The study investigates the impact of human and cultural issues on the success or failure of ERP systems in Arab countries business environment, using a framework of post-implementation use of cross-functional information systems in a longitudinal case study. The study uses focus group teams, composed of business units' end-users, managers and IT specialists (employment cohorts) to elicit the causes of failing ERP services. The study succeeds in developing improved understanding organizational and human factors that can potentially decide the levels of acceptance, usage and utilization of ERP systems in Arab organizations. This is viewed as an opportunity for Arab organizations to be conscious and prepared for managing issues and challenges that could lead to ERP failures. The scope of this research fits within the studies that target the analysis of organizational adoption and use of ERP, using feedback loops analysis method to analyze the complex dynamic socio-technical behavior in enterprises, leading to understand the relationships among the many non-linear variables prevalent in the post implementation stage of the ERP lifecycle.

Chapter 8 examines the challenges, issues and success factors that have been observed by practitioners during the implementation of enterprise systems, specifically focusing on factors that are of importance in ERP implementations in the Arabian Gulf context. The study successfully identifies the main ERP critical success factors (CSF) and highlights the main risks and challenges that threaten the success of ERP projects. The study proposes a structured approach based on CSF to address issues and to overcome challenges and mitigate risks. Such approach if dynamically reviewed and improved could contribute positively to increase the chances of ERP projects success and to provide lessons learned. The study raised some important challenges and factors that influence the preparation of stakeholders in enterprise systems implementation. Although the proposed approach to ensure ERP success can be useful in different contexts, it is essential due to cultural differences between organizations that engage in ERP implementation, its effectiveness is subject to ERP implementation teams to be aware of their context and tailor their processes to their organization, and stakeholders.

Chapter 9 reflects on different factors that are linked to the success of an IT implementation project in a large Egyptian bank. The case highlights leadership actions, as well as other related factors to the effectiveness of IT implementation that are linked to strategic competitiveness and value creation. Findings of the investigation indicate that successful implementation was influenced by the interplay of several management practices, which eventually, had an impact on strategic competitiveness through their impact on some in-house attributes; notably, a dominating constructive cultural pattern leading to higher levels of organizational commitment, and the bank's value chain. This chapter describes the activities and management practices adopted by the bank throughout implementation of two IT-based delivery systems, namely ATMs and Internet Banking, and links the successful outcomes and business gains of the IT implementation to a set of practices and measures that have been adopted. It is anticipated that other organizations of similar cultural settings in the Arab region can benefit from this experience to ensure effective adoption and adaptation of technology that yields tangible business gains and investment returns.

Chapter 10 discusses probably a common but an important factor on why ICT integration fails within organizations in the Arab region. The author places a research lens on a Marketing department in one of the organizations in the United Arab Emirates. The author discusses in details the operation of the department and its implementation of IT applications. Several problems and challenges are highlighted that were factors for unsuccessful implementation. The author aims at emphasising the importance of understanding the components of IS and the interplay between them.

The following five chapters of the book have focused on Internet technologies, applications including e-government, e-learning and e-communities with a focus on the e-readiness of the Saudi Society.

Chapter 11 and Chapter 12 are somewhat related, as both chapters discuss the readiness of an Arab country for adopting ICTs. Chapter 11 focuses on assessing cultural issues that may impede the diffusion of ICT within the Saudi society. The author developed and used an assessment framework of ICT readiness to evaluate 87 organizations in Saudi Arabia. The study revealed seven cultural factors that can have impact on ICT adoption and diffusion. The author also presents a number of recommendations to be considered for enhancing ICT adoption within the Saudi society. Chapter 12 addresses the same subject, but with specific focus on the readiness of Saudi health organizations to adopt electronic health. The study highlights 11 issues related to how ICTs are being integrated in the health organizations. The study revealed through a specifically designed assessment readiness instrument that the health organizations in Saudi Arabia are making initial positive progress towards that endeavour.

Chapter 13 examines the e-learning sphere in the Arab World. The authors discuss the status and quality of e-learning in Arab Universities located in the Middle East. The need for e-learning solutions for the Arab world is essential for bridging the digital divide as there are over 130 million illiterate Arab children with no formal education. The adoption of e-learning systems is of paramount importance as the average yearly population growth in the Arab States is on high increase. Although e-learning systems can help to address the high cost of education, however, according to this chapter such an approach suffers from many different challenges which need to be addressed, if in the long term, e-learning is to be a successful combat to illiteracy. The authors conclude with different solutions and recommendations in order to make a successful match that will result in a better adoption of e-learning technology appropriate to the Arab world environment.

The context of Chapter 14 is not about any specific Arab location but about an Arab virtual space. The chapter examines a set of factors that affect the adoption of emerging technologies such as virtual communities by group of Arab Academics. The author examines a non-profit organization established in North American with members mostly graduate students, have moved from an Arab country to North America to pursue graduate studies at either Canadian and American universities. The organization is a form of socio-academic virtual network aims at integrating and supporting its member living in North America and to introduce Arabic and Islamic culture to the American society, in addition to other interesting values described in the chapter. The chapter provides anecdotal evidence on the motivations and the challenges of the adoption of Internet applications by the society's' members. The success of the network was hampered by explicit and implicit cultural practices. The chapter sheds lights on managerial, cultural and technical changes that are required to improve the adoption of e-services by its members. The presented model has great potential to leverage online academic communities as a mean for social, political, economic and educational reform.

In Chapter 15 the authors examine social software commonly known as Web 2.0 and related implementation in the Western World, it turns focus to the Arab World and puts forward thoughts on the potential, opportunities and challenges for both citizens and governments for embracing this new wave of web-based services. The paper concludes that Web 2.0 applications have great potential to leverage the mission of Arab e-governments through connecting and collaborating with businesses, educational institutions, private and non-government initiatives, and of course individual citizens, many of whom are already beginning to embrace social media. The chapter also highlights the current condition of Arab e-governments and how Web 2.0 can supplement their mission. The authors make a recommendation that before making the paradigm shift and cultural change, governments must make sure they and their citizens are ready for this new channel and need to determine the best way to proceed towards more dynamic governments that can respond and meet the expectations of their supporting communities.

To this end, the primary objective of this book was to bring together diverse views and insightful experiences from researchers and practitioners from around the Arab region to reflect on the nature of ICT adoption at the individual, organizational or country level. In effect, the book offers a unique opportunity to serve multiple purposes to serve researchers, professionals, policy makers, teachers, and students. We expect that this book will be an essential resource for ICT adoption in developing countries.

Salam Abdallah
Abu Dhabi University, UAE

Fayez Albadri
ADMO-OPCO, UAE

Acknowledgment

Completing a book of this magnitude has proven to be a major undertaking. The book could not have come to fruition without the dedication and support of many individuals. We would like to thank the members of the International Advisory Board for their invaluable support throughout the venture. We would also like to acknowledge the great effort by many Arab and Non-Arab academics and practitioners who submitted proposals to participate in the book. However, our appreciation and gratitude is to the distinguished scholarly work of the authors of the books, who have also helped in the review of some chapters. We also extend a special thanks to our dear colleague Dr. Ashraf Khalil, who was one of the editors at the early development stages of the textbook. Our appreciation and gratitude also goes to the College of Research and Graduate Studies at Abu Dhabi University for their grant to support the completion of this book.

We would like to acknowledge the efforts of the following members for contributing valuable advice, authoring and reviewing chapters:

Advisory Board

Edward A. Stohr, *Stevens Institute of Technology, USA*
Khalid Soliman, *Hofstra University, USA*
Ernest Jordan, *Macquarie University, Australia*
Gerald Grant, *Carleton University, Canada*
Mohamed Baka, *Abu Dhabi Water and Electricity Authority, UAE*
Nabeel Al-Qirim, *United Arab Emirates University, UAE*
Rami Gharaibeh, *Jordan University of Science and Technology, Jordan*
Sami Akabawy, *The American University in Cairo, Egypt*
John Venable, *Curtin University of Technology, Australia*
Ezendu Ariwa, *London Metropolitan University, UK*

Authors/Reviewers

Ashraf Khalil, *Abu Dhabi University, UAE*
Sami Akabawi, *The American University of Cairo, Egypt*
Tarek Hatem, *The American University of Cairo, Egypt*
Elham Metwally, *The American University of Cairo, Egypt*
Atef AbuHmaid, *Middle East University, Jordan*

Norita Ahmad, *American University of Sharjah, UAE*
Alanoud Alhaj, *American University of Sharjah, UAE*
Bassam Hammo, *University of Jordan, Jordan*
Basem Saraireh, *University of Jordan, Jordan*
Hamed Al Hinai, *University of Sunderland, UK*
Helen Edwards, *University of Sunderland, UK*
Ali Al Kinani, *King Abdulaziz University for Health Sciences, KSA*
Sonia Kawas, *British Council, Jordan*
Abdelnasser Abdelaal, *Sur College of Applied Sciences, Oman*
Sonda Fakhfakh, *Univeristy of Tunis, Tunisia*
Nasim Matar, *Anglia Ruskin University, UK*
Ziad Hunaiti, *Anglia Ruskin University, UK*
Shahid Halling, *Talal Abu Gazala Group, Jordan*
Šadi Matar, *Ministry of Communications and Transport, Sarajevo*
Muzhir Shaban, *Amman-Arab University for Graduate Studies, Jordan*
Shadi R. Masadeh, *Al-Isra Private University, Jordan*
Issam Wadi, *Trust Technical Services, UAE*
Nabil Jarrar, *ADMA-OPCO, UAE*
Ayman Murad, *UNRWA, Jordan*
Shadi Aljawarneh, *Al-Isra Private University, Jordan*
Mariam Alsaadi, *Abu Dhabi University, UAE*

Chapter 1
A Perspective on ICT Diffusion in the Arab Region

Salam Abdallah
Abu Dhabi University, UAE

Fayez Ahmad Albadri
ADMO-OPCO, UAE

ABSTRACT

Interest in the Information and Communication Technology (ICT) in the Arab world has started to take different strides for many reasons. This growing endorsement of ICT is primarily attributed to economic and social development factors, and is driven by the aim to transform Arab society to a knowledge based society. Such an endeavor could be successfully achieved only through effective creation, adoption and innovation of the technology. This chapter examines the extant literature on ICT in the Arab world in order to glean the level of ICT investment and acceptance, and to attempt to understand the interplay of cultural practices and values on the successful implementations of ICT initiatives. The Arab countries are undergoing major development that is expected to culminate in tangible growth in many areas, attract the attention and interest of both businesses and researchers. This chapter also sets the scene and lays the foundation for the ICT discussions and investigations in the rest of the chapters of the textbook.

INTRODUCTION

The development, dissemination and management of information and knowledge determine the way forward to build an information society, as an enabler for sustainable economic development. Defining the main features of information society, ESCWA (2009) characterizes an information society as "a society that processes information efficiently in its socio-economic development, including the production, exchange, adaptation and use of information for the purpose of development and the enhancement of the quality of life and work environment for all citizens. In order to realize the information society, information and communications technologies (ICTs) need to be used. While ICTs are necessary, they are not sufficient, given that capacity-building must equally be enhanced in knowledge-related areas covering economic, social, legal, educational

DOI: 10.4018/978-1-60960-048-8.ch001

and research." (p. 1). This successful realization of this endeavor is only possible to achieve through a careful consideration of factors that can enhance adoption and innovation of technology and its creation.

For the above reasons, the increasing interest in ICT in the Arab world has compelled various sectors to take different strides primarily for economic and social development with the aim of becoming a knowledge based society. ICT can provide developing countries an opportunity to 'leapfrog' over the industrial stages of development, and to catch up with more developed economies (Murphy, 2006). However, this aim faces many challenges. In the Arab region the process of ICT transfer and adoption is considered low and its infiltration into its societies is hindered due to challenges in IT transfer (Straub et al., 2001). Technology transfer may be defined as a dynamic process, where there is a technology movement from one physical or geographic allocation to another. This process involves acquisition, adaptation, utilization and development (Straub et al., 2001). Rosalined (1985) has argued that technology transfer also include issues such as decision making procedures; software development methodologies planning and organization of the information resource, training and so on. Deep insights into the area of ICT technology transfer, adoption and diffusion in the Arab world is limited due to scarcity in research examining the different perspectives in different contexts.

Like most other developing countries, the Arab region has its own share of ICT project failures. Heeks (2002) argues that ICT project failures in the developing countries is higher than developed countries, possibly due to lack of technical and human infrastructure. Heeks (2000) also highlights that most studies on ICT projects do not provide reasons for the causes of failure, and they fail to take into consideration multiple dimensions. For example using Leavitt's (1965) model of organizational change can be a useful framework to assess the interplay between multiple dimensions

(people, processes, structures and technology), where he suggest that these dimensions need to be aligned to effectively bring about change introduced by the four dimensions.

Our aim in this chapter is to examine the extant literature on ICT in the Arab world in order to glean and understand the level of ICT investment, acceptance and the interaction of cultural practices and values and to set direction for ongoing research in this developing region. Examples from countries will be given to illustrate how they have attempted to benefit from ICT. The chapter also provides a useful reference to the rest of the chapters in this textbook.

ICT IN THE ARAB REGION

The Arab world accounts for 5% of the world's population 2% of the world GDP, which is considered a relatively high number by international standards. So poverty as the main barrier for the digital divide would not be the main reason for the gap. The United Nations Economic and Social Commission for Western Asia (UN-ESCWA), promotes economic and social development of Western Asia through different drivers, and ICT is one of the prime drivers to assist Arab countries in their transformation toward knowledge-based society and global competitiveness in the market. ESCWA as commonly known had developed a set of indicators to assist developing countries to measure the use of ICT and to assist them in formulating polices and ICT-enabled growth strategies and for enhancing social inclusion and cohesion. The indicators are also used for monitoring and evaluating the impact of ICTs on economic and social developments (ESCWA, 2009). These proposed indictors are related to (i) ICT infrastructure and access (ii) access to, and use of, ICT by households and individuals (iii) use of ICT by businesses and (iv) ICT sector and trade in ICT goods. The objective of these indicators is to assist countries who are developing ICT

Table 1. Arab countries internet usage and population statistics (Miniwatts Marketing Group, 2010)

MIDDLE EAST	Population	Usage in	Internet Usage	% Population	User Growth
	(2009 Est.)	Dec-00	Latest Data	(Penetration)	(2000-2009)
Bahrain	728,709	40,000	402,900	55.30%	907.30%
Iraq	28,945,569	12,500	300,000	1.00%	2300.00%
Jordan	6,269,285	127,300	1,500,500	23.90%	1078.70%
Kuwait	2,692,526	150,000	1.000,000	37.10%	566.70%
Lebanon	4,017,095	300,000	945,000	23.50%	215.00%
Oman	3,418,085	90,000	465,000	13.60%	416.70%
Palestine (West Bk.)	2,461,267	35,000	355,500	14.40%	915.70%
Qatar	833,285	30,000	436,000	52.30%	1353.30%
Saudi Arabia	28,686,633	200,000	7,700,000	26.80%	3750.00%
Syria	21,762,978	30,000	3,565,000	16.40%	11783.30%
United Arab Emirates	4,798,491	735,000	2,922,000	60.90%	297.60%
Yemen	22,858,238	15,000	370,000	1.60%	2366.70%

surveys to produce internationally comparable data. Some of these indicators will be discussed throughout this chapter.

Arab countries do vary at the level of the digital divide, examining the Table 1 below, illustrates that the United Arab Emirates, Bahrain and Qatar, have in comparison high internet penetration and this is probably propelled by attracting foreign ICT investment due to the rich oil economy. On the other hand Iraq has the lowest penetration rate due to the existing turmoil. Interestingly to note that Jordan and Lebanon have moderate economies and government support towards ICT, yet still their penetration is high and this might be attributed to the skilled workforce and the openness of their markets (Mezher, 2007). We also note that Saudi Arabia, although is considered as a wealthy country, its penetration is considered low, perhaps due to the conservative nature of the Saudi culture and its skepticism about the role of technology and its impact on a traditionally religious society.

The Madar Research Group also publishes regular ICT use Index taking into consideration four parameters: PC installed base and the number of Internet users, mobile phones and fixed lines. The Index is calculated by adding up the values of these four parameters and dividing the sum by the country's population figure. For the 2008 Index, UAE ranked the highest scoring (2.3) followed by Bahrain, Qatar and Saudi Arabia, and the lowest score recorded was for the Sudan (0.4).

ESCWA also regularly measures the progress made in countries such as Bahrain, Egypt, Iraq, Jordan, Kuwait, Lebanon, Oman, Palestine, Qatar, Saudi Arabia, Syria, the Sudan, United Arab Emirates (UAE), and Yemen. In its 'country status and comparative analysis' report under the section titled, "Regional Profile of the Information Society in Western Asia", ESCWA (2009) confirms that further development in ICT is still required to reduce the existing digital divide. The report provides measures on a number of areas using the above mentioned indicators in order to determine the progress made toward information societies in each country and classified according to four maturity levels. Level 4 is the highest degree of efforts being made by a specific country in order to move into knowledge societies. Table 2 tabulates the different ranking of Arab countries in terms of their progress towards building information society using different indicators. The following

Table 2. Ranking of ESCWA member countries by Maturity level using various indicators (Source, ESCWA, 2009)

Country	Government and All stakeholder Support	ICT infrastructure & Penetration rates	Access to Information & Knowledge	ICT Capacity Building	Building ICT security & Confidence	Adoption of ICT Application	Preservation of Cultural Identify	Total of all levels
United Arab Emirates	4	4	3	4	2	4	3	24
Qatar	4	4	3	3	2	3	3	22
Bahrain	4	4	3	2	1	4	2	20
Saudi Arabia	3	3	2	3	2	3	3	19
Egypt	3	2	3	3	2	2	3	18
Jordan	3	3	2	3	1	3	2	17
Kuwait	3	3	3	2	1	2	3	17
Oman	3	2	3	2	1	3	2	16
Lebanon	3	2	2	3	1	2	2	15
Syrian Arab Republic	2	2	2	2	1	1	3	13
Palestine	1	2	1	2	1	1	2	10
Iraq	1	1	2	1	1	1	1	8
The Sudan	2	1	1	1	1	1	1	8
Yemen	1	1	1	1	1	1	1	7

sections discuss these indicators reflecting on some of the Arab countries.

GOVERNMENT AND ALL STAKEHOLDER SUPPORT

Higher maturity is facilitated when there is positive government support to build the foundation for information society, by developing and adopting policies, strategies and action plans, in addition to the support from private sector and non-government organizations. There is a great difference among the Arab countries in maturity level; some have matured to the highest level, while others only began initiating and developing the ICT industry, and primarily through the support from the private sector and NGOs (ESCWA, 2009). Countries like Bahrain, Qatar and United Arab Emirates have progressed towards level 4 and mainly due to their well developed vision towards moving into knowledge society, along with a clear national ICT policies and strategies, with effective implementation plans supported by governments, public-private partnerships and multi-sector partnerships. Countries at the lower level such Iraq, Palestine and Yemen suffer from political crises and disruptions, which impede governments and the private sector to take actions towards adoption and implementation of national ICT strategies. Countries that are positioned in level 2 and level 3, have ICT polices and strategies but they lack appropriate implementations or slow paced and also lack support from stakeholders. There are other impeding factors such bureaucracy in case of Kuwait and Saudi Arabia or because insufficient resources in case of Egypt and Jordan.

ICT INFRASTRUCTURE AND PENETRATION RATES

The other maturity measure considered by ESCWA (2009) is related to the existing ICT infrastructure in terms of (a) penetration rate of fixed and mobile telephone lines; (b) conducive telecommunication environment to encourage adoption by businesses and individuals, and (c) adequate telecommunication infrastructure.

In this context, we find the Arab countries are witnessing high adoption rates of mobile technologies. For example in Saudi Arabia, United Arab Emirates, Bahrain, Qatar and Oman they exceeded the 100 per cent, while the Sudan and Yemen scored the lowest penetration rate close to 26%. Also we find the spread of internet services are increasing at fast pace, with the United Arab Emirates having the highest rate and Palestine the lowest. The higher penetration is attributed to the promotion of various internet access technologies, such as wireless Local Area Network, 3Gs, and WiMax. The rate of penetration of personal computers is also used to determine the information society maturity level for countries. It has been reported that an average growth rate of 31 per cent during 2007-2008, which is higher than the previous measured period (2005-2006) rated at 12%. Iraq had the highest growth with the rate

of 43% and Palestine scored also low at the rate of 6%. The total installed computers in all of the ESCWA region is about 247,587 million.

At this category, we find again Bahrain, Qatar and United Arab Emirates are leading due to high penetration rate of fixed and mobile telephone lines and quality services that encourage the participation of businesses and individuals. This is supported by having a quality national bandwidth/ backbone for voice and data telecommunication, and also having active related Internet service providers. On the other hand countries like Iraq, the Sudan and Yemen have the lowest level of ICT infrastructure leading to low penetration rates of fixed and mobile telephone lines and low participation in services by businesses and individuals and few Internet related service providers in the market.

These Arab countries had made good progress in developing their ICT infrastructure and growth in the use of the internet, mobiles and personal computers, however, the growth is still offset by high population growth rate and political instability in some countries.

ACCESS TO INFORMATION AND KNOWLEDGE

The availability and ease of accessibility of free public information and knowledge is another indicator that can be measured to determine the progress of a country towards building information society. The indicator also accounts for variables such as the Internet penetration rates, broadband subscription costs as a percentage of income, availability of community public access centers, free flow of information and the quantity of information available on the Internet. The 'Declaration of Principles' made by The World Summit on the Information Society (2003) states that "rich public domain is an essential element for the growth of the Information Society, creating multiple benefits such as an educated public, new jobs, innovation, business opportunities, and the advancement of sciences. Information in the public domain should be easily accessible to support the Information Society, and protected from misappropriation". The declaration has been recognized by most Arab countries, but progress towards that end in some of the countries is slow, due to low internet penetration, and high access costs.

However, some of the Arab countries are beginning to transform themselves into information societies by digitizing and disseminating public information, such as public libraries and archive services, improving local governance through the use of ICT, in addition to the implementation of e-government initiatives with various services. Another related area specific to this sector is the use of Free and Open Sources Software (FOSS), which can provide alternative to commercial software to those who are on limited budget. The adoption of FOSS has also been slow in most countries, except for Egypt, which is showing sign of trends towards building communities of contributors and practitioners. None of the participating Arab countries had made the level four category (see Table 2), the highest level achieved was level 3 by countries like Bahrain, Egypt, Kuwait, Oman, Qatar and United Arab Emirates mainly due to their high Internet penetration rates, widespread broadband Internet usage and low-cost Internet services. Countries at the low maturity level 1 usually have low Internet penetration rates and high Internet costs, and lack e-government services.

ICT CAPACITY BUILDING

Education and training in ICT is another forefront that the Arab world has recognized as important in terms of responding effectively to meet the challenges imposed by the new technologies for the information society. In this context governments have taken national initiatives to provide more ICT capability, by supplying hardware and connectivity followed by conducting training programs to

large sectors of its society. In many cases this was implemented through large international firms such as Microsoft, Cisco and Intel. The initiatives were to raise the levels of ICT literacy, ranged from technology training, curriculum development to integrate technology into the learning process, digitizing course material for e-learning programs, and establishing educational portals.

On the training sides, initiatives include customized ICT programs to meet the needs of qualified professionals such as the introduction of the International Computer Driving (ICDL) and other programs to promote literacy in societies in order to foster better engagement in e-government public services.

The United Arab Emirates is the only country at the maturity level 4, while the rest placed between level 2 and level 3. The United Arab Emirates has well established practices in using ICT in education, especially in e-learning applications, focused IT literacy programs both for the public sector and citizens, and also invest in various R & D projects to advance science and technology. Countries such as Egypt, Jordan, Lebanon, Qatar and Saudi Arabia have extensive use of ICT in schools and universities and IT literacy programs and making certain progress in R&D. While Iraq, the Sudan and Yemen are at the maturity level 1 due to the limited use of ICT in education and training programs.

BUILDING ICT SECURITY AND CONFIDENCE

Threats from the Internet have made it vital for e-service providers to take measures to avoid losing the confidence of the society in using online activities. Several initiatives have been implemented covering network security, privacy and data protection, cyber legislation and international cooperation to minimize malicious attacks. Good example was set by the United Arab Emirates government by introducing the Law of

Electronic transaction and Commerce in 2002, and a secure e-payment method (e-Dirham). Also, in Saudi Arabia they have introduced Public Key Infrastructure (PKI) technology to ensure secure transmission of information and conducting electronic transactions. Oman had also introduced digital signature to facilitate and regulated e-transactions and therefore boosting the public trust in e-services. Countries such as Iraq, Kuwait, Palestine, Lebanon, Qatar and Yemen have not invested in real efforts to build confidence and security in the use of online services such as the use of e-transaction laws, e-signature laws and PKI technologies.

Initiatives to combat cybercrime are also implemented in certain Arab countries to address the technical, legal and policy issues related to network security. For example the establishment of emergency response teams (CERTs) in Oman, Qatar, Saudi Arabia and United Arab Emirates, and the introduction of Anti-SPAM framework specifically in Saudi Arabia to define the role and responsibilities of service providers, to increase awareness and find appropriate countermeasures. Despite these measures, Internet security threats in the Middle East are still at an alarming rate according to Symantec's annual report (2009). The report also indicated that United Arab Emirates had moved from 43 in 2007 to 10 in 2008, in terms of origin of attacks.

Another positive measure to promote confidence and security is through privacy and data protection legislations as seen in Oman, Tunisia and the United Arab emirates. However, investment in IT security and awareness remains at low level due to lack of programs and the absence of virus, spam and spyware protection software, and consequently many of the users and businesses become victims of various malicious attacks. Examining the maturity level of ESCWA countries (see Table 2); we find that these countries are at either level 1 or level 2. Egypt, Qatar, Saudi Arabia and United Arab Emirates are at level 2 and Bahrain, Iraq, Jordan, Kuwait, Lebanon,

Oman, Palestine, the Sudan, Syria and Yemen are ranked at level 1.

ADOPTION OF ICT APPLICATION

The use of ICT at both the public and private sectors is in progress towards creation of national economic welfare as the result of becoming efficient in providing services to citizens, in addition to other values such transparency and accountability. At the government level various applications implementations have taken place in various countries for financial operations, income tax, custom and civil registration. Services through government portals include paying utility bills, traffic fines and taxes, renewal of passports and license. Such applications are aimed at improving the front and the back office operations.

E-government measured in terms of assessment of online presence of governments, ICT infrastructure and skills and knowledge of their citizens. Most of the Arab countries in the ESCWA region have progressed towards a higher maturity level. The United Arab Emirates, Jordan, Bahrain are ranked among the top 50 countries on a global scale rating.

E-government portals are the citizens' gateway to either access information or conduct online transactions. Examination of some of the Arab countries websites indicates high disparity in terms of their level of services. We also note variations within the Arab countries and it is also interesting to note that Web 2.0 tools such as blogs, RSS and forums are still not utilized in most e-governments websites.

ICT applications are increasingly adopted by the business sector in the Arab countries although at varying degrees, from the simple use of the email to conducting sophisticated online transaction such as online shopping, e-banking, e-payment and supply chain management, e-procurement and others. E-banking is quickly being adopted by the banking and financial institutions providing various services such account inquires, and making money transfers between internal accounts. E-commerce is also flourishing and attracting users specifically from the GCC countries with volume estimates exceeding $100 billion and with annual growth of 20% which is within the international e-commerce growth rates. Gulf countries are excelling in this area due to the introduction of various e-payment systems, such as the prepaid internet credit and debit cards, which is a safe approach to conduct business online thus boosting the confidence of users to engage in online transactions, in addition to the option of using the traditional online payment method using online credit cards.

Adoption of ICT application in education is another forefront pursued in the Arab world to reduce the digital divide. Strong national initiatives are being pursued in countries like Jordan, Egypt and Palestine, including the integration of ICT into education at the primary, secondary and higher education. E-learning solutions are also implemented and launched in several of the Arab countries and the most prominent example found in this area is in Syria.

This indicator also measures the extent of improvement in the health care using ICT to better manage day-to-day operations and to extend services such as using telemedicine. In this sector we note innovative implementations, such as mobile health unit project in Egypt, electronic scans and different measurements are exchanged via satellite. Awareness and the need for sharing and integration of information and systems have been recognized across different Arab countries. In-addition, various portals were established to disseminate valuable information to citizens, such as information on medical and health information. We find Bahrain and the United Arab emirates ranked at level 4 since they are making steady progress in different sectors of ICT applications. Countries like Iraq, Palestine, the Sudan, Syria and Yemen are positioned at Level 1 due to poor use of ICT applications in government, weakness or absence of e-commerce and related legislation, and limited usage of ICT in education.

PRESERVATION OF CULTURAL IDENTIFY AND E-CONTENT CREATION

Knowledge societies in the Arab world are also assessed by measuring the amount of digital Arabic content (DAC) as a way to protect its cultural identity by preserving historical documents and manuscript, archiving and indexing resource to promote it to the rest of the world. There have been many positive initiatives towards this endeavor, for example several competitions in the region were initiated to encourage innovation in creating DAC. Another project to promote development of culture, economic and scientific content is the eSyria national blog. The Mohammad bin Rashid Al Maktoum Foundation is also actively engaged in translating and digitizing Arabic books. In Egypt the "Fekr Rama" was established to promote culture for online users.

There is significant progress on the amount of the Arab web content, which has doubled in 2009 with a growth of 43% compared to 2007, Table 3 illustrates the contribution of Arabic web content in some of the countries. Saudi Arabia and United Arab Emirates are leading in this area mainly due to high internet penetration and the attraction of foreign investments.

Most of the Arab countries are placed between Level 1 and Level 3, countries like Egypt, Kuwait, Qatar, Saudi Arabia, Syrian Arab Republic and United Arab Emirates are exerting efforts toward the preservation of cultural identity, heritage, national archives, with Saudi Arabia ranked first in terms of developing online Arabic content. Again, at this category we find the progress of the Sudan and Yemen is hampered due to political and social instability.

Table 3. Country contributions to e-content

Country	E-content Contribution
Bahrain	2%
Egypt	7%
Iraq	0
Jordan	4%
Kuwait	5%
Lebanon	4%
Oman	5%
Palestine	10%
Qatar	2%
Saudi Arabia	39%
The Sudan	0
Syrian Arab Republic	4%
United Arab Emirates	17%
Yemen	0

ENHANCING DIFFUSION SUCCESS RATES

The extant literature of ICT adoption in the Arab countries is evidently lacking compared to what is being published in the Western world. The potential of research in this region is vast and many areas are still untapped. However, it is useful to examine some of the literature in this chapter as a way to evaluate the progress of ICT development and determine better ways of its implementation.

There have been several studies examining the challenges of IT transfer and adoption in the Arab countries. Probably the most cited study is by Straub et. al (2001). Their study had identified that culture is a major factor, which can implicate the successful transfer and implementation of IT.

The argument made by Straub et. al that "Transfer means more than just technology…All too often, new technologies fail in the marketplace because flawed assumptions about considerations totally unrelated to technical merit" (Allan Kuchinsky, 1996). The notion is that cultural and social lives of IT stakeholders influence their attitudes and beliefs toward the use of technology. Their study had illustrated how specific cultural beliefs influence the transference of information technology in the Arab world. The study had focused on a small subset of cultural beliefs such as Arabs sense of time. For example in the Arab business culture long range planning have less value than those in western cultures. Also, the use of Emails should not be seen as a replacement in the Arab world but a supplement to enhance information communication; therefore, Arab countries are not expected to be a mirror image of the Western advancements.

Al-Mabrouk et. al. (2009) had determined major factors that may contribute to the successful transfer of technology in the Arab countries. The study used several research methods to capture data from academic, IT practitioners, government officials, and technology suppliers. The most highly ranked contributing factor was the need to "Formulate flexible government policies, with assistance from private and public corporations, for the selection and introduction of technology" and the second important factor was to "Identify and utilize competitive and high quality suppliers". The study reveals 10 important factors that can be collectively considered by practitioners, policy makers and supplier of technology resources to ensure successful IT transfer in the Arab world. The factors are listed below:

1. Formulate flexible government policies, with assistance from private and public corporations, for the selection and introduction of technology.
2. Identify and utilise competitive and high quality suppliers.
3. Formulate and develop a strategic plan that focuses on the actual IT transfer process and its implementation.
4. Develop open and effective relationships for information sharing between suppliers and acquirers of technology.

5. Measure attitudes towards research and development (R&D) learning and commercialization capabilities.
6. Establish R&D centers to evaluate, promote and encourage technological growth and development in Arab countries.
7. Make use of the consultation services concerning IT transfer and receive support for quick and efficient realization of practical applications for best results.
8. Embrace information technology transfer to improve social lifestyles, without compromising local values, attitudes, beliefs and traditions.
9. Formally evaluate suppliers' performance against organizational requirements.
10. Evaluate effectiveness and quality of candidate technology transfer.

Developing countries who are striving to become competitive states at a global scale, should consider making international alliances and partnerships, to strengthen the development of ICT sector. For example Al-Jaghoub and Westrup (2003) argue that one way for Jordan to succeed in this endeavor is through identifying exemplars to learn from their development processes. The choice of exemplars must have certain similarities with the country under focus. In their study Jordan was compared with Singapore and Ireland and which are regarded as successful in promoting ICT adoption and diffusion through international partnership.

Gahtani (2003) argues that Saudi organizations should allow for a trial use of new ICT products before they are released for use by the general users. Accordingly management can determine any drawbacks, and to plan for successful implementation to the general users, this can save times and costs of rejecting the final system.

Hamade (2009) argues that ICT in the Arab countries are hampered due to limited wealth, poor ICT and English language literacy, weak communications infrastructure, interrupted power supply, high cost of internet services, lack of policies, regulations, and legislations. Accordingly the author suggests a solution of using WiMax to cut down cost and to increase internet penetration.

To increase the internet adoption and the use of ICT in general, Aziz (2009) argues that some of the essential factors that can foster social inclusion is through popularizing the learning of English and to increase Arabic and culturally appropriate content. The author also highlights the need to overcome the issues of IT literacy to increase ICT adoption, and suggests that this can be accomplished through a conducive and holistic environment that involves policies, different tools and approaches to address the various needs of stakeholders.

Organizations in the Arab world are encouraged to build internal knowledge management infrastructure and to participate in establishing and engaging in industry-wide knowledge networks as a way for enhancing productivity, innovation and sustaining competitive advantage. Ahmad and Daghfous (2009), argue in their study with a group of national and multinational companies in the UAE, that the concept of Knowledge management has low priority on the organizations agenda, although they recognize their value in nurturing knowledge building and sharing across organizations. However, the main impeding factor are related the confidentiality of the knowledge and the distrust of the concept of using industry knowledge networks.

Al-Roubaie and Abdul-Wahab (2009) also argue in order to reduce the digital divide, they encourage the use of data mining techniques to discover new knowledge as a way to empowering people with knowledge, leading to innovation and creation of new knowledge and better making decisions.

Al-Mashari and Mudimigh (2003) in their study on ERP implementation in the context of Suadi Arabia, revealed six factors that might be essential for successful implementation. In ERP implementations one of the initial tasks is planning for detailed business process re-engineering (BPR). The first consideration is to avoid scope creep, by focusing on the set of objectives and not to deviate from them. Lack of ownership and transference of knowledge is another factor that is essential to be considered and specifically if the implementation is being carried out by a consultant. In Al-Mashari and Mudigm case study, external consultants were in control of the total project and any decision affecting it. It is essential that ownership to be remained in hands of the organization and at the end of the project transference of all necessary skills must be planned for. The other essential factor is to plan and manage the change by having in-house experienced staff, training and education, user involvement from the beginning to minimize resistance to change. A formal communication strategy must also be placed to keep stakeholders informed about new developments. Measurement metrics must also be devised to provide feedback mechanism to track implementation progress and to identify gaps and deficiencies in performance and to plan for corrective actions. The study concluded with the need for socio-technical perspective to development and to ensure alignment between business strategy and IT strategy. Furthermore, in order to succeeded in ERP implementation companies must reach out to learn from best practice companies with emphasis on total commitment, leadership and persistence within the organization.

Urged by an overwhelming number of ERP project failures, Albadri (2009) developed IPRM, the Integrated Project Risk Model, a structured risk-based approach to managing ERP projects, as a guide for managers and consultants to ensure success of ERP throughout the selection, adoption, assimilation and adaptation stages. The investigation of the influence of ERP issues, concerns, problems and threats on performance in twenty five major Australian and Middle East ERP projects was the main input to developing, testing and refining the structure, components and inter-relationships of the hypothesized risk-based model in a longitudinal research.

With its components, processes, metrics and tools clearly defined and characterized, (IPRM) model is presented as a viable alternative to conventional ERP management approaches. In addressing the question 'Why do IT projects fail?' and examining the suitability and effectiveness of the management practices and tools currently being used in the new complex ERP environment, the research links unfavourable project performance to inadequate or ineffective project governance, managerial competence, project and risk management practices, project control, lack of integration between processes and functions, or software tools. The model integrates components of management and control systems, manager oversight and project governance functions, and is supported by a set of appropriate metrics and tools. In effect IPRM does not only identify the problems, but also answers the question 'How can IT project failures be prevented?'

The components of the IPRM model were the subject of a series of investigations by Albadri and Jordan (2000; 2002; 2003; 2004) including; "Risk Management driven decision-making in ERP projects" (Albadri & Jordan, 2000), "The effective use of software in ERP projects (Albadri & Jordan, 2002), "The Role of controls to minimize risks of ERP projects failures" (Albadri & Jordan, 2003) and "ERP success from governance and oversight perspectives" (Albadri, 2004).

Albadri and Abdallah (2009) shifted the ERP research focus to the post-implementation phase, reflecting on an ERP success story of UAE organization in the effective utilization of ERP applications through a structured approach of 'ERP End-users characterization and competency building'. The findings of the research highlight the criticality of the 'End-Users' factor to the success or failure of the ERP venture. The case study highlights the ineffectiveness of the current approaches to ERP end-users' "training and competency building", that are commonly applied in ERP implementations, and proposes as an alternative, a new structured approach that redefines the traditional role of "ERP Training" from isolated activities concerned with introducing end-users to "how to" use ERP applications to an integral part of a comprehensive "knowledge and change management" strategy that advocates a holistic life-cycle approach to managing ERP Critical Success Factors (CSF). The investigation benefits from the "iceberg competency model", the "training management cycle", and "Kirk-Patrik's evaluation model" as a theoretical basis and context for drawing a comparison of the suitability and effectiveness of training approaches adopted in major ERP implementations in UAE. The research proposed approach, which is built around 'end-user characterization' as the main input into 'competency building' is flexible enough to plug into standard ERP methodologies and may be projected throughout the ERP life-cycle. The end-users characterization and competency building approach is expected to contribute to healthy levels of ERP usage and utilization that leads to positive business gains and return on investment.

It has already been established in the literature that electronic commerce has low uptake in the Arab world for many reasons. Yasin and Yavas (2007) through qualitative observation had identified several culturally-based factors that should be considered to increase the adoption of e-commerce practices in the Arab world. Below is a partial list from Yasin and Yavas (200) article:

- Arab countries favor face-to-face interaction over other mediums.
- Lack of trust over electronic transaction. In the Arab culture trust is a social process that is strengthened overtime.
- English is essential for interacting over the internet.
- Arab culture is high on group and family collectivism, power distance and low on future orientation.

- Arab culture is a slow paced culture, where the method of delivery is much more valued than the efficiently of delivering the message e.g. face-to-face vs. email communication, thus instantaneous delivery of a message through electronic means is not as value as in the Western world.
- Tribal culture controlled and centralized, top ranked are those with information and dispersed on need basis, therefore technologies to enhance dissemination and sharing information is valued.
- Buying and selling in the Arab world is ritual and purchasing decision is made usually on recommendation from family members or close relations rather than gathering and analyzing information from the internet.

Akhter (2007) argues that there are two major factors that impede the adoption of e-commerce in that region, namely the user's educational background and the lack of awareness of using security facilities provided by the website. In return companies selling online must ensure privacy and confidentially is practiced and trust communicated to the user. In another study by Akhter and Kaya (2009) they argue that IT managers in UAE had also concurred that security is the main barrier for the successful e-business adoption by users. Similarly, Hunaiti et al. (2009) argue that Libyan SMEs are willing to use electronic commerce, but they are confronted with several adoption barriers, such as high cost of internet access, lack of security for conducting financial transactions, no legal framework to govern electronic trading, and the postal service is not suitable for handling shipping, in addition to internet filtration.

E-government initiatives in the Arab world are currently at different development stages providing different services level. Chatfield and Althujran (2009) examined 16 different Arab countries websites and portals to carry out comparative analysis of e-government services delivery among Arab countries. The comparative analysis was based on four criteria to evaluate Arab E-government, namely: One-way information flows, Two-way interaction, payment transaction and E-democracy. The outcome of the study did confirm a wide digital divide between Arab countries and the developed countries in addition to the divide within Arab countries with disparity in delivering advanced e-services capabilities. The study had classified these 16 countries into three groups: (a) Leaders: United Arab Emirates, Bahrain and Qatar; (b) Up-and-Comers: the Majority of Arab countries such as Jordan, Lebanon, Kuwait, Egypt, Saudi Arabia, Morocco, Tunisia, Algeria, Oman and Syria; and (c) Laggards: the Sudan, Yemen and Iraq.

The sampled extant literature discussed above can provide us with a useful direction on strategic and tactical matters for both practitioners and academics. This examination also presents a sense of the current state, trends, and how some of the Arab countries are attempting to benefit from ICT despite of the many fraught challenges pertaining to the various stages of its adoption.

CONCLUSION

ICT in the Arab world is attracting the attention of governments and private sectors in order to transform themselves into knowledge-based economies despite of the various challenges being faced such as limited financial resources, technology and skills. Arab countries may be accused of only replicating the technology of the western world, but they are beginning to use technology innovatively especially in the rich oil countries. In this chapter, we have examined various initiatives launched in different countries and some being hailed for their success. However, the endeavor for the Arab world may just have begun; we are yet to see different strategies being planned and implemented in the different categories discussed in the previous sections. Another useful indica-

tor that can be used to determine the successful penetration of ICT in the Arab society is through the examination of the academic research, which is currently lacking in most directions. Further research and development must be also encouraged in the area of ICT adoption in order to determine better ways of planning and implementing successful ICT projects aligned with cultural practices and other challenges.

REFERENCES

Ahmad, N., & Daghfous, A. (2010). Knowledge sharing through inter-organizational knowledge networks: Challenges and opportunities in the United Arab Emirates. *European Business Review*, *22*(2), 153–174. doi:10.1108/09555341011023506

Akhter, F. (2007). A case study: Adoption of information technology in e-Business of United Arab Emirates. *IEEE international Conference on e-Business Engineering,* Hong Kong, China

Akhter, F., & Kaya, L. (2009). Building secure e-business systems: Technology and culture in the UAE. In *Proceedings of the 2008 ACM symposium on Applied computing,* Ceara, Brazil.

Al-Gahtani, S. (2003). Computer technology adoption in Saudi Arabia: Correlates of perceived innovation attributes. *Information Technology for Development, 10,* 57–69. doi:10.1002/itdj.1590100106

Al-Jaghoub, S., & Westrup, C. (2003). Jordan and ICT-led development: towards a completion state? *Information Technology & People, 16*(1), 93–110. doi:10.1108/09593840310463032

Al-Mabrouk, K., & Soar, J. (2009). An analysis of the major issues for successful information technology transfer in Arab countries. *Journal of Enterprise Information Management, 22*(5), 504–522. doi:10.1108/17410390910993518

Al-Mashari, M., & Al-Mudimigh, A. (2003). ERP implementation: lessons from a case study. *Information Technology & People, 16*(1), 21–33. doi:10.1108/09593840310463005

Al-Roubaie, A., & Abdul-Wahab, R. (2009). Data mining and knowledge discovery: An Approach for sustaining development in GCC countries. *International Association of Computer Science and Information Technology* – Spring Conference, Singapore.

Albadri, F. (2009). The Integrated Project Risk Model: A Risk Based Model for Managing Information Technology Projects. In Kidd, T. T. (Ed.), *Handbook of Research on Technology Project Management, Planning, and Operations*. Hershey, PA: IGI Global.

Albadri, F., & Abdallah, S. (2009). Characterization and Competency Building in Context of and Oil & Gas Company. In *Proceedings of the 11th. International Business Information Management Conference*, Cairo, Egypt, Jan 4-6.

Albadri, F., & Jordan, E. (2000). The integrated management model: Risk Management driven decision making in IT Projects. In *Proceedings of the 5th International APDSI Conference, Asia-Pacific Region of Decision Institute*, Tokyo, Japan, July 24-27.

Albadri, F., & Jordan, E. (2002). The integrated Project Risk Management Methodology: The effective use of software in IT Projects. In *Proceedings of the 7th International APDSI Conference, Asia-Pacific Region of Decision Institute*, Bangkok, Thailand, July 24-27.

Albadri, F., & Jordan, E. (2003). The role of Controls to Minimize Risks of Information Technology Projects Failure. In *Proceedings of the 14th Australasian Conference on Information Systems*, Perth, Western Australia, November 26-28.

Albadri, F., & Jordan, E. (2004). Project Success: Project Managers Competence. In *Proceedings of the 3rd International Business Information Management Conference*, Amman, Jordan, July 4-6.

Aziz, S. (2009). Dimensions of IT literacy in an Arab region: A study in Barkha (Oman). *Information and Communication Technologies Development International Conference*. Doha, Qatar.

Chatfield, T., & Althujran, O. (2009). A cross-country comparative Analysis of E-government service delivery among Arab Countries. *Information Technology for Development, 15*(3), 151–170. doi:10.1002/itdj.20124

ESCWA. (2009) *Economic and Social Commission for Western Asia, United Nations Economic and Social Commission for Western Asia publication*. Retrieved April 2010, from http://www.escwa.un.org/information/publications/edit/upload/ictd-09-12.pdf

Hamade, S. (2009). Information and communication technology in Arab countries". *Problems and Solutions. Sixth International Conference on Information Technology:* New Generations. Las Vegas, NV.

Heeks, R. (2002). Information systems and developing countries: Failure success, and local improvisations. *The Information Society, 18*, 101–112. doi:10.1080/01972240290075039

Hunaiti, Z., Masa'deh, R., & Mansour, M. (2009). Electronic commerce adoption barrier in small and medium-sized enterprises (SMEs) in developing countries: The case of Libya, *IBIMA. Business Review (Federal Reserve Bank of Philadelphia), 2*, 37–45.

Ibrahim, R. (1985). Computer usage in developing countries: Case Study Kuwait. *Information & Management, 8*, 103–112. doi:10.1016/0378-7206(85)90038-2

Kuchinsky, A. (1996). Transfer means more than just technology. *Communications of the ACM archive, 39*(9), 28-29.

Leavitt, H. J. (1965). Applying Organizational Change in Industry: Structural, Technological, and Humanistic Approaches. In March, J. G. (Ed.), *Handbook of Organizations*. Chicago: Rand McNally.

Mezher, T. (2007), Information and Communication Technology in the Arab Region in Nazli Choucri, D. Mistree, F. Haghseta, T. Mezher, W.R. Baker, C.I. Ortiz (Eds), on *Mapping Sustainability Alliance for Global Sustainability*, Springer-Verlag New York Inc.

Miniwatts Marketing Group. (2010). *Middle East Internet Usage and Population Statistics*. Retrieved April from http://www.internetworldstats.com/stats5.htm

Murphy, E. (2006). The political impact of information technologies in the gulf Arab states. *Third World Quarterly*, *37*(6), 1059–1083. doi:10.1080/01436590600850376

Straub, D., Lock, K., & Hill, C. (2001). Transfer of information techno to the Arab World: A test of cultural influence modeling. *Journal of Global Information Management*, *9*(4), 6–28.

Symantec EMEA Internet Security Threat Report. (2009). *Symantec EMEA Internet Security Threat Report*. Retrieved April 2010, from http://eval.symantec.com/mktginfo/enterprise/white_papers/b-whitepaper_emea_internet_security_threat_report_04-2009.en-us.pdf

World Summit on the information society. (2003). *Declaration of Principles - Building the Information Society: a global challenge in the new Millennium*. Retrieved April 2010, from http://www.itu.int/wsis/docs/geneva/official/dop.html

Yasin, M., & Yava, U. (2007). An Analysis of e-business practices in the Arab culture: Current inhibitors and future strategies. *Cross Cultural Management: an International Journal, 14*(1), 68–73. doi:10.1108/13527600710718840

Chapter 2
Deciding to Use ICT in the Arab Culture:
The Influence of Obedience to Authority and Collectivism

Sonda Bouattour Fakhfakh
University of Tunis, Tunisia

ABSTRACT

Compared to the Western world, the Arab world endures a lack of ICT use and a scarcity in software production and services. One of the explanations that were advanced in order to determine the reasons of these deficiencies is the Arab world's cultural context. In this chapter, which reports on a doctoral research, we focus on two dominant values which are supposed to be descriptive of the Arab world: obedience to authority and collectivism. We examine their hypothesized influence on ICT use within an Arab context (Tunisia) and at an individual level. Data collection was carried out on 378 Tunisian students in three quasi-experimental settings measuring the degree of obedience, and by a pilot questionnaire determining, among other things, the level of collectivism. While the obedience to the authority figures appears to positively affect the probability of technology use in the first experiment, the rationality seems to influence choices in the second and third experiments, and to deny any pressure of the foreign nationality of the figure. The results of this study also reveal that all participants have a collectivistic tendency. However, the influence of collectivism is found to be mixed.

INTRODUCTION

Since the gate of interpretation (ijtihed) was ended (X[th] century), the Arab world, composed of more than 342 million people, could not retake the lead on the world scene. Neither the oil discovery in the 1930s nor the information and communication technology (ICT) advent allowed the Arab world to ensure its place in the emerging information society. The problem of the Arab world, as far as ICT is concerned, is not related to readiness, given that the access and the infrastructure have relatively been well established (Madar Research Group, 2008). The Arab world really endures a lack of ICT use and a scarcity in software production and services, as compared to the Western world

DOI: 10.4018/978-1-60960-048-8.ch002

(ITU, 2009; UNDP, 2003; UNESCO, 2005). One of the explanations that was advanced to determine the reasons of the existing deficiencies is the Arab world's cultural context (Hill et al., 1998; Straub et al., 2001).

The Arab socio-cultural system has been identified as being complex (Straub et al., 2001) and not homogeneous (Corm, 2007) for its diversity in terms of religion, ethnicity, wealth, resources, history, etc. It has also been described as having ambivalent forces. While the dominant culture emphasises traditional values such as fatalism, imitation, obedience to authority and collectivism, some subcultures do at the same time support modern values like freedom, femininity, independence and innovation (Barakat, 1993).

Shariati (1971) asserts that each society must dip into its own cultural traditions in order to find its own enlightenment. In light of this assumption, we attempt to identify some traditional values that could explain ICT use in Arab world. Many Western models treating ICT acceptance do exist, and were applied on Arab culture. But, as Alterman (2000) suggests, "We need to free our studies of new media and technology in the Middle-East from the straitjacket of western experience" (p. 355). In fact, Arab values may be concomitant with Western values. However, the concepts and their use can be different from Western connotations. So, imposing Western patterns just as they are in the Arab context may condemn our view of present and future. It may also cause researchers or managers -willing to have an idea about the Arab as an individual or as a potential worker- to fall in mistakes.

In this chapter, which reports on a doctoral research, we focus on two dominant values which are supposed to be descriptive of the Arab world (Bouattour & El Louadi, 2004): obedience to authority and collectivism. We examine their hypothesized influence on ICT use in an Arab context (Tunisia) and at an individual level.

We attempt specifically to answer the following question: If an Arab has the choice to use a technology, what technologies will he/she select and how? Do obedience to authority and collectivism influence his/her choice?

The different dimensions of obedience to authority and collectivism are first identified. Then, the relationship between these concepts and ICT use is determined and hypotheses are formulated. Next, we present our quasi-experimental setting followed by results and discussion. We propose some future directions before concluding.

THE ARAB CULTURE AND ICT

Despite the fact that the Arab world has realized a factual progress in terms of mobile phone access and use, the diffusion of Internet seems to be still slowly (ITU, 2009). Overall, the digital gap between industrialized countries and Arab countries -and even inside the Arab countries themselves- remained unchanged between 2002 and 2007 (ITU, 2009). If the economic and technical reasons of this ICT gap could be easily detected, understanding the Arab cultural barriers is more complicated. In fact, the Arab society oscillates between a continuum of homogeneity and heterogeneity of cultural orientations. However, some dominant values could be identified such as obedience to authority and collectivism. Considering these values, the use of ICT might be better adapted to the behavioral patterns of Arabs.

Authority and Obedience

"Those who are older than you, have more wisdom"; "Stand up for your teacher and respect him, the teacher is nearly a prophet[1]"; "Ask an experienced man rather than a doctor"; "The one who does not obey their older, will have problems" are some examples of Arab popular proverbs which emphasise respect and obedience to the authority of the elder, experienced men and teachers. According to these proverbs, authority supposes the existence of a hierarchical structure, an inequality of situations and an unconditional respect to a person that is recognized as superior (Tassin, 1994).

These Arab maxims draw their origins from the Quran and Sunna[2]. In fact, while the Prophet (SAW) has emphasised the authority of scholars "The Ulama are the inheritors of the prophets", the Quran has supported obedience to authority through this Quranic passage: "Oye who believe, obey Allah, and obey the messenger and those charged with authority among you" (S.IV: 58). This verse, as the Hadith, had the aim to apply the recommendations of the Islamic law (Chari'a). But, the selective interpretation used by several politicians to legitimate their power (UNDP, 2004) has made obedience a duty that is inherited from one generation to another (Ibn Khaldoun, 1863), and a socially claimed virtue.

Indeed, religious institutions have tried to teach the respect of the male elder and superior through the concept of family. As noted by the UNDP report (2003), the dominant children's education mode within Arab families is authoritative and over-protected. In spite of the decline of patriarchal roles, especially in the high layer of the Arab society, the father and all other members of agnatic families continue to hold a high-level authority. The representation of this domination inside the family has been extended to the educational structure, the political system and the organizational management (Dwairy et al., 2006; UNDP, 2003). Besides, the combination between obedience and respect has allowed the reproduction of centralized systems (Abdalla & Al-Homoud, 2001) with a paternalistic authoritarian mode (Scouarnec & Silva, 2008). Obedience has also been transmitted to the foreign authority of the West (Elmandjra, 1990) -even after the independence-, encouraging Arabs to be fascinated by what is foreign, and whoever or whatever comes from the West (Ben Zina, 2002; Zogby, 2003).

Thus, attitudes towards authority seem to be influenced by the legacy of the past and by the beliefs and representations acquired in teenage years and confirmed in adulthood. However, it should be noted that being obedient in Arab societies does not have the same connotation as it is in the West. An Arab does not lose his/her individuality. In addition, Quran and Sunna affirm that all humans are equal, but their inequality is only founded on the basis of knowledge and piety. Therefore, Tunisians seem to accept the social inequality if it is based on learning (Smida & Latiri, 2004). Although, it appears that the respect of the scholars (Ahl Al Alm) is decreasing in profit of wealthy figures (UNDP, 2003).

The experiments of Milgram (1974) had shown that the presence of a recognized authority symbolized by a researcher's white blouse can lead an ordinary individual to obey even if the consequences are dangerous[3]. The same results have also been obtained thirty years ago in a replication of Milgram's experiment (Burger, 2009) and even in a virtual environment where the participant was aware of the unreality of the shocks and the learner (Slater et al., 2006).

In consequence, if a figure of authority that is worthy of confidence and is recognized by a community recommends the use of technology, Arabs will be more easily disposed to take his/her advice.

So, the following hypothesis can be put forth:

H1: Obedience to authority positively affects the decision of ICT use.

Collectivism

While an individualist tends to be independent from the group (Hofstede, 1980), a collectivist is ready to be linked to a group (family, friends, colleagues). The collectivist privileges the collective goal over the personal one and gives more importance to the harmony of the group and its values (Triandis, 1989).

According to Hofstede (1980), individualism and collectivism are opposites. For others, and at an individual level, these concepts are independent and can coexist in the same culture (Realo et al., 1997; Rhee et al., 1996). Thus, a member of a collectivistic culture is not necessarily a collectivist. Besides, no culture is exclusively individualistic or exclusively collectivistic (Triandis, 2001). However, Triandis (1989) recognized the influence of the dominant culture on students: belonging to a collectivistic culture increases the likelihood that its members show more collectivistic characteristics.

Being aware that the study of collectivism as a relational process is more relevant than individualism (Gaines et al., 2005) and that the Arab culture is considered as having a collectivistic tendency (Dhaoui, 2000; Hofstede, 1980; Ibn Khaldoun, 1863; Oyserman, 1993), we focus on collectivism. The Arab collectivism seems to vary along a continuum. For example, Saudis are more collectivistic than Lebanese or Egyptians (Hofstede & Hofstede, 2005).

In the Arab world, collectivism is especially endorsed by two elements: (1) ethnic solidarity (Assabiya according to Ibn Khaldoun (1863)) and (2) religious values (Zaket, alms, love and help of parents and others) (Oyserman, 1993). Nevertheless, the power of these two elements seems to have changed. In fact, ethnic solidarity appears to be enhanced in some Arab countries because of the continued persecution and exclusion of minorities within the same society (e.g., Kopts in Egypte, Kurds in Irak or Syria). On the other hand, religious values seem to lose weight (Fitouri, 1991) among, for instance, Tunisians, Algerians (Kabyle), Moroccans or Lebanese, where the irruption of the concept of secularism and the indirect influence of colonizer appear to contribute to the adoption of French ideas. For example, assessment of some young Tunisians on the collectivism/individualism scale showed a tendency to achieve personal success and self-fulfilment, and this, at the same time as taking care of family's solidarity (Smida & Latiri, 2004). These results are in agreement with the statement of Dhaoui (2000) who postulates that among

Maghrebins, "there is a kind of independence in the dependence" (p. 86).

Being collectivist and using technology can be matched. In fact, electronic mail and short messages (SMS) offer the possibility to send message to many people at the same time (especially in festivity, Ramadan and Aïd). Mobile phones, Skype, MSN and satellites allow keeping in touch with distant families, reinforcing the blood relations (Silatou Arrahim) and raising the connection with others from around the world. Moreover, the need to develop friendship and to preserve social ties has pushed Arabs to humanize ICT and to develop new practices. The insertion of smileys, the invention of codes in order to write in Arabic and Tunisian dialect (Zlitni, 2003) and the use of beeping in Morroco[4] (Kriem, 2009) are some examples of collectivistic use of technology. Even technologies that are not based on the physical presence of interlocutors (e.g., instructor in e-learning or members in groupware) seem to require some collectivistic norms such as a predisposition to work in group or a valorisation of social exchanges. Thus, the need for membership and interaction with others can encourage the use of technology.

It is consequently plausible to suggest that:

H2: Collectivism positively affects the decision of ICT use.

EFFECT OF OBEDIENCE TO AUTHORITY AND COLLECTIVISM ON THE CHOICE OF ICT: HYPOTHESES TESTS

Our study explores the relation between two Arab cultural values and the decision to use ICT in the Tunisian context. We used a quasi-experimental design where we manipulated obedience to authority. We also administered a pilot questionnaire to determine, among other things, the collectivism. French was the language used for this study.

We reported the following main results and we discussed them.

Quasi-Experimental Design

The technique chosen to test our hypotheses was quasi-experimentation with a post- type test and a control group.

Participants

The participants were students in Management at the Master's degree level. We conducted three experiments, nearly similar, that were run at universities in the cities of Sfax, Tunis, Manouba and Carthage. Each participant was subject to one experiment only. Data collection was completed in December 2008 for the first experiment and in February, March and April 2009 for the other two experiments. We obtained the responses of 378 students divided into three experiments as showed in Table 1.

Procedure and Material Used

Experiments had paper as support and used a similar procedure. After having obtained the agreement of the instructors, we invited the students to take part in the experiment. The instructions were written and told aloud. Anonymity and confidentiality were also guaranteed. In order to avoid all bias including communication between groups, participants were not informed of the real aim of the experimentation (see Campbell & Stanley, 1963). They were just told that the experiments concerned a doctoral study studying the relationship between Tunisian students and the adoption and use of ICT.

Each experiment was conducted in class and lasted between twenty and twenty five minutes.

Since the number of students at the Master level was rather low, the choice was, first, a toss-up between the experimental and the control groups and then a regular change between the two groups.

The First Experiment

After giving out the instructions above, we informed the participants about the nature of the two documents that we had distributed to them.

The first document contained three pages. The first page constituted of the scenario and reported a discussion between two students (Aymen and Khadija) about their thesis progress and problems. This discussion is guided by a question asked by Khadija: What is the most appropriate technology to be used among three (the mobile phone, the SMS and the electronic mail) to contact a friend (unknown) and to ask him for an English research paper not found in the library.

The answer to this question was collected from three male instructors[5] (fictitious) specialized in information systems and represented by their passport photo[6] in the second page. There was a 27-year-old Tunisian junior lecturer and two 60-year-old professors; one was Tunisian and the other one American[7]. Professors put on suits while the junior lecturer wore a T-shirt.

The American professor's photo was arranged in the first position followed by the Tunisian

Table 1. Descriptive statistics for samples

	Female	Male	Average Age	Experimental Group	Control Group	Total
First Experiment	107	18	24.1	60	65	125
Second Experiment	107	28	25.5	71	64	135
Third Experiment	69	49	26.1	62	56	118

professor's photo and finally the photo of junior lecturer. This order was chosen in a way that it does not support our hypothesis: the participant will follow the answer made by the foreign professor who suggests the least suitable technology. In fact, we believed that with equal ranks, the subject follows the proposal of the American professor even if it is wrong and even if the Tunisian professor is right. In our case, the mobile phone was the most inappropriate technology to use. In fact, Khadija (the actress student) had to formulate her request, to specify the paper's references (title, authors and journal), and to give her electronic address. All these information needed time to be transmitted besides the problems of pronunciation and comprehension, which made the communication expensive. The second best choice, stated by the Tunisian professor, was SMS: writing all these information would be easier, cheaper and safer than calling the friend over the mobile phone. However, using SMS was more expensive and restrictive in terms of the number of characters (160 per message), as compared with electronic mail. Therefore, electronic mail was the best choice to be agreed on by the junior lecturer. Each instructor gave his suggestion without any further explanations.

Being aware of the effect of primacy and assuming that what remained in memory were the last words and images, we had chosen to use the least expensive method (in terms of number of combinations), and to reverse the order of appearance of the three instructors. Then, photos and suggestions were replaced (in order to eliminate the participant memory test) on the top of the third page where the question of the likelihood of the technologies uses was asked, and an example of invalid responses was given[8]. (Table 2)

We collected the first document as soon as the student had finished answering it in order to check out invalid answers. To reduce the Rosenthal[9] effect, we tried to limit our intervention to only drawing their attention to the fact that each technology must have a different answer from the other technologies.

The second document was the pre-questionnaire[10]. It was composed of 14 questions including demographic and situational information about the respondents, their possession, their use of ICT and their experience level. The last question was about collectivism.

This procedure had been applied to experimental groups as well as the control group. The only difference resided within the first document (pages 2 and 3) distributed to the control group: it contained neither the photos of the three instructors nor their ranks and ages. Only their names, their countries and their opinions were specified.

The Second and Third Experiments

It is obvious to ask if the suggestion of the foreign instructor will be adapted de facto because of his nationality. Trying to explore this question, we

Table 2. Experimental manipulations

	Experimental manipulations		
Academic Rank	Professor	Professor	Junior lecturer
Nationality	American	Tunisian	Tunisian
Age	60		27
Look	Suit		T-shirt
Proposed Technology	Mobile Phone	SMS	Electronic Mail
Order of Appearance (page 2)	1	2	3
Order of Appearance (page 3)	3	2	1

reproduced the same first experiment on other students while respecting all the different steps of the procedure. We had only changed the nationality of the instructor. So, in the second experiment, the foreign instructor preserved his rank as professor and proposed the use of SMS, while in the third experiment, the foreign instructor was the junior lecturer and suggested the electronic mail.

Measures

The decision of ICT use was measured by the likelihood of choosing to use each of the three technologies (the mobile phone, the SMS and the electronic mail). This likelihood varies from (1) completely improbable to (5) completely probable. The participant was asked to give a different response to each technology. The obedience was determined by the average of participants' answers that had aligned their opinions on those of the supposed figure of authority.

The authority was measured by the academic rank and age of each instructor, in such a way that the rank and age increase in the same direction. These measures were inspired by Ali & Wahabi (1995), Ben Fadhel (1992) and Wilson (1968).

We also reproduced the ten-item collectivism scale of Yamaguchi (1994) conceived for a collectivistic culture and only referring to individual behavior. We chose friends as a reference group because, at this age, students are closer to their friends than to their family.

Results

Our aim was to highlight the influence of obedience to authority and collectivism on the decision of use of technologies in an Arab context. The collected data was analyzed using SPSS.

The First Experiment

The first hypothesis supposed that the Arab was more likely to obey to highest ranked figure of authority when choosing a technology. Applying the t-test for independent samples, we only found one significant difference between the experimental group and the control group when it comes to the probability to choose the use of the mobile phone, the less suitable choice, which was suggested by the foreign professor (Table 3).

Answers on the collectivism scale (average = 34.73, sd = 6.26) showed a collectivistic tendency. The obtained Cronbach's alpha is higher than 0.74, which is regarded as satisfactory. Data analysis comparing the collectivism level between the two groups revealed that the experimental group was significantly more collectivistic (average = 3.58) than the control group (average = 3.36) (t = 2.007; df = 123; p = 0.047).

The data analysis in relation with our second hypothesis -underlining the existence of one positive relationship between collectivism and the use of ICT-, showed that the correlation between col-

Table 3. Comparison between experimental and control groups and probability of ICT use

	Group	N	Average	t	df	p
Probability (Mobile Phone)	Experimental	60	4.02	2.281	123	0.024*
	Control	65	3.46			
Probability (SMS)	Experimental	60	2.52	- 0.788	123	0.432
	Control	65	2.69			
Probability (Electronic Mail)	Experimental	60	4.02	0.675	123	0.501
	Control	65	3.88			

* Significance level was smaller than 5%.

Table 4. Correlation between collectivism and probability of ICT use

	Probability to choose the use of	r	p
Collectivism	Mobile phone	0.068	0.454
	SMS	- 0.051	0.571
	Electronic mail	- 0.133	0.141

Significance level was smaller than 5%.

Table 5. Comparison between experimental and control groups and probability of ICT use

	Group	N	Average	t	df	p
Probability (Mobile Phone)	Experimental	71	4.00	1.335	133	0.184
	Control	64	3.70			
Probability (SMS)	Experimental	71	2.14	-0.935	133	0.352
	Control	64	2.34			
Probability (Electronic Mail)	Experimental	71	3.77	-0.457	133	0.648
	Control	64	3.88			

Significance level was smaller than 5%.

lectivism and the probability of choosing each of the three technologies was not significant (Table 4).

We should highlight that there was not any significant difference between the two groups that has to do with their age, gender, income (family and individual) and ICT experience level.

The Second Experiment

The aim of this experiment and the next one was to test the influence of the nationality of the authority figure on the participant's obedience.

As shown in Table 5, there were not any significant differences: the fact that the foreign professor suggested the use of SMS as being the suitable technology had no significant effect on the respondent's obedience.

The collectivism scale was one-dimensional (average = 35.29; sd = 6.98) and Cronbach's alpha was equal to 0.78. Pearson correlation between collectivism and the probability of SMS use was negative and significant (Table 6).

There was no significant difference between the two groups neither in terms of age nor gender nor income. The only difference was noted on the level of ICT experience (t = 2.308; df = 129; p = 0.023).

Table 6. Correlation between collectivism and probability of ICT use

	Probability to choose the use of	r	p
Collectivism	Mobile Phone	0.102	0.239
	SMS	- 0.187	0.030 *
	Electronic Mail	0.105	0.224

* Significance level was smaller than 5%.

Table 7. Comparison between experimental and control groups and probability of ICT use

	Group	N	Average	t	df	p
Probability (Mobile Phone)	Experimental	62	4.02	0.700	116	0.485
	Control	56	3.86			
Probability (SMS)	Experimental	62	2.26	0.212	116	0.832
	Control	56	2.21			
Probability (Electronic Mail)	Experimental	62	3.92	0.198	116	0.843
	Control	56	3.88			

Significance level was smaller than 5%.

The Third Experiment

The data illustrated in Table 7 reveal the absence of significant differences between the two groups when choosing ICT.

For the collectivism scale, we had to delete four items in order to reach a Cronbach's alpha of 0.82 (average = 21.98; sd = 4.9). Concerning our second hypothesis, only a positive relationship -which tended to be significant - was revealed between the electronic mail and collectivism (Table 8).

The two groups showed some significant differences. While the control group was older (average = 27.04) than the experimental group (average = 25.25) (t = - 4.489; df = 113; p < 0.01), the last one seemed to belong to a well-off family with a high income (t = 0.227; df = 113; p = 0.003).

Analysis and Interpretation

This study attempts to show how Arabs perceive authority, and how obedience and collectivism could influence their technological choice. Results indicated that ICT use by Tunisian students seem to be, at least partially, consistent with their cultural values.

For the First Hypothesis H1: Obedience to Authority Positively Affects the Decision of ICT Use

The First Experiment

Analysis of the data revealed that the participants aligned their choices on those of the (foreign) professor and opted for the mobile phone. Consequently, the first hypothesis is not rejected.

According to Milgram (1974), the family education mode, the habit of obedience and the extension of the authority system to other hierarchical structures are some pre-requisites for obedience. Besides, and in an experimental context, if the experiment is perceived as scientific and the authority as lawful (in our case, the authority, represented by a suitable-dressed instructors, acts in a student's context), the obedience level will increase.

Table 8. Correlation between collectivism and probability of ICT use

	Probability to choose the use of	r	p
Collectivism	Mobile phone	- 0.119	0.198
	SMS	0.027	0.769
	Electronic mail	0.158	0.087

Significance level was smaller than 5%.

It is consequently obvious that Arabs have acquired an obedient culture and it is understandable that Tunisian participants obey to the authority and respect hierarchy.

But, it appears debatable that the use of technology -which is supposed to reduce hierarchy and increase democracy- could be exacerbated by authority and obedience. Accepting Ibn Khaldoun's view of the role of authority on the reinforcement of change, we agree with Al-Jabri (1994) that the Arab world should not break up with its past, but rather search into its traditions some sources that enable it to turn towards the future.

Being obedient correlates with the work of Hofstede (1980) -who argues for the existence of a high hierarchical distance within Arab countries (at the national level)-, and is supported by findings of other authors (at the organizational level) (Lassoued, 2005; Soyah & Magroum, 2004). Although there is no mechanical determinism of the national or the organizational/familial levels towards the individual, it seems that there is an interplay between these three levels. In fact, Sharabi (1988) and Todd (1983) state that the patriarch Arab figure and the family culture explain the socio-political system. Attiyah (1993), Geertz (2000) and Hofstede & Hofstede (2005) add that, beyond the bottom-up processes, the autocratic political system and the national culture also affect the organisation's structure and the local values and beliefs. Nevertheless, Desjeux (2003) advances that context should be taken into consideration when dealing with macro and micro levels. Besides, Cohen (2007) emphasizes that, "each person defines a situation in light of key values that are important to him or her" (p. 274).

Moreover, the positive impact of obedience to authority on the respondents' choices supports the social influence theory of Fulk et al. (1990). But, it does not deny the rational choice at the moment when following the instructor could be a rational behavior for the participant (Minsky & Marin, 1999).

The Second and Third Experiments

The result obtained in the first experiment showed that, at an equal academic rank, the opinion of the foreign professor was followed. This result had pushed us to explore the pressure of the instructor's nationality on the ICT use.

The results (Tables 5 and 7) show that there was not any influence of the instructor's nationality on the student's choice. However, there was also no support for hypothesis H1.

Milgram (1974) reported that the participant could him/herself feel quite embarrassed when two legitimate authorities gave two different orders. So, he/she applies his/her rationality whether that is correct or not. Baile & Lefièvre (2003) affirmed that although the influence of social pressure may be important in the selection and use of a communication medium, the hypothesis of objectivity could not be discarded. The results shown in Tables 5 and 7 display the fact that there was a near unanimity about the probability average of using each technology, which suggested that there was some rationality. Moreover, the collected post-experiment information revealed that participants who had chosen the mobile phone advanced the argument of rapidity and the necessity of response. On the other hand, students who opted for electronic mail indicated that it was more suitable to ask for information from an unknown person than SMS which was a more of a private technology. So, according to the Rational Choice theory, participants seem to have matched the objective characteristic of the technology with the objective requirement of the communication task.

For the Second Hypothesis H2: Collectivism Positively Affects the Decision of ICT Use

The Yamaguchi measure (1994) showed a collectivistic tendency with respect to the reference group for the three samples. This result goes in the same direction as that of the authors identifying Arabs as collectivistic people, especially those of Ibn Khaldoun (1863) and Hofstede & Hofstede (2005).

The First Experiment

The result shown in Table 4 revealed an absence of significant relation between collectivism and the probability to choose technologies. The second hypothesis is consequently rejected. Therefore, having a collectivistic tendency does not have any influence on the technological choices. However, the relationship between these two variables, although weak, supports the assumption according to which Arabs, at least Tunisians, have a preference for oral speaking (mobile phone, $r = 0.068$) rather than for writing (SMS, $r = -0.051$) and electronic mail ($r = -0.133$) (El Louadi, 2004). Then, the technology use seems to be dependent on its degree of substitution to the face-to-face communication (Hasan & Dista, 1999).

The Second and Third Experiments

The data analysis showed that while the result of the second experiment negatively and partially supports H2, the finding of the third experiment does partially and with a borderline significance support H2.

As shown in Table 6, the relationship between collectivism and the probability of SMS use was negative. It is possible that Tunisian students have considered the friend that will be contacted by Khadija as being an out-of-group member. So, using SMS was unsuitable because SMS seemed to be especially destined to the in-group members such as family and near friends (Desjeux, 2005). In addition, since SMS is considered as a recent technology (launched in 1992), its use may need more time to be integrated in social habits. However, according to Lynn & Gelb (1996), high collectivism is, in general, a barrier to use technologies. For example, the Lebanese, identified as being less collectivistic than the Saudis (Hofstede &

Hofstede, 2005), seem to have the highest SMS rates in the Arab world (Arab Advisors Group, 2005) and among the young, SMS penetration reaches 96% (Arab Advisors Group, 2007).

In the third experiment, the relationship between collectivism and electronic mail use tended to be significant. A preliminary result showed that almost 80% of these participants used electronic mail, at least, once a day. If we know that approximately 10% of the entire Tunisian population are members of the social network Facebook and that the majority of them are between 14 and 34 years-old (Checkfacebook, 2009), it is somewhat plausible to speculate that electronic mail use is enhanced by Facebook connection.

FUTURE RESEARCH DIRECTIONS

Our study is limited to the case of Tunisia and to a convenience sample. Although most behavioral and social science studies use convenience samples, these are not representative of the population. Besides, what could be valid for Tunisia could be different in other Arab countries because of their diversities and traditions. Moreover, what participants said that they would do may not necessarily correspond to what they will really do (Matsumato, 2006). Therefore, results should be replicated in others Arab societies in order to see the presence of similar patterns, especially in the Gulf where religious values, collectivism and foreign authorities could also be more influential. It is also judicious to design a study of other Arab cultural variables in order to be able to determine a framework of explanatory variables of ICT use with Arab lens. In addition, and according to Milgram (1974), complying with the group is more important than obeying to a simple order given by a professor. So, it will be useful to expose students to further experiments to detect the power of the group influence compared to obedience.

CONCLUSION

This study has attempted to explore the influence of obedience to authority and collectivism values on ICT use over three similar quasi-experimentations that were accomplished in an Arab context. While the obedience to the professor's authority appears to positively affect the probability of technology use in the first experiment, the rationality seems to influence choices in the second and third experiments, and to deny any pressure of the foreign nationality of the instructor. Through these results, we can see that authority can be conceived as a constructive value and not only as a destructive one. The results of this study also reveal that all participants have a collectivistic tendency. But, the influence of collectivism is mixed. While the first and third experiments do not show any significant relationship between collectivism and probability of use of ICT, the second experiment illustrates a negative relationship when deciding to use SMS. This result supports the fact that collectivism could contribute to understand the resistance of Arabs, at least Tunisians, to the use of ICT.

However, being collectivistic or having the sense of obedience is not in contradiction with a prospective attitude. The example of Japan which has achieved its progress without the westernization of its society (cultural opening and societal closing) is quite illustrative.

Moreover, it is possible that Arabs found themselves incapable of developing their informational and scientific domains because they are stressed to behave using Western paradigms. In fact, the Arab world seems to perceive science as supporting a capitalistic ideology and a latent atheism (Guessoum, 2006). To change this view, we can refer to the suggestion proposed by Bouraoui & Chastrette (1999) related to modifying students misconceptions. They suggest enlarging the information network of the students by reorganizing the structures of their anterior conceptions and refining them to be adapted to the context.

Similarly, using traditional values as a basis for changing could be a successful strategy.

Our study has the power to recognize, at least at an individual level, one dominant Arab cultural value that could enhance the ICT use among students who will be the future managers or makers of decision.

By accepting the systemic cultural values, Arab institutions should take into account the cultural dimensions in order to redirect their action and improve their social and cultural awareness about technology, at least, in the educational system. The change should come from Arabs. It should not be viewed as imposed. It should be considered as a progressive extension for their values, as Allah had done when the wine had been prohibited.

REFERENCES

Abdalla, I. A., & Al-Homoud, M. A. (2001). Exploring the implicit leadership theory in the Arabian Gulf States. *Applied Psychology: An International Review*, *50*(4), 506–531. doi:10.1111/1464-0597.00071

Al Akremi, A., Nasr, M. I., & Sassi, N. (2007, September). *Impact de la culture nationale sur la confiance interpersonnelle en milieu du travail: Analyse comparative entre la France et la Tunisie.* [Impact of the national culture on interpersonal confidence in work context: Comparative analysis between France and Tunisia]. Paper presented at the 18th congrès annuel de l'Association Francophone de Gestion des Ressources Humaine (AGRH) [Paper presented at the 18th annual congress of the Francphoic Association of Human Resource Management], Fribourg, Switzerland.

Al-Jabri, M. A. (1994). *Introduction à la critique de la raison arabe. Introduction to the criticism of the Arab reason. Translated from Arab by A. Mahfoud & M. Geoffroy.* Paris: La Découverte.

Ali, A., & Wahabi, R. (1995). Managerial value systems in Morocco. *International Studies of Management and Organization, 25*(3), 87–102.

Alterman, J. (2000). Counting nodes and counting noses: Understanding new media in the Middle East. *The Middle East Journal, 54*(3), 355–363.

Arab Advisor Group. (2005). *Only 56% of Arab cellular operators provide the MMS service, while Morocco and Lebanon have the highest SMS rates in the Arab world.* Retrieved October 19, 2009, from http://www.arabadvisors.com/Pressers/presser-170105.htm

Arab Advisor Group. (2007). *Lebanon cellular users survey.* Retrieved October 19, 2009, from http://www.arabadvisors.com/Pressers/presser-161207.htm

Attiyah, H. S. (1993). *Roots of organization and management problems in Arab countries: Cultural or otherwise?* Paper presented at the Arab Management Conference, Bradford.

Baile, S., & Lefièvre, V. (2003, June). *Le succès de l'utilisation de la messagerie électronique: Etude de ses déterminants au sein d'une unité de production aéronautique.* The success of the use of the electronic message: Study of its determinants within a manufacturing unit. Paper presented at the 8th colloque de l'Association Information et Management (AIM). Paper presented at the 8th seminar of the Association of Information and Management, Grenoble.

Barakat, H. (1993). *The Arab world: Society, culture and state.* Berkeley, CA: University of CA Press.

Ben Fadhel, A. (1992). *La dynamique séquentielle culture gestion: Fondements théoriques et analyse empirique du cas tunisien.* Sequential dynamics of management/culture: Theoretical bases and empirical analysis of the Tunisian case. (Unpublished doctoral dissertation), University of Nice, France.

Ben Zina, S. (2002). *Le profil culturel de l'innovateur tunisien: Approche psychologique et comportementale. Cultural profile of the Tunisian innovator: Psychological and behavioral approaches. (Unpublished master's theses)*, University of Tunis. Tunisia: ISG.

Bouattour, S., & El Louadi, M. (2004, October). *Les aspects culturels de l'adoption des technologies de l'information et de la communication dans le monde arabe.* Cultural aspects of the adoption of information and communication technology in the Arab world. Paper presented at the 8th Conférence Internationale de Management des Réseaux (CIMRE), Paper presented at the 8th International Conference of Network Management, Hammamet, Tunisia.

Bouraoui, K., & Chastrette, M. (1999). Conceptions d'élèves et d'étudiants français et tunisiens sur la conduction dans les piles électrochimiques. Designs of French and Tunisian pupils and students on conduction in the electrochemical batteries. *Didaskalia, 14,* 39-60. Retrieved May 12, 2008, from http://documents.irevues.inist.fr/bitstream/handle/2042/23868/DIDASKALIA_1999_14_39.pdf;jsessionid=5E280612129F8D3A4B33ADF7711A4F65?sequence=1

Burger, M. J. (2009). Replicating Milgram: Would people still obey today? *The American Psychologist, 64*(1), 1–11. doi:10.1037/a0010932

Campbell, D. T., & Stanley, J. C. (1963). *Experimental and Quasi-Experimental Designs for Research.* Chicago: Rand McNally.

Checkfacebook (2009). http://www.checkfacebook.com. Retreieved February 11, 2009 from http://www.checkfacebook.com.

Cohen, A. (2007). One nation, many cultures: A cross-cultural study of the relationship between personal cultural values and commitment in the workplace to in-role performance and organizational citizenship behavior. *Cross-Cultural Research, 41,* 273–300.

Corm, G. (2007). *Histoire du Moyen-Orient: De l'antiquité à nos jours* [History of the Middle-East: From antiquity to our days]. Paris: La Découverte.

Credif. (2005) Centre de recherche, d'études, de documentation et d'informations sur la femme. *La femme tunisienne: Acteur de développement regional- approche Empowerment.* [The Tunisian woman: Actor of regional development- Empowerment approach. Desjeux, D. (2003). Les échelles d'observation de la culture. Scales of the culture's observation. *Communication, 22.* Retrieved April 17, 2009, from http://www.argonautes.fr/sections.php?op=viewarticle&artid=383

Desjeux, D. (2005). Usages et enjeux du SMS en Chine, en France et en Pologne. Uses and stakes of SMS in China, France and Poland. *Consommations & sociétés.* Retrieved October 13, 2009, from http://www.argonautes.fr/sections.php?op=viewarticle&artid=353

Dhaoui, H. (2000). *Pour une psychanalyse maghrébine: La personnalité.* For a Maghrebin psychoanalysis: The personality. Paris: L'Harmattan.

Donner, J. (2007). The rules of beeping: Exchanging messages via intentional missed calls on mobile phones. *Journal of Computer-Mediated Communication, 13*(1). Retrieved January 20, 2009, from http://jcmc.indiana.edu/vol13/issue1/donner.html

Dwairy, M., Achoui, M., Abouserie, R., Farah, A., Sakhleh, A., Fayad, M., & Khan, H. K. (2006). Parenting styles in Arab societies: A first cross-regional research study. *Journal of Cross-Cultural Psychology, 37*(3), 230–247. doi:10.1177/0022022106286922

El Louadi, M. (2004). Cultures et communication électronique dans le monde arabe. Cultures and electronic communication in the Arab world. *Revue Systèmes d'Information et Management, 9*(3), 117–143.

Elmandjra, M. (1990, May). *Future of the Islamic world. Future studies: Needs, facts and prospects.* Paper presented at the Future of the Islamic World Symposium, Algiers, Algeria.

Fitouri, C. (1991). L'acculturation, premier facteur du sous-développement. [Acculturation, the first factor of the underdevelopment]. *Série Sociologie, 17,* 47–57.

Fulk, J., Schmitz, J., & Steinfield, C. W. (1990). A social influence model of technology use. In Fulk, J., & Steinfield, C. (Eds.), *Organizations and Communication Technology* (pp. 117–142). Sage publications.

Gaines, S. O., Henderson, M. C., Kim, M., Gilstrap, S., Yi, J., & Rusbult, C. E. (2005). Cultural value orientation, internalized homophobia, and accommodation in romantic relationships. *Journal of Homosexuality, 50*(1), 97–117. doi:10.1300/J082v50n01_05

Geertz, C. (2000). *Interpretation of cultures: Selected essays.* New York: Basic Books.

Guessoum, N. (2006). Science et société dans le contexte arabe. *Science and society in the Arab context.* Retrieved November 13, 2009, from http://science-islam.net/breve.php3?id_breve=546&var_recherche=Guessoum&lang=fr

Hasan, H., & Ditsa, G. (1999). The Impact of culture on adoption of IT: An interpretive study. *Journal of Global Information Management, 7*(1), 5–15.

Hill, C. E., Loch, K. D., Straub, D. W., & El-Sheshai, K. (1998). A qualitative assessment of Arab culture and information technology transfer. *Journal of Global Information Management, 6*(3), 29–38.

Hofstede, G. (1980). *Culture's consequences: International differences in work-related values.* Beverly Hills, CA: Sage.

Hofstede, G., & Hofstede, G. J. (2005). *Cultures and organizations, software of the mind: Intercultural cooperation and its importance for survival.* New York: McGraw-Hill.

Ibn Khaldoun, A. (1863). *Les prolégomènes.* An introduction to history. Translated in French by W. Mac Guckin & B. De Slane. Paris: Librairie orientaliste Paul Geuthner, 1936.

ITU. (2009). *Measuring the information society -The ICT development Index.* New York: International Telecommunication Union.

Kriem, M. S. (2009). Mobile telephony in Morocco: A changing sociality. *Media Culture & Society, 31*(4), 617–632. doi:10.1177/0163443709335729

Lassoued, K. (2005). *Technologies de l'information et culture nationale: Cas de la Tunisie.* Information technology and national culture: The Tunisian case. Retrieved December 12, 2008, from http://medforist.ensias.ma/Contenus/Conference%20Tunisia%20IEBC%202005/papers/June25/43.pdf

Lynn, M., & Gelb, B. D. (1996). Identifying innovative national markets for technical consumer goods. *International Marketing Review, 13,* 43–57. doi:10.1108/02651339610151917

Madar Research Group. (2008). 2007 Arab ICT use Index Growth. *Madar Research Journal, 5*(6). Retrieved April 14, 2009, from http://www.madarresearch.com/

Matsumato, D. (2006). Culture and cultural worldviews: Do verbal descriptions about culture reflect anything other than verbal descriptions of culture? *Culture and Psychology, 12*(1), 33–62. doi:10.1177/1354067X06061592

Milgram, S. (1974). *Obedience to Authority: An Experimental View.* New York: Harper & Row.

Minsky, B. D., & Marin, D. B. (1999). Why faculty members use E-mail: The role of individual differences in channel choice. *Journal of Business Communication, 3*(2), 194–217. doi:10.1177/002194369903600204

Morrisson, C., & Friedrich, S. (2004). La condition des femmes en Inde, Kenya, Soudan et Tunisie. The condition of the women in India, Kenya, Sudan and Tunisia. *Centre de développement de l'OCDE,* document 235. Retrieved September 19, 2005, from http://www.oecd.org/dataoecd/41/48/33645983.pdf

Oyserman, D. (1993). The lens of personhood: Viewing the self and others in a multicultural society. *Journal of Personality and Social Psychology, 65,* 993–1009. doi:10.1037/0022-3514.65.5.993

Realo, A., Allik, J., & Vadi, M. (1997). The hierarchical structure of collectivism. *Journal of Research in Personality, 31*(1), 93–116. doi:10.1006/jrpe.1997.2170

Rhee, E., Uleman, J. S., & Lee, H. K. (1996). Variations in collectivism and individualism by in-groups and culture: Confirmatory factor analysis. *Journal of Personality and Social Psychology, 71*(5), 1037–1054. doi:10.1037/0022-3514.71.5.1037

Scouarnec, A., & Silva, F. (2008). Les pratiques de management en Euro Méditerranée: Proposition d'un cadre d'analyse. Practices of management in Euro Mediterranean: Proposal of an analysis framework. Retrieved April 12, 2009, from www.reims-ms.fr/agrh/docs/actes-agrh/pdf-des-actes/2008scouarnec-silva.pdf

Sharabi, H. (1988). *Neo patriarchy: A theory of distorted change in Arab society.* Oxford, UK: Oxford University Press.

Shariati, A. (1971). *Fatemeh Fatemeh Ast. Fatimah is Fatimah.* Tehran: Husseinieh Ershad.

Slater, M., Antley, A., Davison, A., Swapp, D., Guger, C., Barker, C., et al. (2006). A virtual reprise of the Stanley Milgram obedience experiments. *PLoS ONE, 1*(1): e39. Retrieved July 17, 2009, from http://www.plosone.org/article/info:doi%2F10.1371%2Fjournal.pone.0000039

Smida, A., & Latiri, R. (2004, October). *L'attitude du manager tunisien face à l'avenir*. The attitude of the Tunisian manager toward future. Paper presented at CIDEGEF colloque, University Saint-Joseph, Beirut.

Soyah, T., & Magroun, W. (2004, October). *Influence du contexte culturel tunisien sur l'orientation des systèmes d'information des banques tunisiennes*. Influence of the Tunisian cultural's context on the orientation of the information systems of the Tunisian banks. Paper presented at CIDEGEF colloque, University Saint-Joseph, Beirut.

Straub, D. W., Loch, K. D., & Hill, C. E. (2001). Transfer of information technology to the Arab world: A test of cultural influence modelling. *Journal of Global Information Management, 9*, 6–28.

Tassin, E. (1994). Pouvoir, autorité et violence: La critique arendtienne de la domination. Power, authority and violence: Criticism of the Arendt's domination. In Goddard, J. C., & Mabille, B. (Eds.), *Le Pouvoir* (pp. 266–301). Paris: Vrin.

Todd, E. (1983). *La troisième planète: Structures familiales et systèmes idéologiques. The third planet: Family structures and ideological systems.* Paris: Le Seuil.

Triandis, H. C. (1989). The self and social behavior in different cultural context. *Psychological Review, 96*, 506–520. doi:10.1037/0033-295X.96.3.506

Triandis, H. C. (2001). Individualism-Collectivism and personality. *Journal of Personality, 69*, 907–924. doi:10.1111/1467-6494.696169

UNDP. (2003). *Arab Human Development Report 2003. Building a Knowledge Society*. United Nations Development Programme.

UNDP. (2004). *Arab Human Development Report 2004. Towards Freedom in the Arab World*. United Nations Development Programme.

UNDP. (2005). *Arab Human Development Report 2005. Towards the Rise of Women in the Arab World*. United Nations Development Programme.

UNESCO. (2005). *Unesco science report 2005*. Paris: UNESCO Publishing.

Wilson, P. R. (1968). Perceptual distortion of height as a function of ascribed academic status. *The Journal of Social Psychology, 74*, 97–102. doi:10.1080/00224545.1968.9919806

Yamaguchi, S. (1994). Collectivism among the Japanese: A perspective from the self. In U. Kim, H. C. Triandis, C. Kagitcibasi, S. C., Choi, G. & Yoon (Eds.), *Individualism and Collectivism: Theory, method, and applications* (pp.175-188). Thousand Oaks, CA: Sage.

Zghal, R. (2002). Acquis sociaux de la femme tunisienne et inerties culturelles et institutionnelles: Education, emploi et attitudes envers le statut social de la femme. Social rights of the Tunisian woman and cultural and institutional inertias: Education, employment and attitudes towards the social status of the woman. In Nations Unies (Ed.), *Disparités entre femmes et hommes et culture en Afrique du Nord* (pp. 41-55). Centre de Développement Sous-régional pour l'Afrique du Nord.

Zlitni, S. (2003). Utilisation des NTIC dans les entreprises: Au-delà des usages professionnels. [Use of the new ICT in the companies: Beyond the professional uses]. *Revue Internationale de sociologie et de sciences sociales, Esprit critique, 5*(4). Retrieved May 20, 2006, from http://www.vcampus.univ-perp.fr/espritcritique/0504/esp0504article02.html

Zogby, J. (2003). Why do they hate us? *The Link, 36*(4), 2–13.

ADDITIONAL READING

Al-Krenawi, A., & Graham, J. R. (2003). Principles of social work practice in the Muslim Arab world. *Arab Studies Quarterly*, *25*(4), 75–91.

El Louadi, M. (2008). The Arab world, culture and information technology. In Van Slyke, C. (Ed.), *Communication technologies: Concepts, methodologies, tools and applications* (pp. 151–159). Hershey, PA: Information Science Reference.

El Louadi, M., & Everard, A. (2004, August). *Information technology and the Arab world: A question of culture*. Paper presented at the Americas Conference on Information Systems (AMCIS 2004), New York.

Kloeber, A. L., & Kluckholm, C. (1952). *Culture: A Critical Review of Concepts and Definitions*. New York: Vintage.

Muna, F. (1980). *The Arab Executive*. New York: Mac-Millan.

ENDNOTES

[1] This is a poem verse of Ahmed Chawki in which he venerated the instructor (translated by the author).

[2] Islam is the dominant religion in the Arab world and many Arab countries still maintain strong Islamic values such as Tunisia which is supporting a high level of religiosity (El Akremi et al., 2007). But, even other religions as Judaism or Christianity share some principles with Islam like the respect of the hierarchy (Mullah, Rabbis, Papacy).

[3] In this experiment, the researcher orders the teacher (the participant) to give increasing electric shocks (simulated) to a learner (a confederate) whenever he/she makes a mistake, when associating a pair of words.

[4] Beeping is one form of communication used in developing countries (Donner, 2007). It is usually used to surmount the illiteracy to write SMS or to overcome the cost of communication (Kriem, 2009).

[5] This choice can be explained by the fact that although the image of women work is generally positive and more and more accepted (Zghal, 2002) (for example, women account for 40% of the instructors (Credif, 2005)), some Tunisians seem to be still reticent to admit the exercise of authority by women (Morrisson & Friedrich, 2004). The UNDP Report (2005) underlines a gap between the real transformation of the Arab woman status and her stereotypical images in popular proverbs or in school curriculum, which preserve the patriarchal system. The report adds that the growth of fundamentalist trends and the rise of the influence of conservative channels -particularly in the Arab Gulf public- consolidate the inferiority of the women and emphasise the gender hierarchy.

[6] Slater et al. (2006) have shown that, even in immersive virtual environment, people tend to response to events in a realistic manner and subordinate to authority. On this basis, we suppose that the presence of the instructor's photo will help the participant to get involved into our experiment.

[7] The UNDP (2003) reports that Arabs distinguish consuming knowledge of producing it. So, they import knowledge from foreign countries. Then, we suppose the existence of Westerners' veneration, especially for Americans considered as being the first force, and more trust for their opinion.

[8] We asked the respondent to propose a different probability for each technology. This request will motivate the participant to make a choice.

[9] Called also the Pygmalion effect; for researchers, it consists of unconsciously emitting some indices that push participants to give them the answers that they wish to have.

[10] Before being submitted to the participants, the pre-questionnaire was tested and some modifications were given to make questions easier to understand.

Chapter 3
Embracing ICT by the Jordanian Education System

Atef Abuhmaid
Middle East University, Jordan

ABSTRACT

This chapter discusses the experience of adopting information and communication technologies (ICT) by the Jordanian education system. The discussion tackles the Jordanian Ministry of Education's endeavour to update the education system for the Knowledge Economy by shedding light on its two main recent initiatives - Education Reform for the Knowledge Economy (ERfKE) and the Jordanian Education Initiative (JEI). Both projects included ICT as a key component for upgrading the education system for the Information Age. Jordan's location in the heart of the Middle East as well as the countries scarce resources available for development makes education reform a challenging task for the education system. The chapter discusses the Jordanian education system's approach in reforming itself through the adoption of ICT.

THE JORDANIAN MINISTRY OF EDUCATION: BACKGROUND

The Jordanian Ministry of Education is the central education authority represented by 36 regional directorates of education across Jordan (a small country located in the Middle East). The Ministry plays a crucial role in the society as the country is considered a young society with 38 percent of the population are estimated to be under 15 years of age. Several achievements of the Jordanian education system bear witness to the high priority placed on education in the country.

After independence, the 1950s period witnessed a shift towards *education for all* which was reinforced by the *Knowledge Act No.20* making education compulsory for all children up to grade six. In order to further extend educational provision across the country, the *Education Act No.16* was endorsed in 1964 making education compulsory up to grade nine. Therefore, education provision has expanded rapidly across the country. For

DOI: 10.4018/978-1-60960-048-8.ch003

instance, the number of governmental schools increased from 714 in 1960 to 5,526 schools in 2006 with 1,056,470 students and 58,886 teachers (Ministry of Education, 2007). In addition, the budget of the Ministry of Education reached 12 percent of Jordan's General Budget in 2006 (Ministry of Education, 2006). Currently, both primary and secondary education are provided tuition free for all students across the country. As a result, the literacy rate in Jordan has become one of the highest in the Arab world with 91 percent of the population over 15 years of age can read and write.

The Need for Educational Change

Human capital represents Jordan's main asset. The scarcity of natural resources in Jordan and its weak industry (Al-Sa'd, 2007) made the successive governments of Jordan focus on education as a main driver for human development. As a result, the educational system's graduates have undertaken outstanding role in the development of the Jordanian society as well as other countries in the region due to the good reputation. Therefore, despite the fact that Jordan is a small country with limited natural resources, the education system of Jordan compares favourably with other systems in the Middle East. Accordingly, remittances from the Jordanian workforce abroad are major sources for foreign exchange (Mu'tamen, 2007).

However, global markets have undertaken major transformations placing increasing pressure on education systems worldwide to respond and to undertake reform accordingly in order to be able to survive and compete in the new era. There has been a warning that despite the large spending on education by education systems in the Middle East and North Africa (MENA) region students in the region are not being prepared to participate effectively in the global market (Akkari, 2004; Ministry of Education, 2004; World Bank, 2002; World Bank, 2003). For example, according to Akkari (2004), Arab countries spend over 5 percent of their Gross National Product on education; the highest percentage among all developing countries. Furthermore, despite impressive signs of improving education in term of enrolment and access to public education, *quality* is still a concern. For example, in 1999, Jordan, Tunisia and Morocco participated from the MENA region in the Third International Mathematics and Science Study (TIMSS), the test showed that Jordan was "near the bottom in math and science" compared with other participating countries (World Bank, 1999, p.12) and the *Programme for International Student Assessment* (PISA) report in 2000 reported that Jordan was "statistically significantly below the OECD average" on its science scale (OECD, 2006, p.22). Furthermore, a national assessment found that Jordanian students did not meet the educational system's learning objectives in Arabic, math and science (World Bank, 1999).

The Education System and the Sudden Increase of Student Numbers

The Middle East region is turbulent and frequently shaken by conflicts and wars which have disturbed development and reform across the region. However, the regional disturbances have had severe implications on development projects in Jordan as a small country located in the heart of the region and relies heavily on external assistance in undertaking reform. The unexpected disturbances caused by external, as well as internal, conflicts have frequently shaped reforms in Jordan. Shortly after gaining independence in 1946, Jordan bore the full brunt of the creation of Israel on its western borders engaging in military conflicts and through dealing with large numbers of Palestinian refugees (Chatelard, 2004; Saleh, 1991). The 1967 war with Israel created a wave of an estimated 350,000 displaced Palestinians seeking refuge in Jordan (Khawaja, 2003). Furthermore, other conflicts in the region including the Gulf War I and II and their consequences impacted development projects within the country.

As a result of military conflicts in the region the population of Jordan increased dramatically posing serious challenges for development and reform projects in the country. For instance, a mass influx of refugees and displacement policies accompanied the 1990-1991 Gulf War increased the population of Jordan from 2.8 million in 1987 to 4.2 million in 1995 (Talal, 1998). An estimated one million people arrived to Jordan voluntarily or as a result of displacement policies from countries on each side of the conflict (Hear, 1995). Likewise, as a result of the 2003 Gulf war, approximately 750,000 Iraqis settled in Jordan straining the country's infrastructure (Gavlak, 2007) and adding to the already mounting pressure on the country's economy.

The sudden growth of the population has posed a major challenge to the country and to the educational system in particular. Frequently, the Jordanian government has voiced concern about the pressure of the refugees issue on the country's limited resources especially on the health-care and education systems (Gavlak, 2007). The 1991 Gulf War "created a surge in student numbers; currency pressures; increased budgetary; constraints; and delays in construction, technical assistance, and the delivery of supplies" (World Bank, 2000, p.3). Typically, in the face of a new wave of refugees resources are diverted to where they are mostly needed and in this case to providing the refugees with the basics. In the educational context, when the system faces a sudden increase of its student population it is expected to act swiftly to provide students with classrooms, textbooks, desks and other basics. Understandably, the diversion of resources and efforts to accommodate the unexpected situation comes at the cost of the planned development and reform projects which was the case with the 1990s reform projects aimed to update the education system through the integration of ICT as the reform fell short of its rhetoric (Hasan, 2000). Similarly, the aftermath of the

2003 Gulf war diverted some aspects of reform projects which focused on updating the education system for the knowledge-based economy. At the beginning of the schooling year 2007/2008, the education system received 50,000 Iraqi students mainly in the already crowded schools in Amman, the capital (Alkhawaja, 2007).

Additionally, internal student immigration from private to public education due to the increasing economic pressure adds to the mounting pressure on the education system to accommodate unexpected waves of students. For instance, 7,000 students moved unexpectedly from private education to public education during the 2007/2008 schooling year (Alkhawaja, 2007). The significance of this number can be recognized considering that 20 percent of Jordanian students study in overcrowded schools (World Bank, 2003) and the planning for the schooling year had not considered this extra number (Abuhmaid, 2009).

Moreover, Jordan relies heavily on international aid especially in undertaking development and reform. Therefore, any disturbance happens to the external sources of funding development projects directly impacts the projects on the ground. For instance, the year 1982 witnessed a sudden drop in oil prices critically reducing aid received by Jordan from the Gulf States considering that foreign aid comprised 40 percent of Jordan's state revenue in 1980 (Moore, 2004). Furthermore, as a result of the two Gulf Wars, the country faced economic problems which hampered development projects caused basically by a decline in foreign aid and remittances (Abuhmaid, 2009; Moore, 2004).

ICT-Related Initiatives in Jordan

Over the last three decades, the education system of Jordan had several ICT-related initiatives. In 1987, the *National Conference for Educational Development* became a milestone for Jordan's education reform. In particular, the Conference highlighted the need for integrating computers across the education system (Jaradat, 1992). Subsequently, the Directorate of Educational Computing was established in 1988. Nevertheless, the lack of equipment, the high cost involved, and the lack of trained teachers were obstacles to integrating ICT across the education system (Tawalbeh, 2001). In addition, the reform efforts at that stage focused primarily on providing schools with computers, and minimal attention was paid to other crucial factors that are believed to be essential for the success of ICT integration, such as upgrading, maintenance, or training in schools. Instead, the integration of computers across the education system targeted two main areas (Jaradat, 1992):

- Improving access to technologies.
- Introduce computers as learning aids and target skills required by students to benefit from computers.

Furthermore, the unexpected mass return of Jordanians after the 1990-1991 Gulf War as well as a new wave of refugees (Hear, 1995; Seijaparova & Pellekaan, 2004) disturbed the initial initiatives for ICT adoption by the education system as significant resources had to be redirected to accommodate the sudden increase of student numbers. During the 1990s, schools continued to receive computer labs although their use was confined to computer subjects.

The turning of the 21st century witnessed concrete steps taken to incorporate ICT throughout the Jordanian education system. His Majesty King Abdullah II shortly after he ascended to the throne in 1999 envisioned that "Jordan will become an IT hub for the region." Consequently, the Ministry of Education committed itself to reforming the education system envisioning that "the Ministry of Education will be the single largest user of IT in Jordan" (Ministry of Education, 2001, p.3). In addition, Jordan's e-strategy was launched in 2000 drawing on data from other countries in order to identify the country's comparative strengths and weaknesses. Nevertheless, according to the World Bank (2005), the e-strategy was weak in implementation, monitoring and evaluation so,

it did not achieve its goals. Hence, the e-strategy was updated in 2004 to become more realistic by identifying achievable goals; subsequently, the initial goals were lowered substantially.

In 2002, the Ministry of Education developed its "Vision Form for the Future of Education in Jordan" setting the scene for education reform for the 21st century (World Bank, 2003). Concerned with the access issue, the Ministry focused on improving the ratio of students to computers to become 8/1 (Heresh & Twal, 2002; World Bank, 2003) as well as connecting schools countrywide to the Internet and intranet. During the academic year 2002-2003, the Administration of Curricula and School Textbooks within the Ministry was preparing new curricula in which ICT was being integrated and new software for Arabic and English was developed (Ministry of Education, 2004). In 2001, Queen Rania Al-Abdallah Center (QRC) was established in order to develop and sustain education in Jordan through using modern technologies. The Center's goal was to support e-learning and teaching practices by providing the education system with all the necessary resources for students, teachers and schools. Remarkably, the Center has been a stimulus to ICT infusion and management across the education system. Moreover, in 2003 His Majesty the King stressed that Jordan was moving in the direction of reforming the educational system and that he had been "closely following this initiative (the e-learning strategy) … simply because it is the single most important element in our modernisation drive" and that the modernisation of the education system was not "just for Jordan, if we are successful here, we'll be doing it for the rest of the Middle East" (Qadamani, 2003).

REFORM FOR THE KNOWLEDGE-BASED ECONOMY

Realizing the pressing needs for change, the Ministry of Education initiated two major ICT-enriched initiatives aimed to integrate ICT across the education system: the Education Reform for the Knowledge Economy (ERfKE) and Jordan Education Initiative (JEI). The two initiatives are detailed and discussed next.

Education Reform for the Knowledge Economy (ERfKE)

In 2003, the Ministry of Education launched an ambitious large-scale educational reform named Education Reform for the Knowledge Economy (ERfKE). The initiative is the largest reform project ever funded by the World Bank (Cisco Learning Institute, 2004). The aim of the project was to create a comprehensive transformation of the educational system in order to equip students with knowledge, skills, attitudes, and competences required for the knowledge-based economy (National Center for Human Resources Development, 2005). The project identified four objectives for the reform:

a. Reorienting education policy objectives and strategy through governance and administrative reform.
b. Transforming education programmes and practices for the knowledge-based economy by:
    ◦ Preparing curriculum and assessments for the knowledge economy.
    ◦ Providing professional development for personnel.
    ◦ Providing required resources to support effective learning.
c. Supporting provision of quality physical learning environments.
d. Promoting readiness for learning through Early Childhood Education.

Several ICT professional development programmes (e.g. International Computer Driving Licence (ICDL), Intel Teach to the Future, World Links, and Thinking Tools) were adopted in order to prepare teachers to deal with ICT. 60,000 teach-ers received training under the ICDL, 4,500 in the Intel Programme and 2,570 in the World Links. In addition, resources were secured to providing schools with computer labs and connection to the Internet and Intranet. Furthermore, considerable efforts and resources focused on developing complementary e-contents. Accordingly, the Ministry launched its electronic portal on the Internet, the EduWave, in order to provide teachers and students with e-courses and activities that support teaching and learning. Mentors were also mobilised to support teachers and provide them with assistance and guidance while using ICT.

Jordan Education Initiative (JEI)

In order to provide ERfKE with a rapid prototype model for discovering best practices and lessons learned during its implementation in Jordanian schools, the Ministry of Education adopted the JEI and the Discovery Schools initiatives (McKinsey & Company, 2005). The JEI is an ICT-enriched initiative functioned as an "accelerator" for ERfKE by concentrating on the performance of Discovery Schools before ideas and programmes were to be rolled out nationally under the ERfKE. One hundred Discovery Schools located in Amman, the capital of Jordan, were the main focus of integrating ICT in Jordanian schools.

The objectives of the JEI were organized in three tracks:

• **Track 1**: New approaches to learning and teaching. This track incorporated three sub-tracks which focused on ICT integration in schools through:
    ◦ In-classroom Technology, which aimed to infuse ICT in everyday practices by providing teachers with training, ICT tools (e.g. laptops and projectors) and internet connection (Cisco Learning Institute, 2004).
    ◦ e-Curricula development, which aimed at developing rich digital con-

tent in new outcomes-based curricula from Grades 1-12 in multiple subjects (McKinsey & Company, 2005, p.8).

- ○ Training, which aimed to develop and provide ICT professional development covering basic ICT skills, pedagogical use of ICT, adoption of new e-Curricula, and change management and leadership.
- **Track 2**: Lifelong Learning.
- **Track 3**: ICT industry Development.

REFORMING EDUCATION THROUGH PARTNERSHIP

For developing countries, like Jordan, initiating and implementing large-scale reform projects is an immense task due to the scarcity of resources and limited capacity within these countries for undertaking costly projects. Therefore, developing countries usually seek financial assistance and expertise from external international donors and lending agencies.

Due to the high cost of education provision, scarcity of resources, and the growing numbers of students, the Education system of Jordan has built partnership with external partners at both local and international. Partnership has been in place between the education system with the World Bank, UNICEIF as well as other international organizations since mid 1980. For example, the Seventh Education Project in 1988 was funded by international partners as the World Bank financed school construction while the European Union provided $2.5 million to support technical assistance and fellowship program (World Bank, 2000). In addition, in 1990s, the Ministry of Education implemented a project called the Global Education Program, which was offered and sponsored by UNICEF. Additionally, in partnership with the World Bank, Japan Overseas Economic Cooperation Fund and other international organizations, the Ministry of Education implemented *The First and*

Second Human Resources Development Sector Investment Loan I and II (HRDSIL II and HRDSIL II) projects between 1990 and 2000. Most recently, in 2003, the World Bank was the major partner, along with other international organizations, for the education system in implementing the Education Reform for the Knowledge Economy project (ERfKE). Figures show that foreign aid constitutes a great deal of the total spending on education in Jordan. For example, in 2005 the contribution of Arab and international organizations was as high as 51 percent of the total spending on education in Jordan (Ministry of Education, 2007).

Conspicuously, since the end of the colonial era large international private companies mainly from developed countries often penetrated local markets through local private affiliates or under aid proposed to economically poorer nations. A growing number of large international enterprises have established development or aid affiliates, which act as a means for these enterprises to penetrate new markets through aid projects claiming greater access to developing countries (e.g. Cisco Systems has established *Cisco Institute* and Intel has established *Intel Education Initiative*). This new trend of involving the affiliates might project a new strategy in expanding businesses and influences globally under the cover of aid and development.

Recently, "public-private" partnership has been added as another dimension to implementing ICT-related initiatives in the Jordanian education system. This type of partnership was stressed by the UN Millennium Development Goals of the 2000 citing building public-private global partnership in order to tackle development problems in developing countries and to make available the benefits of ICT to these countries (United Nations, 2005). Understandably, the limited capacity of the Jordanian local private sector along with the immense task of reforming education through ICT made it inevitable for the education system to seek assistance from the international private sector. The involvement of the international private sector in ERfKE and the JEI was substantial in training, financing, and the development of digital materials.

International institutions have launched numerous aid programs in order to support technological development in poor nations, however, the motivation behind such programs has been debated. While developed countries argue that such aid is motivated by moral values aiming to assist poorer countries to prosper, participate and benefit from the knowledge-based economy (UNDP, 1999), critics are sceptical, arguing that in practice this has often served to benefit the industrial world (Castells, 1999; Forman, 1995; Norris, 2001). According to Fischer (2006) it is "simplistic to maintain that rich states offer generous and compassionate aid devoid of mercenary ulterior motives" (p.177). Wilson (2004) argued that the reason behind developed countries' concern about developing countries' embrace of ICT is commercial, political, and social. In this sense, developing countries are increasingly becoming large markets for Western industries, and a significant part of the global market. Chossudovsky (2003) explains what she calls *"policing countries through loan conditionalities"* saying that international financial institutions oblige developing countries through the so-called conditionalities attached to loan agreements which can "appropriately redirect their [developing countries] macro-economic policy in accordance with the interests of the official and commercial creditors." (p.35)

Partnership in ERfKE and the JEI

The education system of Jordan, as many other education systems in the developing world, relies heavily on international partners in financing reform projects. For phase one of ERfKE, which was implemented between 2003 and 2008, resources were secured through partnership between the Jordanian Ministry of Education and external partners; such as: The World Bank, Canadian International Development Agency (CIDA), USAID, UNICEF, National Council for Family Affairs (NCFA), Japan International Cooperation Agency (JICA), Arab Fund for Economic and Social, and European Investment Bank (EIB). Foreign contribution of in the ERfKE was substantial as the World Bank alone provided 32 percent of the total spending on the project, and the government of Jordan sharing constituting 35 percent as it is clear from Table 1.

The *motto* of the JEI was "a new model of public-private partnership" stressing the role of private sector in public education reform projects. Apparently, the JEI, along with the ERfKE, show substantial shift toward direct involvement of the private sector in public education reform in Jordan. Similar to the ERfKE, the share of both local and global private sectors in financing the JEI was 56 percent and 32 percent provided by international donors compared with only 11 percent the share of the Jordanian government (McKinsey & Company, 2005). The digital materials for mathematics were developed by Cisco Learning Institute

Table 1. Proposed Education Spending Under ERfKE (US$), (Adapted from: World Bank, 2003, p.45). (IBRD = International Bank for Reconstruction and Development, the World Bank)

	2003	2004	2005	2006	2007	2008	Total
Total Financing Required US$M							
IBRD	5	20	30	28	27	10	120
Government	13	41	24	22	20	10	130
Others	-	-	32	35	34	19	120
Total Program Financing	18	61	86	85	81	39	370

and Rubicon, a Jordanian local private company and sponsored by Cisco Systems. Twenty-two partners from the private sector and seven NGOs participated in the JEI illustrating the diversity of partners from multiple sectors involved in the project (see Table 2).

THE EFFECT OF EXTERNAL PARTNERSHIPS IN EDUCATION REFORM

ERfKE and the JEI, as well as earlier education reform initiatives in Jordan, illustrate the country's heavy reliance on international partners in undertaking education reforms especially ICT-related initiatives. While the diversity can enrich reform by providing expertise and resources, sustainability and the quality of the reform might be compromised due to the involvement of diverse partners.

Sustainability of ICT Integration

Clearly, integrating ICT into education systems is a long process starts by providing schools with technologies. However, while this step is a prerequisite for all the following steps, it requires other supporting factors in order to reap the benefits of ICT integration including maintenance, technical support, training, upgrade, new software, and replacement of old technologies. For instance, no matter how advanced the computer is, it might be useless if the software or hardware were inadequate. Likewise, software frequently

Table 2. Key partners in the Jordan Education Initiative. (Based on McKinsey & Company, 2005)

	Private sector	**Public Sector**			
		Governmental	**NGOs/Foundations**	**Academics**	
Global	Cisco Systems Computer Associates Corning Cable Systems Dell DHL France Telecom / Jordan Telecom HP Intel Microsoft Pearsons Vector Capital / CommercialWare	US government – USAID – MEPI (US State Dept.) British Council North Virginia Technology Council	Cisco Learning Institute Krach Family Foundation Reuters Digital Vision Foundation	University of Keele	
					19
Local	Aramex e-Dimensions Fastlink Intaj ITG Menhaj MobileCom Razorview Rubicon STS Synttex	Jordanian government – Ministry of Education – Ministry of ICT – Ministry of Planning	Int@j Young Entrepreneurs Association of Jordan NetcorpsJordan		
	22	4	7	1	15
Total					34

becomes obsolete requiring replacement. Thus, ICT integration is a dynamic process and whenever it becomes static it loses one of its key features and becomes outdated. In this sense, sustaining the initial efforts in ICT integration is an integral part in its success.

The dilemma of implementing ICT projects in developing countries, which rely heavily on external aid, is over emphasizing the launch stage of the projects with less attention paid to the sustainability issue. Usually, making a decision regarding ICT-related reform is bonded to the consent of the external aid partners to support and fund the project. However, any disruption or discontinuation of the external fund impacts the whole reform project leaving the initial investment in disarray. Remarkably, the Jordanian education system has been successful in attracting a great deal of international support for its projects aiming to update itself for the knowledge-based economy. Many partners supported the Ministry's plans for reforming the education system through supporting ERfKE and the JEI (see Table 1 and Table 2). Clearly, the wide range involvement of these partners in both projects illustrate the country's success in promoting national reform projects and securing international funding critically needed for undertaking national projects. The involvement of aid partners in ERfKE and the JEI are mainly involved under the umbrella of 'aid.' Therefore, the Jordanian education system had to agree to the *conditionalities* attached to the agreements with aid partners. For instance, the Ministry had to reduce the proposed time for ERfKE from 10 to 5 years, because the World Bank did not consent to the 10-year deadline for the project completion. In addition, dispute emerged between the Ministry of Education and the private company, which was responsible for developing the digital materials, during the implementation of the e-content regarding the copyright of these materials. The company wanted to retain the copyright of the materials in order to distribute them in other countries. Another dispute emerged around the

responsible party for updating the digital materials (Abuhmaid, 2009) repeatedly causing delay during the implementation process. Obviously, the education system did not have complete control over the timing and conditions in which external partners became involved in its reform projects.

Moreover, education systems in developing countries are warned of the "shaver-and-blades" strategy while adopting ICT-related education initiatives. In this strategy, while the hardware ("shaver") is sold to these systems at a relatively low price, or even provided free, by companies in order to capture market share, and then the software ("blades") is sold at a high price. Globalisation has paved the way for large companies to expand and to find new markets, and therefore they might use such strategy to gain ground in new markets. Understandably, private companies may find it profitable to fund and assist projects that can be financially promising in the short or long term. For instance, a computer company is likely to be interested in supporting ICT integration into schools, as these computers and the related technologies will continuously need upgrade, maintenance, replacement and software. Therefore, if a company provides a given educational context with computers (the shaver) it can capture a share in the recurrent spending on the provided machines through maintenance and replacement.

Quality of ICT Integration

The fragile nature of education landscape necessitates extra caution when attempts are made to build public-private partnerships. In a large and complex reform projects, profitable companies are willing to provide financial assistance to parts of the reform which serve their interests. However, supporting certain elements in the reform might create imbalance within ICT-related reforms, such as ERfKE and the JEI. The imbalance can occur when support is directed towards accelerating certain components of the reform project while others remain lagging behind because they do not match the interests of external partners. Consequently, the success of the whole reform could be at stake considering the importance of the holistic approach in education reform. Therefore, balance has to be struck between the agendas of the education system and those of the external partners in order not to skew the reform towards what benefits profitable companies while moving at a snail's pace in areas that fail to excite aid partners.

As indicated earlier, Cisco Systems along with its local partner, Rubicon, developed the e-contents for mathematics. However, according to McKinsey and Company (2005), this investment was aligned with Cisco Systems' long-term commercial interest which was motivated by the opportunity to develop e-content to be useable and trialed within a real school context. In this way, other education systems, especially those of other Arab countries, might then be convinced to adopt the developed 'product' as it would have been already trialed in a similar education system (Abuhmaid, 2009; McKinsey & Company, 2005). Similarly, in 2003 as a result of public-private partnership between the Ministry of Education, Menhaj, and JICA bilingual digital material was developed in English and Arabic, which might suggest that the material was developed for the Jordanian education system with an eye on the larger regional and international markets.

Clearly, the involvement of different kinds of donating partners can impact the effectiveness of education reform projects. Inevitably, external sources of funding national reform projects have never been value-free; rather, they are politically or commercially bonded. Usually, the range of partners (e.g. politicians, development bureaucracies, none government organizations, implementers and participants) means that various objectives are sought (Appleby, 2005). While the announced objective of aid is development, other unannounced objectives might be designed to support the political, strategic, military, and commercial objectives of donor countries and lending agencies (Escobar, 2004). Thus, the reliance on external

aid for funding major national reform projects including ICT integration with scarce resources and limited expertise has its implications to the ICT reform projects in Jordan. For instance, Jordan's political standpoint in the 1990/1991 Gulf War had implications for the country's relationships with other countries held different political stances regarding the war. Consequently, aid from these countries was critically affected, leaving ICT integration initiatives, as well as other development projects in the country, at that stage in disarray.

Built-In Capacity of the Educational System

The built-in capacity within an organization is a decisive factor in the success or failure of implementing new initiatives. This factor is crucial to be considered while adopting and implementing reform projects in order to avoid superficial reform. Therefore, organizations are warned of adopting reform projects which are beyond their own capacities, either individually or organizationally, to put into practice (Fullan, 2001). While building capacity was addressed in the initial planning for the ERfKE project, the implementation was not within the scope of the Ministry of Education either in terms of quality or quantity. During the implementation phase of the ERfKE, the Ministry of Education found itself under overwhelming pressure to hold all elements of the reform together and to maintain its holistic nature. However, the limited capacity of the education system diverted efforts and attention to isolated achievements as efforts had to be shifted several times from one component of the reform to another in order to contend with pressures and deadlines. For example, after the e-contents were developed and became available online, the Ministry realised that teachers were not ready or prepared to utilise them. Accordingly, training programs for teachers were accelerated and new ones were introduced especially programs that deal with the pedagogical utilization of ICT.

Nevertheless, the acceleration of training came at the cost of other components of the project especially follow-up and on-going support by mentors. That is, during the first year of implementing the ERfKE and the JEI, especially in the Discovery Schools, mentors used to visit teachers in these schools daily or four times a week, however, these visits dropped down dramatically afterwards. As mentors were shifted to work on developing new curricula, their visits to schools became barely once or twice a year (Abuhmaid, 2009).

Moreover, the launch and implementation phase of ERfKE and JEI experienced delay. For example, there was a delay caused by the failure of private companies to comply with deadlines for developing e-contents, the dispute over the copyright of the e-contents, and the responsible party for upgrading the e-contents in the future. Additionally, the complexity of the implementation phase and lack of resources slowed the deployment of in-classroom technologies as part of the JEI. According to McKinsey and Company (2005), out of the hundred Discovery Schools only 13 schools had received Math labs equipment, laptops, projectors, wireless devices and connectivity to the Internet, which clearly illustrates the gap between what had been planned and the actual achievement during the implementation stage.

Clearly, the private sector moves at a relatively faster pace than the public sector institutions. In a highly centralized system, like the education system of Jordan, changes and development usually advance slowly, therefore, the system might find it a difficult task to adapt to a fast paced and complex reform. Rondinelli *et al.* (1990) stressed that implementing complex projects poses serious challenges to developing countries. While the need for rapid change might pressure education systems for reform, the implementation of overly ambitious projects creates equally serious problems, as the education system has to shift attention continuously from one component to another. In some cases, involving aid in educational reform would create extra pressure on the education system as aid partners have their deadlines and expectations from the reform which overloads the system and might fragment the reform which are the main enemies of large-scale reforms (Fullan, 2000a; Fullan, 2000b).

The variety of aid partners, mainly from the developed world, involved in national reform projects poses a real contextual challenge to the Jordanian education system. A gap might evolve between the defined goals and the actual achievements as a result of ignoring the content of the reform and the institutional context in which it is implemented (Johnson, 2000). The gap might be generated due to the concern with the alignment of the objectives of the reform with the conditionalities attached to the aid. During the planning stage for ERfKE, the Ministry had to negotiate its plans with the World Bank as the major fund partner in the project. As the World Bank did not agree on the timeline drawn by the education system for the reform, the Ministry has to reschedule its plans to be more aligned with the conditions of its partners in order to gain their approval. Furthermore, it is worthwhile considering that relying on external aid means that the education system might not have complete control over the flow of funds in order to sustain all stages of the reform especially when these partners are involved in the project under the umbrella of 'aid'. For instance, as part of the JEI, teachers in the Discovery Schools received donated laptops from partners in the reform. The laptops were used ones, so, the majority of them became dysfunctional and needed repair or replacement, however, as they were "donation" and were received under the umbrella of "aid" they were not guaranteed, and therefore, they were simply withdrawn from schools without replacement (Abuhmaid, 2009). Furthermore, aid partners might manoeuvre ICT reform in certain directions according to their interests. It seemed that the education system faced a dilemma: while it had to maintain its focus on educational outcomes from ICT integration across the educational system, it was consciously or subconsciously mindful of the requirements and demands of aid partners. Often, this consciousness took the form of overemphasizing, or even manipulating some achievements in order to maintain the support of aid partners. For example, an earlier study, revealed that a senior official in the Ministry requesting certain modifications to a statistical report about the numbers of computers in schools, because the true figures were not representable (Abuhmaid, 2009).

SOLUTIONS AND RECOMMENDATIONS

It is not a straightforward task to search for solutions in a highly complex situation. It is as daunting as it sounds to homogenise the diverse factors involved in local reform projects including the built-in capacity, uniqueness of the local context, scarcity of resources available for development projects, the need for external assistance, and the diversity of aid partners in national reform projects make the search for a formula to solution a task. As the education system of Jordan is trapped in poverty and economic challenges, it will continue to seek assistance from external aid partners while trying to undertake reform projects such as ICT adoption. Therefore, this complexity requires extra care to coordinate and manage all the associated factors in order to keep them educationally driven and to prevent them from clashing due to their self-interests.

Understandably, external aid partners can accelerate reform projects in developing countries through providing fund and expertise. Yet, such sources of fund are neither stable nor indefinite. In addition, the involvement of external partners in national reform projects has to consider the local context where ICT integration is implemented because an experience based on research conducted in a particular context is not necessarily applicable to another. For countries like Jordan, the investment in ICT necessitates additional planning prior to adopting educational reform initiatives considering the scarcity of resources and the high dependence on external aid. Hence, the failure of such initiatives in achieving their objectives implies wasting precious resources badly needed in many sectors across the education system especially the resources are basically loans or grants. Moreover, the education system has to be cautious of the conditionalities attached to foreign assistance and to keep them educationally oriented.

Furthermore, education systems have to be aware of their built-in capacity when they adopt ICT-related reform initiatives. Therefore, ICT-related initiatives should embrace a holistic approach which pays enough attention to all levels of the system in order to avoid achieving limited *patches of success*. Strategies for building capacities need to be considered systematically as effective classrooms require supportive instructional materials, schools, school districts and related legislations (Massell, 1998) considering the centralized nature of the Jordanian education system. Conceivably, the performance at one level is interlinked and influenced by the performance at all other levels. For instance, while developing educational materials and textbooks that embrace and integrate ICT and providing teacher professional development are crucial ingredients for ICT integration, yet, this should be considered in light of the larger contexts of schools and the educational system (e.g. infrastructure, maintenance, follow-up, and school cultures).

Additionally, substantial efforts and resources should be directed to building capacity within all levels of the education system. It is essential for the education system, while drawing reform strategies and adopting initiatives, to consider its organizational development and building its human capital. Securing funding for ICT integration projects through international aid organizations and building partnerships with the private sector might be inevitable for the system, however, parallel strategies should focus on capacity building within the organization. Therefore, a proportion of aid might be dedicated to efforts that focus on strategies to help the system getting rid of external aid in the future. Thus, it is essential for the education system to determine what to adopt of the external aid and when this aid is needed as well as framing the whole involvement of aid partners to keep it aligned with the system's educational objectives.

FUTURE RESEARCH DIRECTIONS

Education reform is a challenging task to any education system and it is even more challenging to systems with limited resources as they are pressured to seek and accept external assistance. This chapter has illustrated the case of the Jordanian education system's attempt for reform while involving many external aid partners. It became clear that such involvement adds more challenges to the education reform. A main challenge is posed by managing the involvement of the many of aid partners therefore, more intensive and grounded research is needed to determine and identify strategies for managing various partners and efforts in education reform. In particular, research might focus on particular cases and analyse the failure or success of involving external aid partners and the private sector in national education reform and their impact on the educational agendas.

CONCLUSION

In this chapter I have argued that Jordan, as many other developing countries worldwide, faces serious challenges while it is trying to update its education system for the 21st century. It became clear that the manoeuvre towards ICT adoption by the education system while it relies heavily on external aid makes the reform fragile as the education system attempts to strike balance between the local educational context including its most pressing needs and its built-in capacity to undertake large-scale reform projects. The

chapter suggests that the education system to conserve sizable efforts to managing external involvement and to undertake reform projects which are within its own capacity rather than in the minds of external aid partners. Although the Jordanian context is unique, some of the issues identified above are shared with other countries in the MENA region and with other developing countries in many other parts of the world where poverty and limited capacity of education systems force these systems to seek financial assistance and expertise from external partners mainly from the developed world.

REFERENCES

Abuhmaid, A. (2009). *ICT Integration across Education Systems: The Experience of Jordan in Educational Reform*. Saarbrücken, Germany: VDM Verlag Dr. Müller.

Akkari, A. (2004). Education in the Middle East and North Africa: The Current Situation and Future Challenges. *International Education Journal*, 5(2), 144–153.

Al-Sa'd, A. (2007). *Evaluation of Students' Attitudes Towards Vocational Education in Jordan*. *Lärarutbildningen*. Malmohogskola.

AlKhawaja, K. (2007, August 15). Seven Thousand Students Move from Private to Governmental Schools. *ALRAI*, P.1.

Appleby, R. (2005). *The Spatiality of English Language Teaching, Gender and Context*. Unpublished doctoral dissertation, University of Technology, Sydney.

Castells, M. (1999). Flows, Networks, and Identities: A Critical Theory of the Informational Society. In Castells, M., Flecha, R., Freire, P., Giroux, H. A., Macedo, D., & Willis, P. (Eds.), *Critical Education in the New Information Age* (pp. 37–64). New York: Rowman & Littlefield Publishers, Inc.

Chatelard, G. (2004). *Jordan: A Refugee Haven, Migration Policy Institute.* Retrieved February 19, 2007, from http://www.migrationinformation.org/Profiles/display.cfm?ID=236

Chossudovsky, M. (2003). *The Globalization of Poverty and the New World Order* (2nd ed.). Pincourt, Canada: Global Research.

Cisco Learning Institute. (2004). *Case Study: Jordan Education Initiative.* Retrieved July 28, 2008, from http://www.cisco.com/web/about/ac48/pdf/EU_Case_Study_JEI.pdf

Escobar, A. (2004). Development, Violence and the New Imperial Order. *Development, 47*(1), 15–21. doi:10.1057/palgrave.development.1100014

European SchoolNet. (2005). *Assessment Schemes for Teachers' Ict Competence- a Policy Analysis.* Retrieved August 20, 2006, from http://www.eun.org/insight-pdf/special_reports/PIC_Report_Assessment%20schemes_insightn.pdf

Fischer, H. (2006). *Digital Shock: Confronting the New Reality* (Mullins, R., Trans.). Montreal: McGill-Queen's University Press.

Forman, R. (1994). Australia in Laos: In-Country Delivery of University Programs. In Thomas, P. (Ed), *Teaching for Development: An international Review of Australian Formal and Non-formal Education for Asia and the Pacific.* (pp 98-104). Canberra, ACT.

Fullan, M. (2000a). The Return of Large-Scale Reform. *Journal of Educational Change, 1*(1), 5–28. doi:10.1023/A:1010068703786

Fullan, M. (2000b). The Three Stories of Education Reform. *Phi Delta Kappa International, 81*(8), 581–584.

Fullan, M. (2001). *The New Meaning of Educational Change* (3rd ed.). New York: RoutledgeFalmer.

Gavlak, D. (2007, July 27). Jordan Appeals for Help in Dealing with Iraqi Refugees. *Washington Post (Washington, D.C.)*, A16.

Hasan, O. (2000). Improving the Quality of Learning: Global Education as a Vehicle for School Reform. *Theory into Practice, 39*(2), 97–103. doi:10.1207/s15430421tip3902_6

Hear, N. V. (1995). The Impact of the Involuntary Mass 'Return' to Jordan in the Wake of the Gulf Crisis. *The International Migration Review, 29*(2), 352. doi:10.2307/2546785

Heresh, J., & Twal, N. (2002, August 30). IT Awareness and e-Education Making Progress. *Jordan Times*, p. 18. Amman.

Jaradat, E. (1992). The Philosophy of Educational Development in Jordan. *Risalat Al Muallim, 33*(3), 5–62.

Johnson, B. L. (2000). Great Expectations but Politics as Usual: The Rise and Fall of a State-Level Evaluation Initiative. *Journal of Personnel Evaluation in Education, 13*(4), 361–381. doi:10.1023/A:1008157300684

Khawaja, M. (2003). Migration and the Reproduction of Poverty: The Refugee Camps in Jordan. *International Migration (Geneva, Switzerland), 41*(2), 29–57. doi:10.1111/1468-2435.00234

Massell, D. (1998). *State Strategies for Building Capacity in Education: Progress and Continuing Challenges.* Paper presented at the Consortium for Policy Research in Education, Philadelphia, PA.

McKinsey & Company. (2005). *Building Effective Public-Private Partnerships: Lessons Learned from the Jordan Education Initiative.* Retrieved August 22, 2007, from http://www.jei.org.jo/KnowledgeCenterfiles/McKinsey%20Final%20Report_May%2005.pdf

Ministry of Education. (2001). *E-Learning: A Strategic Framework.* Amman.

Ministry of Education. (2004, September). *The Development of Education: National Report of the Hashemite Kingdom of Jordan.* Paper presented at the International Conference on Education, Geneva.

Ministry of Education. (2006). *The National Strategy for Education.* Amman: The Department of Educational Research and Development.

Ministry of Education. (2007). *A Brief About the Ministry of Education.* Retrieved January 9, 2008 from http://www.moe.gov.jo/glm.pdf.

Moore, P. (2004). *Doing Business in the Middle East: Politics and Economic Crisis in Jordan and Kuwait.* Cambridge, New York: Cambridge University Press. doi:10.1017/CBO9780511492242

Mu'tamen, M. (2007). The Role of the Jordanian Educational System in Development Towards the Knowledge Economy. *Risalat Al Muallim, 43*(1), 12–21.

National Centre for Human Resources Development. (2005). *Terms of References for the Formative Evaluation Study of School Social & National Education Curriculum and Its Implementation.* Retrieved May 25, 2007, from http://www.nchrd.gov.jo/TORSNE.pdf.

Norris, P. (2001). *Digital Divide: Civic engagement, information poverty, and the Internet worldwide.* Cambridge: Cambridge University Press.

OECD. (2006). *Programme for International Student Assessment.* Paris: OECD.

Qadamani, E. (2003, September 29). King Launches 1st E-Learning Forum. *Alrai,* P.3. Amman.

Rondinelli, D. A., Middleton, J., & Verspoor, A. (1990). *Planning Education Reforms in Developing Countries: The Contingency Approach.* Durham: Duke University Press.

Saleh, H. (1991). Water Resources and Food Production in Jordan. In Wilson, R. (Ed.), *Politics and the Economy in Jordan* (pp. 14–54). London: Routledge.

Seijaparova, D., & Pellekaan, J. W. H. (2004). *Jordan: An Evaluation of World Bank Assistance for Poverty Reduction, Health and Education.* Washington, DC: The World Bank Operations Evaluation Department, World Bank.

Talal, A. B. (1998, September). *Teacher Education and School Reform in a Changing World.* Paper presented at The 43rd World Assembly, International Council of Education for Teaching, Amman.

Tawalbeh, M. (2001). The Policy and Management of Information Technology in Jordanian Schools. *British Journal of Educational Technology, 32*(2), 133–140. doi:10.1111/1467-8535.00184

UNDP. (1999). *Human Development Report 1999.* New York, Oxford: Oxford University Press.

United Nations. (2005). *UN Millennium Development Goals.* Retrieved May 25, 2007, from http://www.un.org/millenniumgoals

Wilson, E. J. (2004). *The Information Revolution and Developing Countries.* Cambridge, Massachusetts: MIT Press.

World Bank. (1999). *Education in the Middle East & North Africa: A Strategy Towards Learning for Development.* Retrieved February 22, 2005, from http://www.worldbank.org/education/strategy/MENA-E.pdf.

World Bank. (2000). *Partnership for Education in Jordan.* Retrieved January 8, 2005, from http://lnweb90.worldbank.org/oed/oeddoclib.nsf/24cc3bb1f94ae11c85256808006a0046/ca2a4e7ab293853c852568a8004fc84b/$FILE/193precis.pdf.

World Bank. (2002). *Lifelong Learning in the Global Knowledge Economy: Challenges for Developing Countries.* Washington, DC: World Bank.

World Bank. (2003). *Project Appraisal Document on a Proposed Loan to the Hashemite Kingdom of Jordan for an Education Reform for Knowledge Economy I Program.* Retrieved September 22, 2005, from http://www-wds.worldbank.org/servlet/WDSContentServer/WDSP/IB/2003/05/10/000094946_03043004015982/Rendered/PDF/multi0page.pdf

World Bank. (2005). *Monitoring and Evaluation Toolkit for E-Strategies Results.* Washington, DC: Global Information and Communication Technologies Department, World Bank.

ADDITIONAL READING

Abuhmaid, A. (2007, October). *Sustainability of ICT Integration in Developing Countries: An Example from Jordan.* Paper presented at the MIT LINC Conference: Technology-Enabled Education: A Catalyst for Positive Change, Amman.

Alomari, A. (2009). Investigating online learning environments in a web-based math course in Jordan. *International Journal of Education and Development using Information and Communication Technology, 5(3).*

Bannayan, H. (2007, October). *Jordan Education Initiative.* Paper presented at the MIT LINC Conference: Technology-Enabled Education: A Catalyst for Positive Change, Amman.

El-Berr, S. (2004). *Non-German Donor Activities in Education in Jordan Center for Development Research, University of Bonn.* Retrieved January 11, 2005, from http://www.zef.de/fileadmin/webfiles/downloads/projects/el-mikawy/jordan_non-german-donor.pdf.

El-Berr, S., & El-Mikawy, N. (2004). *Regional Perspectives on Education Reform in the Arab Countries: ZEF.* Retrieved January 11, 2005, from http://www.zef.de/fileadmin/webfiles/downloads/projects/el-mikawy/study_regional_perspectives.pdf

Emaduldeen, M. M. (2007)... *The Role of the Jordanian Educational System in the Development Towards the Knowledge Economy Risalat Al Muallim, 43*(1), 12–21.

enGauge. (2003). *enGauge 21st Century Skills: Literacy in the Digital Age.* North Central Regional Educational Laboratory and the Metiri Group. Retrieved September 18, 2004 from, http://www.ncrel.org/engauge/skills/engauge21st.pdf.

Fullan, M. (2000a). The Return of Large-Scale Reform. *Journal of Educational Change, 1*(1), 5–28. doi:10.1023/A:1010068703786

Hallinger, P., & Heck, R. H. (1998). Exploring the Principal's Contribution to School Effectiveness: 1980-1995. *School Effectiveness and School Improvement, 9*(2), 157. doi:10.1080/0924345980090203

Hepp, P., Hinostroza, E., Laval, E., & Rehbein, L. (2004). Technology in Schools: Education, ICT and the Knowledge Society. Retrieved June 18, 2006, from http://www-wds.worldbank.org/external/default/main?pagePK=6419302 7&piPK=64187937&theSitePK=523679&menuPK=6 4187510&searchMenuPK =64187283&theSite PK=523679&entityID=000160016_200501 10162933&searchMenuPK=64187283&the SitePK=523679

Hill, P. W., & Crevola, C. A. (1999). The Role of Standards of Educational Reform for the 21st Century. In Marsh, D. D. (Ed.), *Preparing Our Schools for the 21st Century* (pp. 117–142). Alexandria, Virginia: Association for Supervision and Curriculum Development.

Holland, R. (2002). *Fedral Law Will Require Research-Based Programs* (pp. 1–3). School Reform News.

Iskandarani, M. Z. (2008). Effect of Information and Communication Technologies (Ict) on Non-Industrial Countries-Digital Divide Model. *Journal of Computer Science, 4*(4), 315–319. doi:10.3844/jcssp.2008.315.319

Johnson, B. L. (2000). Great Expectations but Politics as Usual: The Rise and Fall of a State-Level Evaluation Initiative. *Journal of Personnel Evaluation in Education, 13*(4), 361–381. doi:10.1023/A:1008157300684

Kozma, R. B. (2005). National Policies That Connect ICT-Based Education Reform To Economic and Social Development. *Human Technology, 1*(2), 117–156.

Kozma, R. B. (2006). *Contributions of Technology and Teacher Training to Education Reform: Evaluation of the World Links Arab Region in Jordan.* Retrieved July 02, 2007, from http://www.wlar.org/pdfs/WLAR_Jordan_Evaluation.pdf.

Leithwood, K., & Jantzi, D. (2006). Transformational School Leadership for Large-Scale Reform: Effects on students, teachers, and their classroom practices. *School Effectiveness and School Improvement, 7*(2), 201–227. doi:10.1080/09243450600565829

Leithwood, K., Jantzi, D., & Mascall, B. (2002). A Framework For Research On Large-Scale Reform. *Journal of Educational Change, 3*, 7–33. doi:10.1023/A:1016527421742

Maddux, C., & Cummings, R. (2004). Fad, Fashion, and the Weak Role of Theory and Research in Information Technology in Education. *Journal of Technology and Teacher Education, 12*(4), 511–533.

Mansell, R., & Wehn, U. (1998). *Knowledge Societies: Information Technology for Sustainable Development.* Oxford: Oxford University Press.

McConnell International. (2002). *The National E-Readiness of the Hashemite Kingdom of Jordan: A Global View of Jordan's Competitive Advantages.* Washington, DC: USAID.

Ministry of Education. (2002). *Vision Forum for the Future of Education in Jordan.* Amman. Retrieved January 05, 2005, from http://www.moe.gov.jo/WeB/FORUM%20FINAL%20RE-PORT.doc.

Mu'tamen, M. (2002). *Towards a Vision for a New Education System.* Amman: Ministry of Education.

Nadler, L., & Nadler, Z. (1993). *Sculpting the Learning Organization.* San Francisco: Jossey-Bass.

OECD. (2001). *Schooling for Tomorrow Learning to Change. Education and Skills.* Paris: OECD.

Pedrelli, M. srl, P., & Emilia, R. (2001). *Developing Countries and the ICT Revolution.* Luxembourg: European Parliament Directorate General for Research.

Plomp, T., Anderson, R. E., & Kontogiannopoulou-Polydorides, G. (Eds.). (1996). *Cross National Policies and Practices on Computers in Education.* Dordrecht, The Netherlands: Kluwer Academic Publishers. doi:10.1007/978-0-585-32767-9

Rogers, E. M. (2003). *Diffusion of Innovations* (5th ed.). New York: Free Press.

Sahlberg, P. (2006). Education Reform For Raising Economic Competitiveness. *Journal of Educational Change, 7*, 259–287. doi:10.1007/s10833-005-4884-6

Servon, L. J. (2002). *Bridging the Digital Divide, Technology, Community, and Public Policy.* Oxford: Blackwell Publishing. doi:10.1002/9780470773529

Todaro, M. (1985). *Economic Development in the Third World* (3rd ed.). New York: Longman.

Tyack, D., & Tobin, W. (1994). The "Grammar" of Schooling: Why Has It Been So Hard to Change? *American Educational Research Journal, 31*(3), 453–479.

Underwood, J. (2004). Research into Information and Communications Technologies: where now? *Technology, Pedagogy and Education, 13*(2), 135–143. doi:10.1080/14759390400200176

United Nations Economic and Social Commission for Western Asia. (2005). *National Profile for the Information Society in Jordan.* Retrieved May 11, 2006, from http://www.escwa.org.lb/wsis/reports/docs/Jordan_2005-E.pdf.

Wallace, M. (2003). Managing the Unmanageable: Coping with complex educational change. *Educational Management & Administration, 31*(1), 9–29. doi:10.1177/0263211X030311002

World Economic Forum. (2004). *Jordan Education Initiative.* Amman. Retrieved August 20, 2006, from http://www.moict.gov.jo/moict/downloads/JEI_Track_Update_Document.pdf

Chapter 4
Integrating ICT in Education:
Impact on Teachers' Beliefs and Practices

Bassam Hammo
University of Jordan, Jordan

Basem Saraireh
University of Jordan, Jordan

ABSTRACT

In this chapter, we present the findings of a study carried out at the University of Jordan to evaluate ICTE, an accredited, one-year high diploma training program in ICT Education. The ICTE program aims at preparing Jordanian teachers to use information and communication technology in education. In this study, we attempt to examine the program in terms of its social impact on teachers' lives and how it changed their pedagogical beliefs and practices in teaching. We adopt an evaluation methodology based on empirical analysis of the program modules and students profiling. This study used a combination of quantitative and qualitative tools including questionnaire interviews and document analysis. We focus on problems experienced by students, and highlight issues and recommendations to be considered for the success of the program. During the past three years, over 860 participants graduated from the ICTE program after they have been thoroughly trained on cutting-edge software packages to be used through the school curricula in Jordan.

INTRODUCTION

The government of the Hashemite Kingdom of Jordan recognized that the investment in the ICT sector is a key element for creating sustainable economic development and improving the quality of life of its citizens. To reach this goal, the Ministry of Education (MoE) of Jordan commenced an e-learning initiative to deliver educational resources and in-service training to engage students and teachers with online activities to catch up with the fast evolving world of ICTs. The e-learning initiative also aimed at preparing the new generation of students to compete in the global marketplace of the 21st century (MOE, 2003; MoE, 2009).

During the past decade, MoE has exerted significant education reform efforts, by offering Jordanian teachers the opportunity to participate

DOI: 10.4018/978-1-60690-048-8.ch004

in professional development programs, designed to help teachers to learn how to integrate technology effectively and efficiently in the classroom. The main focus was on how to prepare Jordanian students to acquire the key skills of the 21ˢᵗ century, such as digital literacy, problem solving, critical thinking and collaboration.

The e-Learning initiative has been scheduled to achieve high returns within a short time frame. Accordingly, over the past 10 years, the MoE of Jordan had exerted efforts to integrate ICT into the curricula, connect school labs and invest on software and hardware solutions and teacher training programs. To cope with the rapid global changes, the MOE has initiated a nationwide training program to spread computer literacy, and to encourage the use of computers among teachers. Training programs, such as ICDL®, Intel®, and EduWave® were launched for the purpose of assisting teachers to use ICT more effectively and efficiently in the school curricula (Al-jaghoub & Westrup, 2003).

Despite the extensive resources dedicated to the e-Learning project, it has become clear that the initiative was not able to deliver its major intended goals. Many of the computer labs at schools were left locked and unused because of lack of trained technicians to operate them. Many of the students were mostly playing non-educational games on computers for hours instead of using them for learning. Unfortunately, there were no clear plans to guide the transformation of using ICTs. The ICT training programs were not delivered concurrently with the reform of the curricula and there were limited training for teachers on how to use these technologies. Most teachers were not adequately prepared to use technology effectively, and the benefits from the training programs were not realized. (Al-jaghoub & Westrup, 2003; Hammo et al., 2007; Mofleh et al., 2008).

Integrating advanced ICT in the reformed curricula was challenged by many Jordanian teachers who lacked the necessary ICT training. Many of these teachers required practical training on how to integrate new technologies into their traditional teaching style and to transform their passive traditional classrooms into collaborative learning environments. It became evident that not much progress would be achieved in the teaching/learning process unless teachers have acquired the knowledge of using and integrating computer technology in their curricula (AACTE, 2008).

Recent advancements in computers, Internet, networking and mobile technologies forced many Jordanian teachers to recognize the importance of learning and the need to use educational technology effectively. This requires teachers to have ample time to learn and develop their skills outside the regular school hours. They require a well-planned professional development program based on a theoretical model that can be linked to the curriculum objectives, in addition to having access to support staff. These requirements are essential if teachers are to use technology effectively to improve learning practices and accomplish tangible returns. (Brand, 1997).

In this chapter, we present our findings from a case study carried out at the University of Jordan to investigate and evaluate a training program, aimed at preparing Jordanian teachers to use information and communication technology in education. In this study, we attempt to examine the program in terms of its social impact on teachers' lives and how it transformed their pedagogical beliefs and teaching practices.

The findings of the study are supported through empirical analysis, using surveys, interviews and documents analysis. In this chapter we also discuss the ICTE program and its modules and our experience in developing and sustaining the program during the past three years.

BACKGROUND

The rise of the global economy and the advent of the technology compel students to acquire and master new skills in their academic and life activities. Students of today need to learn how to collaborate, apply critical thinking, and how to find better ways to solve problems to meet the challenges of everyday life (Jaber, 1997).

In developing countries, ICT can offer economical, social and cultural opportunities for people (UNCDF, 2006). Using ICT in education allows teachers and students to participate in a world in which activities are increasingly transformed by accessing the Internet and other technologies. By using the Internet, for example, teachers and students can access large amounts of data at a very low cost. In addition, teachers and students can exchange opinions, share news and information through discussion groups and forums. Accordingly, it is expected that the new generation of Jordanian students will significantly contribute to the development of a knowledge based society.

Several research studies identified the role of teachers as being the most important factor to the successful integration of ICTs in education (Loveless, 1996; Vannatta, 2000; Sandholtz, 2001; Bitner & Bitner, 2002; Zhao et al., 2002; Conlon & Simpson; 2003; Guha, 2003).

A number of studies also revealed that teachers who are socially aware and know where to go for support are more likely to successfully integrate ICT in their classroom. Good ICT support environment usually includes technicians, administrators and knowledgeable peers, who can provide assistance when needed (Hruskocy et al., 2000; Schmid et al., 2001; Bitner & Bitner, 2002; Mouza, 2002; Guha, 2003). Other studies in the literature identified some major barriers to the integration of ICT in schools and the factors that hindered the readiness in using new media in classrooms (BECTA, 2004; Totter et al., 2006; Tella et al., 2007). These factors were reported in BECTA (2004) and they include:

1. Lack of time to learn new technologies and attend training sessions to learn computer basics (Fabry & Higgs, 1997; Manternach-

Wigans et al., 1999; Preston et al., 2000; Vannatta, 2000; Beyerbach et al., 2001).

2. Lack of confidence in using ICT equipment (Pelgrum, 2001).

3. Lack of knowledge in using hardware and software effectively (Snoeyink & Ertmer, 2001).

4. Teachers require assistance to overcome their fears, concerns, and anxiety (Russell & Bradley, 1997).

5. Lack of sufficient amount of computers in the classroom (Preston et al., 2000).

6. Lack of support from administration, technical services and colleagues to resolve technical problems as they occur (Bitner & Bitner, 2002).

7. Lack of models and motivational factors that allow for experimentation without fear of failure (Snoeyink & Ertmer 2001).

THE ICTE PROGRAM: HILIGHTS AND PRACTICES

Helping Jordanian Teachers to Bring Technology to the Classroom

Part of its mission, the University of Jordan (UJ) is committed to serve and participate in the development of the local community. In 2006, UJ signed a partnership agreement with Ohio University, (USA), and Al-Faisal International Academy (Saudi Arabia), to develop a high diploma program in Information and Communication Technology in Education referred to as ICTE Program (UJ Connect, 2007; Franklin, 2007).

The ICTE program has been developed because of the urgent need to train teachers on using ICT in education. Its main objective is to train and prepare Jordanian teachers for mobilizing the necessary technologies, skills, and educational resources to prepare their students to face the challenges of using new technologies.

Hereinafter in this chapter, the term "*participants*" is used to refer to K-12 teachers joining the ICTE program. During the past three years over 860 participants, from both public and private schools, have been trained on cutting-edge software packages to be used in teaching the reformed school curricula in Jordan. Field trip visits to follow up with participants who graduated from the ICTE program during the past three years showed that participants were implementing what they have been trained on in their classrooms. Most participants reported that their students were more motivated when the Internet and computers were used efficiently in classrooms.

The ICTE Study Program

The ICTE program is modeled after the Ohio College of Education's nationally and internationally accredited Instructional Technology graduate program. Table 1, shows the courses of the program, classified into two main categories:

1. *Educational courses* (4 courses): These courses concentrate on the integration of technology, constructivist practices, classroom management when technology is used, and emerging technologies for improving teaching and learning. The *Assessment Techniques* module allows teachers to practice the use of educational assessments during the academic school year.

2. *Information Technology courses* (5 courses): These courses represent the core of the training program. They intended for training the participants on using the cutting-edge software packages usually used in integrating ICTs in the classroom curricula.

The Objectives of the ICTE Program

The following objectives were determined when the ICTE high diploma program was initially developed (Franklin & Franklin, 2007), they include:

1. To prepare teachers who are committed to the pedagogy and who understand the role of technology in preparing the new generation of Jordanian students to compete effectively and efficiently in an evolving information society.

2. To upgrade teachers' knowledge/skills in using new technology in the teaching/learning process.

3. To develop bi-lingual (Arabic/English) training courses for the effective use by teachers whose mother tongue is Arabic.

4. To disseminate the developed courses in printed format, electronic material delivered through Blackboard™, whiteboards,

Table 1. ICTE study plan

Courses in Education			
1	Principles of School Curriculum	2	Learning and Teaching Strategies
3	Instructional Design	4	Assessment Techniques
Courses in ICT			
Course Title		**Tools**	
1	Technology Applications in Education	Word®, Excel®, Access®, PowerPoint®, Publisher®	
2	Internet Applications in Education	Adobe Dreamweaver®	
3	Instructional Multimedia	Macromedia Flash® & Authorware®	
4	Graphics Design for Visual Media	Adobe Photoshop®	
5	Programming Concepts for Teachers	VB.Net®	

PowerPoint presentations, and in the nearest future on CD-ROMs and DVDs.

5. To develop the program at the local and regional levels.
6. To reach decision makers and teachers around the country within their schools to introduce the program and its benefits.

Teaching Methodologies

Throughout the training program, the participants were encouraged to commit and apply to the following strategies:

1. To identify their technical deficiencies and seek assistance when needed.
2. To share their knowledge of computers and Internet with colleagues.
3. To seek mentors and colleagues for assistance.
4. To integrate what they have learned in the curriculum and get feedback from their students.
5. To discover what their students know about technology and ask them for assistance.
6. To identify weekly goals regarding familiarity with the technology.
7. To start using e-mail and to collaborate with others having similar experiences.

8. To discuss issues, ideas, and suggestions with others in the same field.

Additionally, hands-on exercises were encouraged through practical tasks in every lecture at a ratio of 50% of the lecture time.

Demographics of the ICTE Participants

The ICTE program has grown rapidly and now it is approaching its fourth year. Since its initiation in 2006, the program has admitted (300) students every year. It attracts teachers from major cities in Jordan. The program will eventually be refined and exported to neighboring countries. Figure 1, shows that around (60%) of the participants are from government public schools, and sponsored by the Ministry of Education. The other 40% are from private schools and other self-funded. The cost of the program is around 2000 JD ($2800). Figure 2, shows the distribution of the participants based on the years of experience in the field of education. The majority of the participants have from 5-10 years of teaching experience. However, there is disparity in the participant's technology and English speaking skills. This inconsistency was problematic for both the faculty and students. The limited English speaking ability of the students

Figure 1. Distribution of participants based on work sector

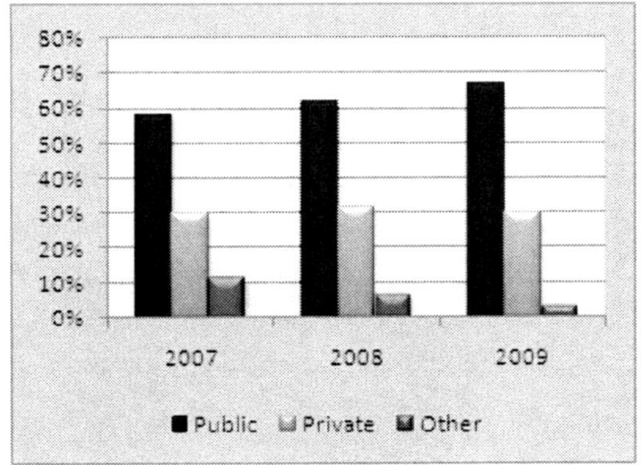

Figure 2. Distribution of participants based on years of experiences

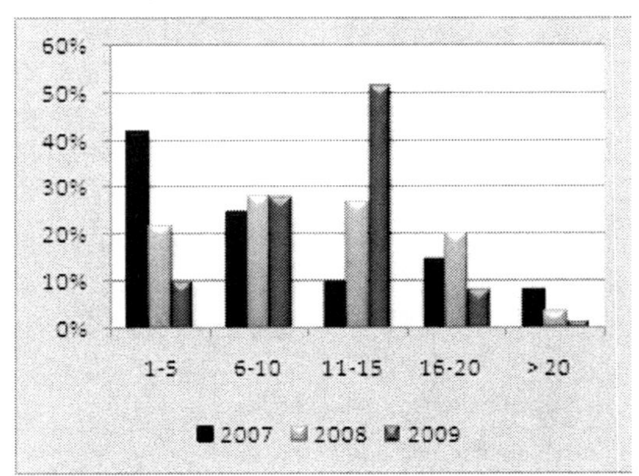

represented an additional burden of translation for the instructors. A remediation plan for students with poor technical and English speaking skills was also activated.

Figure 3, depicts the distribution of the participants according to their gender. The chart shows that the female participants represent (60%) of the students.

The training time of the program varies according to individual's previous knowledge of computers, age, among other factors. Time is usually set according to the needs of the participants. This will ensure they will acquire the knowledge and skills to use computers as a tool for both personal and educational purposes.

The distribution of the participants according to their majors and teaching levels is depicted in Table 2. The study sample includes (60) participants admitted to the ICTE program on the first semester for the academic years 2008-2009. Figure 4, shows the participants' previous knowledge in using computers. Around (83%) of the participants owned either personal computers, laptops or both. Participants teaching science, mathematics, physics, biology and computer science are the most intensive users of ICTs.

Concerning their use and knowledge of the Internet, Figure 5 shows the previous knowledge of the participants in using the Internet. Less than

(53%) of the participants connect frequently to the Internet, while (25%) of the participants never connect to the Internet. However, only (40%) of the participants know how to seek information directly from the Internet sites or through search engines such as Google. The majority of the participants were passive users; where they only know how to navigate through the web pages.

Program Evaluation

As stated earlier, the ICTE program is approaching its fourth year. Since it was launched, the ICTE program has undergone a comprehensive yearly assessment for the past three years (2007, 2008

Figure 4. Distribution of participants based on computer knowledge

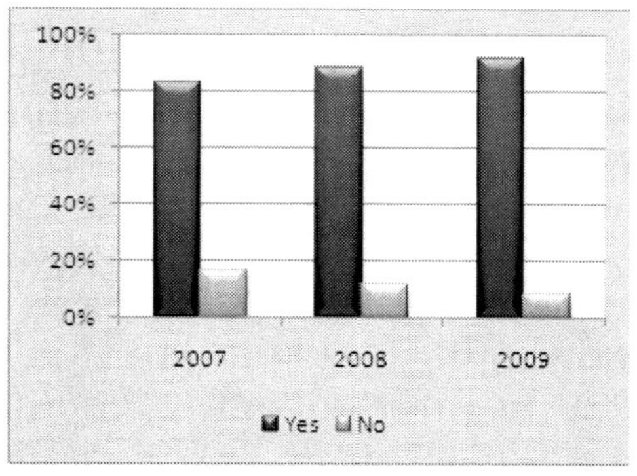

Figure 3. Distribution of participants based on gender

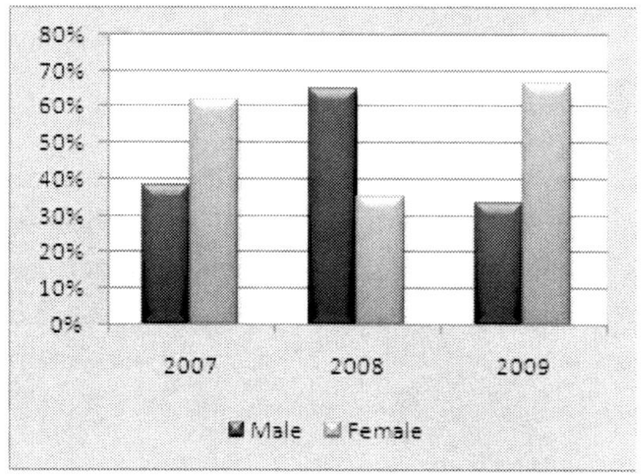

Figure 5. Distribution of participants based on Internet usage

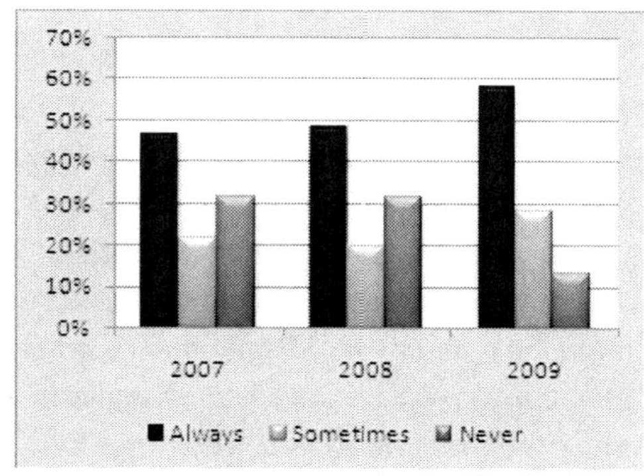

Table 2. Distribution of participants based on their majors, teaching levels and years of experiences

Major	2007	2008	2009
Arabic Language	4	3	5
English Language	7	4	6
Social Studies	5	3	4
Math	11	9	4
Science	10	15	9
Islamic Studies	4	3	9
Art	2	0	0
Physical Studies	1	0	4
IT Major & Business	16	23	19

Level	2007	2008	2009
Kindergarten	3	1	1
Grade 1-3	5	8	14
Grade 4-10	18	14	24
Grade 11-12	14	20	12
Staff	6	8	6
Others	14	9	3

& 2009). The program assessors were specialists from the Ohio University. As a partner in the development and the implementation of the program, Ohio University needed to ensure that the program was up to the same level of its twin program being offered at Ohio University in the United States. Other goals of the assessment included; providing the basis to guide the ICTE program improvements and gaining the support of the policy makers in Jordan to support the integration of ICT in education. Most of the surveyed participants insured that their interest and enthusiasm for teaching had increased vastly under the ICTE program. The following section contains the qualitative data collected from interviews and the quantitative data collected through the Student Satisfaction questionnaire administered by Ohio University (Franklin & Franklin, 2007; 2008; 2009).

Evaluation Methodology

To evaluate the ICTE program from the participants' perspective, a questionnaire was emailed to the participants. The questionnaire was designed to collect data concerning the ability to meet the mission, goals, objectives and satisfaction with the program. Interviews with the students were conducted by assessors from Ohio University. The interviews were recorded, analyzed and used to make recommendations concerning the ICTE

program. In the following sections, we focus on the main parts of the ICTE program that has been assessed during the past three years.

A) *Statistical Analysis of Student Satisfaction Questionnaire:* in its first year, 2007, sixty participants out of 150 responded to a survey submitted through email. The return rate was 38% which may be indicative of the lack of Internet access away from the University of Jordan campus. The questionnaire examined the overall satisfaction of students with the ICTE program. The survey, conducted by Ohio University, used a Likert 4-point Scale in which a "4" indicates "very satisfied" and "1" indicates "very dissatisfied" (Franklin & Franklin, 2007; 2008; 2009). In the following years, 2008 and 2009, a sample of 60 students was selected randomly to conduct the survey.

Findings

Qualitative and quantitative data were used to examine the issue of students' satisfaction with the ICTE Program. The overall satisfaction by participants rating for the years 2007, 2008 and 2009 was (3.11, 2.95, 3.02) respectively. Each of these ratings suggests satisfaction by the partici-

Table 3. Overall satisfaction with the ICTE program

Criteria	Scale 1 to 4		
	2007	**2008**	**2009**
The ICTE high diploma Program's reputation	3.11	2.95	2.95
The ICTE high diploma Program's ability to keep pace with recent developments in the field of educational technology	3.02	3.03	2.85
The effectiveness of the services and facilities which the program offers	2.58	2.54	2.69
Quality of computer support	3.07	2.56	2.63
Quality of Blackboard site used for content delivery	2.89	2.68	2.73
Quality of Internet access	3.21	2.57	2.57
Quality of email connectivity	2.81	2.52	2.29
Relationships and interactions with other participants in the ICTE High Diploma Program	3.43	3.19	3.11
The number of computers available in the training lab	3.12	3.15	3.14
My overall experience as a student in the ICTE high diploma Program	3.11	2.95	3.02

pants concerning their experiences in the ICTE program as indicated in Table 3.

In terms of cooperating with other colleagues, very positive statements were received, suggesting that participants value the opportunity to discuss and collaborate with colleagues within the program. Such relationships can assist to prevent participants from feeling isolated in their schools. The overall comments of the participants were very optimistic and thoughtful. The program has great impact on the participants' social lives, on their schools and their families.

B) *Usefulness of Education and Information Technology courses:* The participants' perception of the usefulness of the Information Technology courses is depicted in Table 4. Only the course, *"Programming Concepts for Teachers"* using *"Visual Basic"* was not considered useful. A Likert 4-point scale with "1 = less important" and "4 = most important" was used in this questionnaire.

Findings

The low rate assigned to the *"Instructional Design"* module was strongly supported by student comments. Students suggested that the material was not practical to be taught in the classroom and it should be removed. Overwhelmingly, students suggested the alternative software "Flash" as substitute for "Visual Basic". Other findings are reported in Table 4.

C) *The importance of the ICTE program to the participants:* The ICTE program has been evaluated based on data collected from the participants to compare the importance of the subjects presented in the courses offered in the ICTE program to their work as instructors in the K-12 schools. Findings indicate that the ratings were positive concerning the content of the ICTE and its application to the work of teachers in the K-12 schools.

The Impact of the ICTE Program on the Participants

There is considerable evidence of the impact of using ICT in education for participants who joined the ICTE program in the past three years, indicated by the following:

Table 4. Usefulness of the ICTE program courses to participants

Course Topic	Scale 1 to 4		
	2007	2008	2009
Technology Applications in Education	3.19	3.00	3.05
Internet Application in Education	3.28	3.10	3.19
Instructional Multimedia	2.16	2.78	2.74
Graphic Design for Visual Communication	2.72	2.94	2.97
Programming Concepts for Teachers	2.80	2.80	2.75
Principles of School Curriculum	2.54	2.83	2.68
Learning and Teaching Strategies	2.15	2.71	2.68
Assessment Techniques	2.57	3.02	3.08
Instructional Design	2.20	2.63	2.61

1. There is strong evidence of immediate increase in applying ICTs in the day-to-day work of teachers. This included an increased efficiency in planning and preparation of work due to more collaboration between teachers.

2. Teachers gained positive attitude towards ICT and considerably increased their confidence in using ICTs. There relations with students have been improved dramatically.

3. The overall comments of the participants were very optimistic and thoughtful. The program has great impact on the participants' social lives, on their schools and their families.

4. The program has successfully overcome the teachers' isolation, by connecting them to colleagues, mentors and curriculum experts on continuous basis.

5. After they have integrated the knowledge and skills gained from the ICTE program, many teachers reported that they observed more confidence in their students' performance. Students were able to handle more complex assignments and achieve more using higher-order skills because of the capabilities provided by the technology.

Challenges Facing the ICTE Program and Recommendations

The following recommendations were derived from in-depth discussions with the key stakeholders of the ICTE program. Feedback obtained from faculty members who are teaching in the program, administrators, participants in the ICTE program and partners from Ohio University, the University of Jordan and Al-Faisal International Academy. The following recommendations are essential to the continuous success of the program:

1. Participants in the ICTE program must be provided with laptop computers, to increase their positive attitudes and confidence in integrating ICT in teaching.

2. Participants should be provided with all the required software, including virus protection, and Internet connectivity in their homes to facilitate the development of on-line instructions and to accelerate acquiring skills and knowledge.

3. The ICTE program should continue to provide courses to reflect emerging educational teaching practices and encourage teachers to upgrade their ICT skills and gain more pedagogical knowledge.

4. Teamwork on projects and collaborative learning are powerful learning strategies and should be encouraged.

5. Since most of educational context have been changed to adopt the new technologies, teachers should be encouraged to include new competencies in the curricula and ways of assessing them should be explored.

6. Participants complained about the lack of time to integrate ICT in education. It is unrealistic to expect a teacher to work from 7AM to 4PM teaching and then spend four hours completing course assignments. We recommend to leverage on teamwork, group projects and provide Internet access from home.

7. Follow-up on graduates to ensure that they implement what they have learned.

8. Remain updated with global ICT trends; particularly those pertaining to educational systems.

9. Recruitment of qualified female instructors is essential to mentor female participants in the ICTE program.

FUTURE RESEARCH DIRECTIONS

In this chapter, we presented a case study focusing on training Jordanian teachers on using ICT in education.

Findings from our study shows that teachers had widely perceived ICT as having positive impact on the students' engagement in learning activities and it enables them to meet the skills for the new century. However, the ICTE program still needs improvement to meet its stated goals and to instill it in the teaching culture. As another outcome of this study, the following key suggestions should be considered for sustainable future of ICT research directions:

1. To explore ways to establish an education communication network to facilitate the sharing of knowledge, skills and experiences among teachers who attend the ICTE program. The network can be also used to share modern curricula and to improve the quality of teaching/learning process.

2. To explore ways to encourage the local ICT market, the academic sector and the government to investment more in research and development (R&D) related to ICT and to partner together in a focused research to remove barriers to adopting ICTs in education.

3. To conduct rigorous studies to measure the impacts of ICTs on the learning outcomes of students. Also to conduct follow up studies on the consequences of integrating ICT in the curricula and its impact on the behavioral skills, jobs and social life of graduated students.

4. To conduct studies in the Arab world to measure the impact of using ICT on the lives of Arab citizens with focus on economic growth, employment, education, politics, health care and social aspects.

CONCLUSION

In this chapter, we have described the challenges facing the teachers and the students in an evolving information-based global society. To cope with the new technology, it is not enough to equip schools with cutting-edge technology. The new era requires changes in the way we teach and learn and hence, it requires changing the teachers' pedagogical beliefs and practices. Therefore, teachers should not be left behind and they should be trained on how to use and integrate the new technology in the curricula they teach. We presented the results of our investigation and evaluation of a high diploma training program named ICTE, established at the University of Jordan to train teachers on the best practices to adopt and use ICTs in education. The program is running into its fourth year successfully. After the program has been evaluated by experts in ICT and yielded positive feedback, we are now very optimistic that this program will act as a blueprint for future teacher training programs in Jordan and the region. Graduates from this program will facilitate the integration of technology through curricula and encourage their colleagues to adopt ICT in education.

ACKNOWLEDGMENT

We would like to thank from Ohio University, Dr. Teresa Franklin/College of Education and Dr. Douglas Franklin/College of Health and Human Services for their great efforts in evaluating the ICTE program for the past three years. Also we would like to thank the steering committee overseeing the program and the anonymous faculty members and students in the program.

REFERENCES

AACTE Committee on Innovation and Technology (Ed.). (2008). *Handbook of Technological Pedagogical Content Knowledge (TPCK) for Educators*. NY: Routledge.

Al-Jaghoub, S., & Westrup, C. (2003). Jordan and ICT-led development: towards a competition state. *Information Technology & People*, *16*(1), 93–110. doi:10.1108/09593840310463032

BECTA. (2004) A review of the research literature on barriers to the uptake of ICT by teachers. British Educational Communications and Technology Agency. Retrieved February 10, 2010, from http://partners.becta.org.uk/page_documents/research/barriers.pdf

Beyerbach, B., Walsh, C., & Vannatta, R. (2001). From teaching technology to using technology to enhance student learning: pre-service teachers' changing perceptions of technology infusion. *Journal of Technology and Teacher Education*, *9*(1), 105–127.

Bitner, N., & Bitner, J. (2002). Integrating technology into the classroom: eight keys to success. *Journal of Technology and Teacher Education*, *10*(1), 95–100.

Brand, G. (1997). What research says: training teachers for using technology. *Journal of Staff Development*, *19*(1), 10–13.

Conlon, T., & Simpson, M. (2003). Silicon Valley verses Silicon Glen: the impact of computers upon teaching and learning: a comparative study. *British Journal of Educational Technology*, *34*(2), 137–150. doi:10.1111/1467-8535.00316

Connect, U. J. (2007). *Information and Communication Technology in Education Diploma*. Retrieved November 10, 2009, from http://www.ju.edu.jo/alumni/UJConnect12/topics/ICT.htm

Fabry, D., & Higgs, J. (1997). Barriers to the effective use of technology in education. *Journal of Educational Computing*, *17*, 385–395. doi:10.2190/C770-AWA1-CMQR-YTYV

Franklin, T. (2007). *A Cultural Bridge: Instructional Technology Program Goes Global*. Retrieved February 10, 2010, from http://www.coe.ohiou.edu/news-events/jordanicte.htm

Franklin, T., & Franklin, D. (2007). *ICTE Final Report*. (Unpublished report), University of Jordan, Jordan.

Franklin, T., & Franklin, D. (2008). *ICTE Year 2 Evaluation Report*. (Unpublished report), University of Jordan, Jordan

Franklin, T., & Franklin, D. (2009). *ICTE Year 3 Evaluation Report*. (Unpublished report), University of Jordan, Jordan

Guha, S. (2003). Are we all technically prepared? Teachers' perspective on the causes of comfort or discomfort in using computers at elementary grade teaching. *Information Technology in Childhood Education Annual*, 317–349.

Hammo, B., Hudaib, A., Huneiti, A., & Al-Zoubi, M. B. (2007). *Preparing Jordanian Teachers to Face the Challenges in Applying Information and Communication Technology in Education.* In the Fourth Annual Conference of Learning International Networks Consortium, MIT-LINC 2007, October 28-30, Amman, Jordan.

Hruskocy, C., Cennamo, K., Ertmer, P., & Johnson, T. (2000). Creating a community of technology users: students become technology experts for teachers and peers. *Journal of Technology and Teacher Education, 8*(1), 69–84.

Jaber, W. (1997). A survey of factors which influence teachers' use of computer-based technology. Dissertation Virginia Polytechnic Institute and State University.

Loveless, T. (1996). Why aren't computers used more in schools? *Educational Policy, 10*(4), 448–467. doi:10.1177/0895904896010004002

Manternach-Wigans, L., & Bender, C. L., & Maushak, N. J. (1999) Technology integration in Iowa high schools: perceptions of teachers and students. A paper presented at *the American Education Communications & Technology Conference,* San Antonio, TX.

MOE. (2003). Incorporating information and communication technology (ICT) in technology learning process. *Teacher's Journal, 42*(1).

MOE. (2009). Jordan Ministry of Education's e-Learning Initiative. Retrieved October, 15, 2009, from http://www.moe.gov.jo

Mofleh, S., Wanous, M., & Strachan, P. (2008). Developing Countries & ICT Initiatives: Lessons Learnt for Jordan's Experience. [EJISDC]. *Electronic Journal on Information Systems in Developing Countries, 34*(5), 1–17.

Mouza, C. (2002). Learning to teach with new technology: implications for professional development. *Journal of Research on Technology in Education, 35*(2), 272–289.

News, U. J. (2009). *About UJ.* Retrieved November 10, 2009, from http://www.ju.edu.jo/Pages/AboutUJ.aspx

Pelgrum, W. J. (2001). Obstacles to the integration of ICT in education: results from a worldwide educational assessment. *Computers & Education, 37,* 163–178. doi:10.1016/S0360-1315(01)00045-8

Preston, C., Cox, M., & Cox, K. (2000). *Teachers as innovators: an evaluation of the motivation of teachers to use Information and Communications Technology.* London: MirandaNet.

Russell, G., & Bradley, G. (1997). Teachers' computer anxiety: implications for professional development. *Education and Information Technologies, 2,* 17–30. doi:10.1023/A:1018680322904

Sandholtz, J. H. (2001). Learning to teach with technology: A comparison of teacher development programs. *Journal of Technology and Teacher Education, 9*(3), 349–374.

Schmid, R., Fesmire, M., & Lisner, M. (2001). Riding up the learning curve- elementary technology lessons. *Learning and Leading with Technology, 28*(7), 36–39.

Snoeyink, R., & Ertmer, P. (2001). Thrust into technology: how veteran teachers respond. *Journal of Educational Technology Systems, 30,* 85–111. doi:10.2190/YDL7-XH09-RLJ6-MTP1

Tella, A., Tella, A., Toyobo, O. M., Adika, L. O., & Adeyinka, A. A. (2007). An Assessment of Secondary School Teachers Uses of ICTs: Implications for Further Development of ICT's Use in Nigerian Secondary Schools. *The Turkish Online Journal of Educational Technology, 6*(3), 12.

Totter, A., Stutz, D., & Grote, G. (2006). ICT and Schools: Identification of Factors Influencing the use of new Media in Vocational Training Schools. *The Electronic Journal of e-Learning, 1*(4), 95-102.

UNCDF. (2006). *UNCDF*. Retrieved January 10, 2010, from http://www.uncdf.org/english/local_development/documents_and_reports/thematic_papers/devcom/200611_bridge/

Vannatta, R. (2000). Evaluation to planning: technology integration in a school of education. *Journal of Technology and Teacher Education, 8*(3), 231–246.

Zhao, Y., Pugh, K., Sheldon, S., & Byers, J. (2002). Conditions for classroom technology innovations. *Teachers College Record, 104*(3), 482–515. doi:10.1111/1467-9620.00170

ADDITIONAL READING

Ascough, R. S. (2002). Designing for online distance education: Putting pedagogy before technology. *Teaching Theology and Religion, 5*(1), 17–29. doi:10.1111/1467-9647.00114

Bauer, J., & Kenton, J. (2005). Toward technology integration in the schools: Why it isn't happening. *Journal of Technology and Teacher Education, 13*(4), 519–546.

Bawden, D. (2001). Information and digital literacy: a review of concepts. *The Journal of Documentation, 57*(2), 218–259. doi:10.1108/EUM0000000007083

Clark, D. (2002). Psychological myths in e-learning. *Medical Teacher, 24*(6), 598–604. doi:10.1080/0142159021000063916

Cuban, L., Kirkpatrick, H., & Peck, C. (2001). High access and low use of technologies in high school classrooms: explaining an apparent paradox. *American Educational Research Journal, 38*, 813–834. doi:10.3102/00028312038004813

Davis, N. (2003). Technology in teacher education in USA: What makes for sustainable good practice? *Technology, Pedagogy and Education, 12*(1), 59–84. doi:10.1080/14759390300200146

Dillon, C., Hodgkinson, L., Needham, G., & Paker, J. (2002). *Developing and assessing information literacy key skills*. Paper presented at the International conference on IT and information literacy, 20th-22nd March 2002, Glasgow, Scotland.

Diochon, M. C., & A. F. Cameron (2001). Technology based interactive learning: designing. *International student research project, 2*(2) 114-127.

Donlevy, J. (2003). Online learning in virtual high school. *International Journal of Instructional Media, 30*(2), 117–122.

Garrison, D. R., Cleveland-Inns, M., & Fung, T. (2004). Student role adjustment in online communities of inquiry: Model and instrument validation. *Journal of Asynchronous Learning Networks, 8*(2), 61–74.

Guillen, M. (2001). Is globalization civilizing, destructive or feeble?: A critique of five key debates in the social science literature. *Annual Review of Sociology, 27*, 235–260. doi:10.1146/annurev.soc.27.1.235

Heines, J. M. (2000). Evaluating the effect of a course web site on student performance. *Journal of Computing in Higher Education, 12*, 57–83. doi:10.1007/BF03032714

Housego, S., & Freeman, M. (2000). Case studies: Integrating the use of web based learning systems into student learning. *Australian Journal of Educational Technology, 16*(3), 258–282.

Howland, J. L., & Moore, J. (2002). Student perceptions as distance learners in Internet-based courses. *Distance Education, 23*(2), 183–196. doi:10.1080/0158791022000009196

Loch, K., Straub, D., & Kamel, S. (2003). Diffusing the Internet in the Arab World: the role of social norms and technological culturaltion. *IEEE Transactions on Engineering Management, 5*(1), 45–63. doi:10.1109/TEM.2002.808257

Lokken, S., L., Cheek, W., K., & Hastings, S., W. (2003). The Impact of technology training on family and consumer sciences teacher attitudes towards using computers as an instructional medium. *Journal of Family and Consumer Sciences Education, 21*, 18–32.

McCartin, M., & Feid, P. (2001). Information literacy for undergraduates: where have we been and where are we going? *Advances in Librarianship, 25*, 1–27. doi:10.1016/S0065-2830(01)80018-0

Muirhead, W. D. (2000). Online education in school. *International Journal of Educational Management, 14*(7), 315–324. doi:10.1108/09513540010378969

Mumtaz, S. (2000). Factors affecting teachers' use of information and communications technology: a review of the literature. *Journal of Information Technology for Teacher Education, 9*, 319–341.

Newhouse, P. (2001). Development and use of an instrument for computer-supported learning environments. *Learning Environments Research: An International Journal, 2*(2), 115–138. doi:10.1023/A:1012411332666

Nieuwenhuysen, P. (2000). Information literacy courses for university students: some experiments and some experience. *Campus-Wide Information Systems, 17*(5), 167–173. doi:10.1108/10650740010360539

Pearson, J. (2003). Information and communication technologies and teacher education in Australia. *Technology, Pedagogy and Education, 12*(1), 39–58. doi:10.1080/14759390300200145

Petrides, L. A. (2002). Web-based technologies for distributed (or distance) learning: Creating learner-centered. Educational experiences in the higher education classroom. *International Journal of Instructional Media, 29*(1), 69–77.

Phillips, J. A. (2000). Teaching a graduate course using an on-site and on-line approach. In *Proceedings of the International Conference on ICT in Education*. University Putra Malaysia. Kuala Lumpur. 123-130.

Poole, D. M. (2002). Student participation in a discussion-oriented online course: A case study. *Journal of Research on Computing in Education, 33*(2), 162–177.

Ronteltap, F., & Eurelings, A. (2002). Activity and interaction of students in an electronic learning environment for problem-based learning. *Distance Education, 23*(1), 11–22. doi:10.1080/01587910220123955

Rosie, A. (2002). Online pedagogies and the promotion of deep learning. *Information Services & Use, 20*(2/3), 109–116.

Saunders, G., & Klemmif, F. (2003). Integrating technology into a traditional learning environment: Reasons for and risks of success. *Active Learning in Higher Education, 4*(1), 74–86. doi:10.1177/1469787403004001006

Savolainen, R. (2002). Network competence and information seeking on the Internet: from definitions towards a social cognitive model. *The Journal of Documentation, 58*(2), 211–226. doi:10.1108/00220410210425467

Selinger, M., & Austin, R. (2003). A comparison of the influence government policy in information and communications technology for teacher training in England and Northern Ireland. *Technology, Pedagogy and Education, 12*(1), 19–38. doi:10.1080/14759390300200144

Song, L., Singleton, E. S., Hill, J. R., & Koh, M. H. (2004). Improving online learning: student perceptions of useful and challenging characteristics. *The Internet and Higher Education, 7*, 59–70. doi:10.1016/j.iheduc.2003.11.003

Thurmond, V. A., Wambach, K., Connors, H. R., & Frey, B. B. (2002). Evaluation of Student Satisfaction: Determining the Impact of a Web-Based Environment by Controlling for Student Characteristics. *American Journal of Distance Education, 16*(3), 169–189. doi:10.1207/S15389286AJDE1603_4

Vannatta, R., & O'Bannon, B. (2002). Beginning to put the pieces together: a technology infusion model for teacher education. *Journal of Computing in Teacher Education, 18*(4), 112–123.

Volery, T. (2000). Critical success factors in online education. *International Journal of Educational Management, 14*(5), 216–223. doi:10.1108/09513540010344731

Vonderwell, S. (2003). An examination of asynchronous communication experiences and perspectives of students in an online course: A case study. *The Internet and Higher Education, 6*(1), 77–90. doi:10.1016/S1096-7516(02)00164-1

Watson, G. (2001). Professional development that engage with teachers' hearts and mind. *Information Technology for Teacher Education, 10*(1/2), 17–19.

Yuen, A., & MA, W. (2002). Gender differences in teacher computer acceptance. *Journal of Technology and Teacher Education, 10*, 365–382.

Chapter 5
ICT in Education:
Culture, Practice and Involvement

Sonia Kawas
British Council, Jordan

ABSTRACT

This chapter is based on work carried out as part of the scope of the ICT in Education project jointly sponsored by Jordan's Ministry of Education and the British Council, with the participation of four other Middle Eastern countries (Egypt, Syria, Lebanon and the Palestinian Territories). The author's main empirical findings and comments relate to the duration of the project (viz. between 2003 and 2008), showing how ICT in education brought positive impact on the delivery of teaching in class, and also how online forums can be utilized as opportunities to exchange invaluable information and knowledge in this sector. Current status, challenges, solutions and recommendations are based on the author's findings and experience whilst managing and working on the project.

INTRODUCTION

Introduced in 2003 by the British Council, the *ICT in Education* project had rolled out by March 2004 a regional British Council pilot project in Near East and North Africa (NENA). Running in parallel in five regional countries, the project aimed at developing a range of strategies for improving school-level ICT, to support national initiatives of introducing IT throughout the schooling systems in Jordan and the regional countries involved. Training activities and workshops for all teachers, supervisors and principals were core elements of this project.

This case study looks at both the early stages of the project, and how it evolved to include strategic leadership in ICT, together with the technicalities and infrastructures, culture, outcomes and challenges.

Lessons learned from Jordan's experience and the author's recommendations for ICT in Near East and North Africa (NENA) are all based on her personal experience in the field on this project.

DOI: 10.4018/978-1-60960-048-8.ch005

Though this study focuses mainly on Jordan, certain issues may also reflect the region as a whole.

BACKGROUND

The first issues realised of the literature review is the positive results around introducing ICT in the education system, and how it enables distant learners to benefit from education, with the technologies available. A study by Shirazi et al (2009) reflected on the expectations of ICT to bring to the economic freedom in several countries in the Middle East, with comparison in technology users, digital divide, regulations and economic growth. On the other hand, the effects of introducing ICT in primary education in the 'midst of uncertainty' as Kiridis et al (2006) described, states as a fact that, though the use of technology is becoming a familiar scene in many countries around the world, ministries need to consider perceptions and attitudes of teachers towards the use of ICT. Kiridis continues to argue that there is a need to convince teachers in the opportunities of embedding ICT in primary education approaches and tools.

Organization Background

The British Council is the United Kingdom's principal agency for cultural relations with other countries. A charity organisation registered in the UK, its purpose is to enhance the reputation of the UK in the world as a valued partner. It promotes the UK in its entirety, reflecting and celebrating its cultural, ethnic and political diversity.

Working with ordinary people as well as with governments and decision-makers in 110 countries, the British Council builds relationships and creates opportunities. It targets specific groups of people through programmes in education, English language teaching, libraries and information, the arts, science and technology, and governance and human rights.

Project Background

Jordan is one of the leading countries in the region in terms of its investment in educational systems reform, led by HM King Abdullah II, which all started when Jordan's *Education Initiative* was launched in 2003. Amongst many issues of concern mentioned in the *Education Initiative* were:

- Teacher training
- Curriculum reform
- Adopting of ICT as a tool for learning
- Improving ICT infrastructures

This initiative in fact supported a larger, five-year *Education Reform for the Knowledge Economy* project (ERfKE), which came in two phases, the first phase focusing on the above-mentioned areas, with a government investment of approximately US$400m. Introducing ICT in schools was a high priority on the government's agenda, and the British Council's *ICT in Education* training was included in the ministry's Professional Development Plan, reflecting the degree to which the ministry gave this British Council sponsored project high importance.

The British Council has engaged in this project since 2004, working with its stakeholders (ministries of education) to develop the Knowledge Economy in Jordan (and regional countries) through ICT and its usage in education, policies and visions.

The project background is referred to the United Nations Development Programme (UNDP) Arab Human Development Report 2003, which underlined the importance of knowledge to Arab countries as a powerful driver of economic growth, and how effective participation in the knowledge economy required creating knowledge societies in the Arab countries, with the dissemination of high quality education to all, thus suggesting that the need for reform of the Arab educational systems, with new educational methods should be given a high priority in the region.

The British Council formed a team of UK consultants, who visited Jordan, Syria, Lebanon, Egypt and Palestine, and identified the need to develop educational methods that were more student-centered, through teams and projects, and more self-evaluation in all dimensions of learning.

In order to address these areas, the following core competencies were evidently required:

- Knowledge and understanding of ICT as an enabler, as well as a subject
- Development of a strategic vision and policies for ICT in education
- New, learner-centered pedagogies
- The need to develop and support capacity in human resources.

The situation of these countries at that time was profiled as follows:

Jordan was benefiting from considerable World Bank support through the *Education Reform for the Knowledge Economy* initiative, in addition to other projects which were providing the means for Jordan to adopt educational reform through curriculum development, teacher training and improved technological infrastructure. Syria, however, lacked international aid, which was an obstacle in achieving the recognised need for educational reform and for the development of the country's technological infrastructure. The political situation in Palestine made national approaches to educational reform very difficult. Lebanon and Egypt, on the other hand, were the only countries in the region with national e-strategies, but there was clear evidence indicating that there were significant gaps between the theory and the delivery.

All these countries shared important common similarities, such as language, culture, the traditional values of their education systems, low internet connectivity, a shortage of computers, a shortage of educational resources and the availability of qualified, skilled teachers.

The British Council developed its NENA Education Strategy to direct its regional work in education, identifying a specific audience which would benefit from the British Council's work in the region and its focus on educational programmes. The strategy clearly identified three audience types to be targeted:

1. **A Target Group level 1 (T1):** which includes ministers of education, secretaries general, and other high officials of similar ranking in the educational sector and/or organisations, who deal only with matters at a superficial level
2. **A Target Group level 2 (T2):** which includes the main persons active on this project, ranging from ministry officials, head teachers, school teachers and educational supervisors (of science, mathematics and English) to educational decision-makers
3. **A Target Group level 3 (T3):** which includes students, who were not specifically targeted on this project, but who would, nevertheless, be influenced by the change and development this project would bring to the educational system.

Though the project mainly focused on the development of teachers and head teachers, through face-to-face events and interactions, it also created a regional, online forum to expand this interaction further to a regional level, with an attempt to sustain this online community for the purpose of sharing knowledge and expertise.

Some of the main corporate outcomes the strategy was expected to culminate include the following:

- Long-term relationships with ministers of education and ICT of the countries involved in the project. This was done through several approaches such as:
  - Establishing and maintaining groups of 'think-tanks'
  - Creating groups of core trainers on the project

 ○ Signing memoranda of understanding with related ministries.
- **Positive partnerships for social change:** Working together with those schools involved in the project, through educational supervisors, teachers and head teachers, on improving the quality and approach to education.
- **Self-development:** Teachers developing their skills in using ICT in class, thus resulting in positive outcomes for students and education.

The project involved training sessions which the British Council organised for members of the different participating groups, in addition to many meetings for middle and high-level Ministry of Education officials in the UK and their NENA counterparts, all aiming to raise awareness of the benefits of ICT in education, and to enable learning from the UK's experience of the professional development, skills knowledge-sharing, and new technologies used.

Key stakeholders on the project were:

1. Ministries of Education (MoE)
2. Ministries of Higher Education
3. Ministries of ICT
4. Centre for Educational and Research Development (Lebanon)
5. United Nations Development Programme (UNDP)
6. Birzeit University
7. UK educational specialists

An online Community of Practice (*Online Blackboard*) allowed networking of teachers from the region with their counterparts, supported and sustained by each of the British Council's Learning Centres' facilities and e-resources. This was a modern, high-tech, 'new' online tool which most of the participants were not familiar with and which they therefore lacked confidence in using. The online community works by networking teachers on the project to produce a model for supporting and sustaining teachers' skills and confidence in using ICT in the classroom by:

- Providing access and support from regional and UK practitioners
- Supporting teachers through the British Council Knowledge and Learning Centres (in each country) by providing them with internet access and ICT facilities.

The project began with a pilot teacher-training session, for teachers of Science, Mathematics and English, with selection criteria agreed upon with the MoE. Emphasis was laid upon the additional necessity of involving participants' head teachers on these training sessions, to ensure implementation, and supervision of learning points, in class.

PROJECT ISSUES, CONTROVERSIES AND PROBLEMS

Issues: ICT in Education Project, Jordan

With the importance of introducing ICT in education to the Ministry's agenda in mind, the British Council arranged a major Jordan-UK *ICT in Education* conference in March 2004. The conference outcomes clearly defined precise, practical, and prioritised recommendations related to curriculum, access, content, professional development, support and evaluation. It identified the key needs of each country, as well as niche areas for the British Council within the region and within individual countries, mainly in educational leadership, capacity building and enhancement, best practice, frameworks for e-learning resource creation and sharing, and English teaching with new technologies.

The project began with training teachers on ICT and how to use technology to facilitate learning, examining and grading. Though participants undertook ICDL training prior to their enrolment, this project's training further enhanced their ICT skills and how to implement skills practically in class, together with their methodologies. To further the participants' experience, the British Council introduced an online forum/community component to the project, which was a relatively new facility for all participants, and enabled them to share and learn from other experiences in the region. The project required a sufficient amount of involvement on online forums, which was quite challenging for participants due to their vague knowledge of how to use such a communication tool and to the restricted use of ICT facilities. Nevertheless, as the project progressed, with frequent training sessions and workshops, the participants were assigned tasks and duties which required interaction with their counterparts on the forum, thus strengthening their confidence. Naturally, this caused an increase in usage levels and more online engagement which, in turn, created a culture of knowledge-sharing amongst participants.

During the three years of the project, training on using ICT in the classroom was therefore cascaded to these teachers, and generally speaking, this training was of great benefit to trainee (cascaded) teachers, who indicated in their feedback that the skills they learnt on this training were used in preparing and presenting lessons, and also in developing their student exam material.

As head teachers' and educational supervisors' involvement on the first part of this project was to a lesser extent than their teachers, a need for training leaders in education (i.e. head teachers, school principals, and educational supervisors) emerged as a vital request. This was made evident during a conference held to define the further stages of ICT in education and the progress of ministries of education in this area thus far.

Training workshops were therefore designed and cascaded, targeting leaders in education, and so the *Strategic Leadership in Education* (SLICT) phase of the project came into being. The strategic leadership workshops were delivered based on a

SLICT curriculum, which was jointly developed by the British Educational Communications and Technology Agency (Becta) and the National College of School Leadership, UK. It was developed as a training course to enhance the ICT confidence of school leaders in taking the leadership on strategic changes towards improving the quality of education in their schools.

To cascade the training to further educational leaders, the training content was translated and adjusted to local needs, but copyright issues around the original material restricted its further usage.

The majority of participants on the SLICT training workshops showed a positive attitude towards utilising ICT in education and leading on such a change in their school's classroom teaching. On the other hand, utilising ICT for administrative and communicative purposes in the school, planning for and promoting the usage of ICT, and demanding more ICT facilities and equipment were major concerns with regard to ICT in education.

Amongst the several online resources available, teachers were encouraged to use wikis and blogs, and to use the Ministry's intranet (electronic platform) *'Eduwave'* and the project's online forum *Moodle*™ more frequently. Of particular note is that science teachers used lab simulation software to demonstrate experiments to their students, and this was observed by their subject supervisors.

The project ran between 2003 and 2008 with Jordan's Ministry of Education and was one of the major leading projects between the British Council and the Ministry. It also ran parallel in 4 other regional countries and their ministries, and was a pilot for regional British Council projects in the NENA region.

Controversies and Problems

Technology Components and Concerns: At the early stages of the project, the British Council recruited an online platform for the project *(Online Blackboard)* as a communication and knowledge-sharing facility between local teachers on the project. As the number of teachers was limited, and since they all worked on the same curriculum and accessed more or less the same educational and teachers' resources, the platform did not, unfortunately, yield the expected outcomes and benefits.

Online Blackboard lacked a user-friendly format, bearing in mind our participants were new to such a tool and to the concept of knowledge-sharing, especially online. In addition, ICT skills being at a relatively average level created a lack of confidence in this area, exacerbated by restrictions/limited access to computers and IT equipment in the workplace (at that time). The online forum users were not very comfortable using this platform, and thus usage was close to nil.

After careful evaluation, the British Council considered changing their online platform and did, in fact, replace it with *Moodle*™. *Moodle* was a far more extensive platform, with various facilities and tools which encouraged users to participate and interact together, and in a user-friendly format, with many language support features. (This was a major concern of participants, as the majority of participants were not proficient in communicating in English.)

The platform was structured based on participants' requirements for each country, i.e. each country had its subject divisions, communication tools, participants' names and profiles, etc., where they could communicate together on a local level. For those wishing to communicate with their counterparts in other countries, they were able to easily navigate between country pages and divisions, and communicate or take part in any discussion forum threads.

The introduction of *Moodle* was greatly appreciated by participants on the project, and it introduced them to new communication and resource tools they were not previously aware of, e.g. blogs, wiki, discussion threads and forums. What also made the platform more personalized was the ability of its users to upload photos, which

put faces to names. Though not, perhaps, of much importance in the IT world, considering our users and the availability of such a platform, this feature seems to have motivated their participation. What's also worth mentioning is that the capacity of the online platform was crucial in enabling participants to upload/download resources and lesson plans across the five countries.

At a later stage, the project's electronic platform underwent further major changes. The project moved its online platform from an external host to the British Council's hosted servers, and this was then the third change made to the online platform from the project's inception.

Participants of other British Council projects with teacher training components running in any of the five countries, were also encouraged to use *Moodle* as an online platform, but this created confusion amongst those participants of the *ICT in Education* project and the other projects. Online discussion projects were, in certain cases, irrelevant to the *ICT in Education* project, so this created a somewhat unnecessary burden on the platform's capacity. With regard to the number of users, this of course resulted in an increase in numbers, but not necessarily in active participation on the platform. What should have also been considered was the necessity of other projects' participants receiving proper training on the platform, in order to understand the various facilities of the platform, and how to employ such a platform effectively for their skills and knowledge development.

To ensure a frequent use of the online platform, online assignments and regular discussion threads could have been created, as these encourage frequent usage. However, this was neither clearly defined on the *ICT in Education* project, nor was there a person delegated such a responsibility. This was also absent from certain other British Council projects which benefited from the platform's presence (other than the *ICT in Education* project).

The manner in which the new platform was structured cancelled out the country divisions, and reduced the structure to only subject matter forums (mathematics, science and English), in addition to a general forum relating to utilising ICT in education. This allowed for more subject matter discussions.

Moodle represented an important platform for the users and the project itself; however, most participants pointed out the fact that it was not used as much as they had expected. An assessment of the usage of theplatform, carried out in April 2008, revealed that an average of 8.7 individual users visited the platform daily from all five participating countries, and amongst them only 2.6 were Jordanian visitors.

Other problems which limited the usage of the platform were identified: there was no designated area and/or forum for head teachers and educational supervisors to enable them to discuss issues related to the management and administration issues of schools; the limited use of the platform caused frustration amongst teachers, and thus it became less used and, in turn, less useful.

Management and Organisational Concerns: The online forum enabled stakeholders to connect for online communications, debates and the exchange of resources. *Moodle*™ was used by some teachers as well. Although the number of participants on the project was remarkably large, the number of active users on the online forum was relatively small in comparison. Such a difference was one of the major challenges on this project.

Other important outcomes included identifying the following:

- The need to provide a forum for exploring issues of capacity building and strategic educational leadership for ICT in schools
- The need to consolidate regional relationships with a view to informing future regional and UK collaboration in this area
- The need to share the outcomes of the NENA *ICT in Education* project with key stakeholders beyond the region.

Current Challenges facing the organization: Many of the challenges mentioned were experienced during the period of the project's lifespan (2003 – 2008), and although many were overcome, it is nevertheless worth expanding a further on such challenges and their relation to ICT:

IT infrastructure: usually MoE venues have some connectivity and/or infrastructure issues, specifically in connectivity speed during peak hours, which was due to the workload on their intranet *Eduwave*. In addition, there were security access issues: for instance it was not possible to open "unrecognised" sites, leading to a serious problem of restricting use of the online community (users were unable to communicate, or exchange knowledge with their counterparts, locally or regionally), plus the fact that downloading was extremely difficult and time-consuming.

Bureaucracy in procedures: because of the sheer value of the IT resources and equipment, certain institutions had quite rigid procedures for allowing their staff to access and use such facilities, so accessibility for teachers was not what was expected. The teacher(s) had to go through the process of requesting and justifying what they would be using of facilities for, the time, duration, websites accessed, etc. Of course, this varied between institutions, but generally speaking, this proved to be an issue.

Lack of availability of ICT: some schools did not have any IT equipment, and some had only a limited amount, whilst others had either outdated technology or connectivity was not available, even though the physical structure of their computer lab was complete and ready. Again, a delay in providing connectivity, IT equipment, or any ICT services to venues was due to internal processes, or a long 'waiting list'.

Capacity of the online community: with an expanding number of participants on the online forum, and quite a high flow of uploads, the online community exceeded its capacity limits, which created complications when participants were communicating together or exchanging documents. Half-way through the project, the British Council had to consider expanding the forum's memory capacity, as this did not allow for as quick a communication speed as expected, and therefore created confusion, especially as the project was running locally over remote geographical areas, in addition to the fact that an online forum was a new tool for some. This all resulted in participants "turning away" from frequent usage of the forum.

E-content monitoring: this was quite a challenge, as the project involved five countries working together on the online community forum, and naturally participants found it a good opportunity to showcase their own experience/schools/country, and so on. This resulted in various uploaded content which, in many cases, was irrelevant or indeed not related to education at all. An urgent need to monitor the e-content and restrict uploading was called for – a decision which was welcomed as a solution to lessening the forum's high burden on its capacity. Therefore, it was important to assign persons from the three subject areas (science, mathematics and English) to take the responsibility of monitoring and filtering their local e-content for relevancy, importance and benefit to the local curricula, and indeed worth sharing with counterparts in other countries.

Student vs. teacher IT knowledge: a surprising issue – unexpectedly, students' ICT knowledge and skills far exceeded their teachers', which was quite a challenge in class and needed wise control. In many schools and grades, students were not only more knowledgeable, but also more confident on ICT, which resulted either in their demotivation (sometimes distracting them from the lesson), or their being ahead of their teacher. This field challenge therefore requires careful consideration – the amount of time teachers designated for developing their ICT skills was far less than that of their students (due to personal commitments, duties, etc.), whilst students were more receptive to new technologies, and devoted more time and willingness to learn and develop

which, all combined, stretched the ICT knowledge gap between teacher and student.

Facilitating online communities: as this was a new tool for participants, it required a considerable amount of facilitation to motivate them to take part in all e-exchanges and communication, and particularly in subject areas where not many resources were available, such as mathematics and where participants were not confident or frequent users of ICT. It was difficult to find suitably motivating online facilitators with these subject backgrounds, who were also confident in ICT and able to "keep the ball rolling" through the e-discussions and e-content exchange. At certain stages, the subject forum fell behind, resulting in losing participants' collaboration and input on the forum.

SOLUTIONS AND RECOMMENDATIONS

IT infrastructure: fiber-optic communication infrastructure is being piloted in some schools and this is very promising with regard to connectivity quality, speed and technology. The author learned from participants on the project that this is one of the country's future plans for technology. This could bring highly positive outcomes for any ICT-related projects, with benefits reflecting on both teachers and students.

Technologies: it is worth mentioning the introduction of handset "gadget" technologies (amongst them the so-called One laptop per child), which the author expects to flood the market in sudden and massive quantities, and to therefore become easily available and more frequently used amongst students. This would cause an increase in involvement and interaction with many online resources and e-communities, which in turn would further students' ICT skills and confidence. Unless teachers are trained on and updated with the various changes in the ICT field, the student vs. teacher ICT skills gap will continue to expand further.

Sustainable e-interaction: it is important to plan how to guarantee sustainability of an online community, as it is at high risk of becoming little-used, out of date, or even obsolete. This comes through content – ensuring the quality and relevancy of content is well-maintained and updated, which will encourage participants to use the forum content bank continuously as a reference. Furthermore, the changing culture of knowledge sharing and encouraging participants to exchange information and knowledge will help secure a sustainable and wide usage – and naturally, 'word-of-mouth' recommendation plays an active role in spreading interest to newcomers to take part in online forums.

Language barrier and diversity: it is very important to enable communication to take place in various languages within online forums, even if they are located within Arabic-speaking countries only – it is important to bear in mind the diversity of the region: some of the resident population are of different origins (Kurds, Caucasians, Armenians, etc.), and they may be more comfortable communicating in their own language which, in turn, encourages the survival of such languages. The author noticed some participants on the project communicated with each other in Kurdish and Armenian, which was interesting, and reflected how well the forum supported diversity. Furthermore, limiting the forum to one language only may create a barrier, as not all participants may be confident in communicating in/using this language, which would restrict their access to and interaction with such forums.

Competing e-communities: we are all aware of emerging online communication tools (e.g. Facebook, blogs, Twitter, etc.), and how organisations are introducing these tools on their pages as channels to connect with their end-users. Online community users now have a wide variety of resources and information easily available and accessible. Such tools may have an effect on the level of participants' engagement with any online project communities and forums. The author ex-

pects that if such communication tools had been as widespread during the project's life cycle as they are now, this could have brought about a sharp drop in the number of participants engaging with and interacting on the forum.

To avoid such results in future ICT projects, it is important to look at online facilities from the perspective of:

- What input it would bring to the project
- How to secure participants' interaction and motivation
- How to build participants' confidence
- How to maintain up-to-the-minute content
- The facilitators' role and input
- How to stand before any competing forums and maintain position.

Students on e-communities: whilst managing the online forum, the author considered the possibility of involving students on the project's online community and forums. Though *Moodle*™ had the infrastructure to involve students, it was not part of the project's initial plan and there was no possibility for further alteration to the project plans. The author strongly recommends involving students with their teachers on online forums and communities in future projects. This would enhance the (educational) communication channels between them, motivate transparency and decrease the student vs. teacher ICT skills gap. It would give teachers the opportunity to be part of their students' "online world", thus building trust between the teacher and his/her students. However, it is important to consider a time allocation for teachers for such a duty, and also the availability of ICT infrastructure and equipment.

ICT favourability: it is very important to establish and, above all, very strict selection criteria when choosing participants for the project and all training events and workshops. Personal interference from internal decision-makers may lead to a downgrade of participants' quality and on participation on the project. It is crucial to ensure favourability does not play a part, or at least, does not negatively affect the desired outcomes.

Finally, it is worthwhile mentioning that, although Jordan is ahead of other countries in the region in terms of using ICT in education, nevertheless the Ministry still benefited from the project through its supporting of, and even guiding, the Ministry's professional development activities.

FUTURE RESEARCH DIRECTIONS

Future trends in ICT in education go towards school linking, exchanging and developing curricula between schools in different countries. *Connecting Classrooms*, a British Council project expected to build on the accomplishments of *ICT in Education. Connecting Classrooms* is a global programme that creates partnership between clusters of schools in the UK and others around the world. This partnership is expected to bring an international dimension of young people's learning, to improve their knowledge and skills in understanding other cultures. It is also expected to further develop teachers' skills and schools' curricula. Potential future research opportunity within this domain is worthwhile evaluating the impact *Connecting Classrooms* has brought on education and the facilitation of ICT in this project.

CONCLUSION

This study showed there were significant positive elements around introducing ICT in education, and the expansion ICT is taking place in NENA countries. However, certain restrictions and challenges remain present, be it with technicalities, infrastructures or skills, countries that have limitations in expanding their ICT infrastructure, as regulations and intervention hinders the grown and development of their ICT infrastructure, need to consider carefully changing their procedures, as technology is advancing at a speedy pace. Of course, this is expected to reflect on the educational reform, curricula and teachers' development.

REFERENCES

Kiridis, A., Drossos, V., & Tsakiridou, H. (2006). Teachers facing information and communication technology (ICT): the case study of Greece. *Journal of Technology and Teacher Education, 14*(1), 75–96.

Shirazi, F., Gholami, R., & Higon, D. (2009). The impact of information and communication technology (ICT), education and regulation on economic freedom in Islamic middle eastern countries. *Information & Management, 46*, 426–433. doi:10.1016/j.im.2009.08.003

ADDITIONAL READING

Doornekamp, G. (2002). A comparative study on ICT as a tool for the evaluation of the policies on ICT in education. *Studies in Educational Evaluation*, (28): 253–271. doi:10.1016/S0191-491X(02)80008-6

Kiridis, A., Drossos, V., & Tsakiridou, H. (2006). Teachers facing information and communication technology (ICT): the case study of Greece. *Journal of Technology and Teacher Education, 14*(1), 75–96.

Naumanen, M., & Tukiainen, M. (2009, October). *Guided Participation in ICT-education for Seniors: Motivation and Social Support*. Paper presented at the 39th ASEE/IEEE Frontiers in Education Conference, San Antonio, TX.

Pelgrum, W. (2001). Obstacles to the integration of ICT in education: results from a worldwide educational assessment. *Computers & Education, 37*, 163–178. doi:10.1016/S0360-1315(01)00045-8

Shirazi, F., Gholami, R., & Higon, D. (2009). The impact of information and communication technology (ICT), education and regulation on economic freedom in Islamic middle eastern countries. *Information & Management, 46*, 426–433. doi:10.1016/j.im.2009.08.003

Smeets, E. (2005). Does ICT contribute to powerful learning environments in primary education? *Computers & Education, 44*, 343–355. doi:10.1016/j.compedu.2004.04.003

Tondeur, J., Van Braak, J., & Valcke, M. (2007). Curricula and the use of ICT in education: two worlds apart? *British Journal of Educational Technology, 38*(6), 962–976. doi:10.1111/j.1467-8535.2006.00680.x

Wang, T. (2009). *Rethinking Teaching: How ICTs Can Positively Impact Education in Architecture.* Paper presented at the 2009 International Conference on Information Management and Engineering, USA.

Chapter 6
Strategic Information Systems Planning in United Arab Emirates Organizations

Fayez Ahmad Albadri
ADMO-OPCO, UAE

Salam Abdallah
Abu Dhabi University, UAE

ABSTRACT

This exploratory research focuses on SISP approaches in UAE organizations and their potential impacts on IS&T investment. The study uses a survey and simple model to characterize, classify and examine SISP approaches in 17 UAE organizations. Four typical cases deemed representative of the main SISP categories were examined further to determine specific strengths and weaknesses, and to appraise the suitability and effectiveness of each category considering the organization culture and business environment. The study concludes with emphasis on the importance of SISP to UAE organizations' ability to achieve their business goals. It also suggests that the proposed SISP classification model can be used by UAE and Arab organizations to identify any inadequacies in their existing SISP approaches, and understand the cultural shift needed to adjust these approaches to guide effective investment in IS&T that is aligned with business strategy and focused on business objectives and goals.

INTRODUCTION

In a competition-intensive business environment with increasing internal and external pressures, especially as the negative impact of the recent global financial crisis is surpassing the most gloomy outlooks, it has become critical for organizations to ensure efficiency and effectiveness of their planning, management and control of

projects and other initiatives geared towards the realization of their business goals and objectives. The wide recognition of Information Systems and Technology (IS&T) as business enabling tools, and their proven benefits to the business in modern organizations, has justified the investment of huge funds in a wide range of IS&T infrastructure and applications. Strategic Information Systems Planning (SISP) provides organizations with the processes to guide effective investment in IS&T, and to ensure that the adoption and assimilation

DOI: 10.4018/978-1-60960-048-8.ch006

of such technologies will culminate in tangible business gains and investment returns that can translate into competitive advantages.

(SISP) is a concern for both Information Technology and business managers. SISP is described as "a cyclic process to identify a range of information systems applications and related infrastructure to act as an enabler to drive business strategies for value creation" (Lederer & Sethi, 1988, Philip, 2007; Pricewaterhouse, 2008). The importance of IT strategic planning is indicated by its ranking in the in the top 10 IT management concerns for the past two decades (Philip, 2007).

When examining the IS literature we find little evidence on the use and influence of SISP in the Arab world, and virtually nothing on the influence of culture on adopted approaches. Therefore, research question that arises is of two parts: "How do organizations go about SISP, particularly those in the United Arab Emirates (UAE)?" and "Are there cultural influences on organizations' selection and adoption of SISP approaches?". This research is focused on the UAE, which is part of the Oil-rich Arabian Gulf region, and is becoming highly industrialized with GDP reaching US$ 190 billion. According to the research firm IDC, IT spending in the Gulf region was expected to exceed US$ 9 billion in 2008, with the United Arab Emirates being the second largest market for IT spending after Saudi Arabia (IDC n.d).

With investment of this size, it is important for managers to know if such investment is creating value for the stakeholders and to have some indicators of anticipated investment returns. We contend that IT planning generally and SISP in particular, is the critical stage that may decide how IT should be spent to drive value creation for competitive advantages in modern organizations.

This exploratory research investigates SISP approaches in UAE organizations and their potential impacts on IS&T investment. This builds on the findings of a case study that examined the approach of a major UAE Utility (Abdallah and Albdari, 2009).

A model that was developed and tested previously is used to evaluate SISP approaches in 17 UAE organizations who participated in a survey associated with this study, and to characterize and classify these approaches. Four organizations selected as typical cases that represent the main SISP categories are investigated in more detail to obtain a better understanding of specific strengths and weaknesses, and to examine the appropriateness and effectiveness of each approach category considering the business environment, organization culture and priorities.

The study concludes with a discussion of the selected group of cases and a number of recommendations to help UAE organizations understand to what category their approaches belong and to take necessary action to adjust their SISP approaches to guide suitable and effective IS&T investment that is aligned to business strategy and focused on the realization of business objectives and goals.

BACKGROUND

The concept behind Strategic Information Systems Planning is about using technology to support the organization's business strategy, with the intention of leveraging existing infrastructure for effective deployment and to guide future IT acquisitions (Boynton et al., 1987; Earl, 1993, Mentzas, 1997). Earl (1993, p.1) argues that SISP should address the following areas: a) Aligning investment in IS with business goals; b) Exploiting IT for competitive advantage c) Directing efficient and effective management of Information Systems resources d) Developing technology policies and architectures.

Earl argues that the first two areas are concerned with information systems strategy, the third with information management and the fourth with information technology strategy. Lederer et al. (1988) has also argued that SISP is a way to determine a portfolio of IS&T infrastructure to meet a business's objectives as declared in their business strategy. The process of planning is a complex task involving a high degree of dependency between the major business processes and technologies. Mentzas (1997) argues that although there are many formal methodologies to assist in planning, problems still exist in three areas: a) the need for integrating SISP with the overall corporate strategy; b) the moderate practical utility of existing SISP methodologies; and c) limited management involvement and commitment to SISP. Mentzas (1997) has proposed a generalized strategy planning approach as a foundation for developing SISP, and as a way to overcome these problems. The approach consists of five phases, namely: a) strategic awareness; b) situation analysis; c) strategy conception; d) selecting strategy and e) planning strategy implementation. Another commonly used approach to planning is proposed by Premkumar (1991), which is based on the input of information and resources that is later translated to a set of strategic plans and, when implemented, produces certain outcomes. Regardless of the methodology, Synnott and Gruber (1981) argue that the outcome of the SISP endeavor is usually a plan that includes: a) IS mission statement; b) IS objectives; c) linkage of the IS objectives to organization goals; d) IS action plan for achieving IS objectives; e) assignment of specific tasks to specific individuals or units; and f) mechanism(s) for management control, feedback and reporting. Although SISP should be comprehensive to be effective, Newkirk et al (2003) argue that both too much and too little implementation planning could have an opposite effect on SISP success. Newkirk also reveals that at the implementation phases, there is a need for more planning than the early stages.

Developing strategies is not enough; Teo et al. (2001) argue that one of the major problems in IS planning is translating strategies to action plans. Hartono et al. (2003) argue that managers should plan for implementation by explicitly including actions in their plans in order to avoid wasting valuable resources. The plans should as a minimum include sections pertinent to: a)

identifying resources; b) change management and c) organizational structure (Baker, 1995). Furthermore, Brown (2008) has also argued the need for the consideration of not only internal factors, but also external factors which may have the greatest impact and make "a well formulated plan obsolete and impractical". Brown et al. (2008) is referring to the external business environment and external IT environment.

SISP CLASSIFICATION MODEL

To investigate the issues and challenges that need to be addressed by UAE organizations in terms of adopting suitable and effective SISP approaches as a basis for their investment, acquisition and adoption of ICT technology, this study is primarily concerned with developing reliable tools and a mechanism that can be used to indicate the type of SISP approach that is being used by the concerned organization, and to pin-point the specific weaknesses and strengths of these approaches in terms of effectiveness and formalization, which can be followed by a corrective action for improvement.

The SISP classification model used in this investigation is based on the main SISP principles, components, features and attributes as described in the literature and discussed earlier. The study benefits from twelve SISP features that were extracted and validated by the authors as part of a previous case investigation (Abdallah and Albadri, 2009). For the purpose of this study the following features are deemed adequate to characterize a formal and effective strategic planning process: 1) approach-structured, 2) business-aligned, 3) management-sponsored, 4) technology-based, 5) long-term, 6) control-documented, 7) process-phased (initiation, formulation, implementation, and assessment), 8) team-dedicated, 9) analysis-oriented, 10) risk-managed, 11) Non-routine and 12) awareness-paralleled. These features were mapped to the two following key SISP attributes:

Table 1. SISP Classification features related to SISP attributes

SISP Attribute	SISP Feature
B. Effectiveness	1. Approach-structured 2. Business-aligned 3. Long-term 4. Analysis-oriented 5. Risk-managed 6. Technology-based 7. Awareness-paralleled
B. Formalization	1. Process-phased 2. Non-routine 3. Document-controlled 4. Management-sponsored 5. Team-dedicated

A) formalization and B) effectiveness (as shown in Table 1).

This classification model used in the investigation was intended to be used for characterizing and classifying SISP approaches adopted by different UAE medium to large size organizations who participated in the study survey. Questions corresponding to each of the SISP features outlined in Table 1 were given to the participating organizations' representative. With the help of a simple rating system, the answers of the participants were rated and consolidated to gauge the level of formalization and the level of effectiveness. The combined rating of the SISP formalization and effectiveness attributes were projected into a four-quadrant classification grid, by plotting the results into the quadrant to which the case is best fitted as shown in the Classification Matrix (Figure 1). All the participating organizations (cases) were represented in the Matrix based on their SISP profile, reflecting the combination of their SISP formalization and effectiveness ratings.

Category A (Formal-Effective) Profile

Category A (Quadrant 1) represents SISP approaches that are effective and formal. This category of SISP is highly disciplined in its approach to strategic planning and IS&T investment, and usually has a dedicated function, well defined

Figure 1. SISP classification matrix

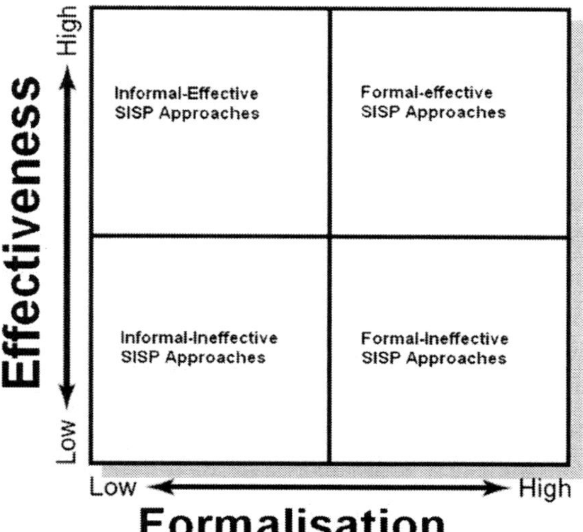

processes and a team with clear objectives and mandate to plan and manage an organizations' IS&T strategic plans (usually with undivided support from senior management).

As a result, this category is effective in analysis of the strengths and weakness of the organization, and capable of identifying opportunities and threats associated with the dynamics of the external environment and technology trends. It is common to have well defined short, medium and long term strategies and plans which include prioritized IS&T projects and initiatives supported by evidence of investment returns and risk mitigation plans.

IS&T investments pertinent to this approach are usually justified and should have a fairly high success rate in terms of realizing their objectives leading to business gains that contribute to the overall company goals and objectives.

Category B (Formal-Ineffective) Profile

Category B (Quadrant 2) represents SISP approaches that are formal but not adequately effective. This category of SISP may be structured, have the right processes in place and may even have a dedicated team and functions for SISP. However, the approach itself is not adequate as a result of not having the right or competent resources capable of accomplishing the sought objectives. In some cases the lack of alignment of IS&T investment with business strategy renders the planning unrealistic and not serving the company. Other reasons relate to not using appropriate analysis tools or not being aware of the dynamic of the external environment, especially those related to technologies.

This category, as a result will end up with an ineffective selection of technologies that can contribute to the business, and can culminate in lesser priority projects taking too many resources. Failed and ignored projects and initiatives are associated with this approach. This leads to triggering a 'blaming' cultural along with conflicts across functions and divisions. It is thought that the main cause in such cases is related to problems with vertical channels connecting middle and senior management on one hand and also horizontal disconnections across business and technology functions on the other.

Category C (Informal-Effective) Profile

Category C (Quadrant 3) represents SISP approaches that although informal are fairly effective in pursuing strategic planning. Although this category of approaches may not have clearly structured functions or processes (or even dedicated teams to be involved in strategic planning), competent individuals may take it upon themselves (sometimes with management support or encouragement) to do what is needed in terms of analysis and planning that forms the basis for producing strategies and plans. This category usually works fine until the disappearance of key people, management or personnel who were individually involved. The SISP that belongs to

this category is constantly threatened because of the lack of formality and documentation.

Category D (Informal-Ineffective) Profile

Category D (Quadrant 4) represents ad-hoc approaches to IS&T investment where there are no clear processes, functions or dedicated teams involved. There is no forward planning involved and the decisions for investment in IS&T are based on management hunch and impressions. It is not uncommon in such cases to see such organizations wasting resources and funds on many failed projects and unutilized applications. It is thought that the root cause of this problem is related to the business itself and the lack of clear vision by senior management. This category is common among companies that have no clear control over spending and typically survive due to government support and subsidy. In the private sector many organizations of this profile may vanish in no time.

THE CASE STUDIES

The study requested medium to large size UAE organizations to complete a survey questionnaire that includes a number of questions related to the SISP approaches adopted by the organization, specifically in the context of IS&T investment. Out of thirty (30) organizations that were approached for participation, only seventeen (17) consented to take part. The completion of the questionnaires was made flexible at the convenience of the participating parties, by allowing questionnaires to be completed either in face to face interviews, over phone or by email. The approached organizations were of diversified background, and include the main UAE industries of Oil & Gas, Utility, Banking, Manufacturing and Real Estate. A non-disclosure agreement with the participants has prohibited the inclusion of the names of the participating organizations and has limited the use of the collected information to only academic research purposes.

The distribution of the participating organizations by industry is illustrated in Figure 2.

The classification process has resulted in assigning different ratings to each of the 17 participating organizations as illustrated in the classification matrix (Figure 3).

Guided by the scores of the organizations' SISP approaches within each category, we have selected the following 4 different cases as representative approaches:

- Case B (combined score: 84%) to represent Category A (Formal-Effective)
- Case N (combined score: 64%) to represent Category B (Informal-Effective)
- Case Q (combined score: 37%) to represent Category C (Formal-Ineffective)
- Case K (combined score: 14%) to represent Category D (Informal-Ineffective)

Case B (Utility)

Case B is related to a major UAE utility organization. The investigation indicates clearly an adoption of an SISP approach and processes that are formal and effective. The organization gives great importance to the role of SISP, and has based all important decisions pertinent to IS&T investment on a three-year strategy and five-year plan which were developed and prioritized by a dedicated competent team who had open channels with business functions and senior management. The organization has managed in record time to rejuvenate its network and hardware infrastructure, created a data center, replaced its legacy systems by an integrated ERP system that supports all functional areas, has implemented important E-procurement and E-registration online functions and is currently exploring new state-of-the art technologies including RFID, Business Intelligence and Data warehousing. The organization has wisely implemented awareness campaigns in conjunction with new IS&T initiatives and projects. There are also some positive indications of investment returns which are contributing to gains in more senior management confidence and support.

Case N (Real Estate)

Case N is related to a leading Real Estate and Property developer. The investigation indicates that although the SISP approach is not formal, the investment in IS&T was generally effective. This could be attributed to the fact that the company embarked two years ago on a program to acquire

Figure 2. Participating organizations by industry

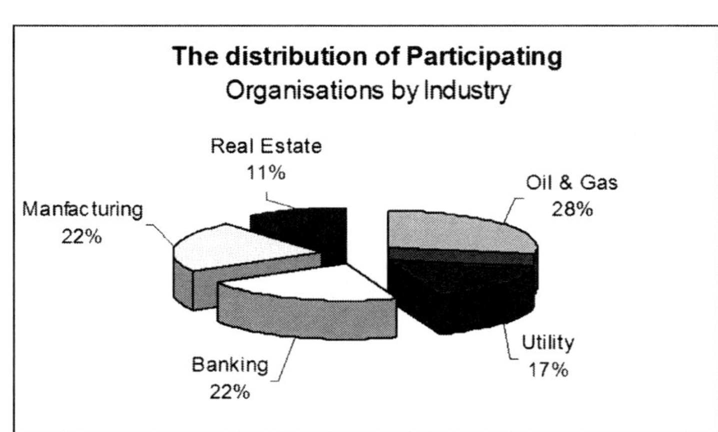

Figure 3. Survey results SISP classification matrix

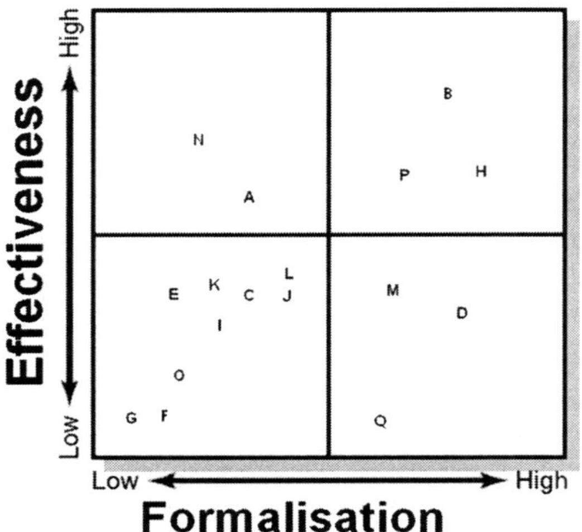

ISO certification and this has somehow allowed IS&T investment planning to be included in the scope of the preparations for the quality assurance certification. Although there are no formal SISP processes with dedicated teams and resources, the quality assurance control seems to have provided the mechanism to informally formulate, analyze and implement the selection process of suitable technologies. However, there is a major concern that after the company was ISO certified, the control mechanisms may be loosened and the badly needed continuous review of the IS&T options and priorities in a rapidly changing dynamic environment would be drastically compromised.

Case Q (Banking)

Case Q is related to a local bank. The investigation shows that the SISP approach adopted by the bank is formal in the sense that the IT department has within it a section that is dedicated for IS planning and projects, and the planning processes are documented. However, it was conveyed very clearly that the majority of IS&T investment is done by senior management and the role of the IT department is marginalized (especially in the case of major initiatives). This was evident in the management selection of the internet online banking and the back-office systems supporting the finance, accounting and loans departments. Management seems to view the role of the IT department as routine support for PCs and standard operating environments, and not for strategic planning and investment decisions. A close view of the main systems utilized indicates there is clear evidence of hiccups and failures and excessive hidden costs in post-implementation, as well as user acceptance problems and an attitude to over customize systems to reflect management's views.

Case F (Energy)

Case F represents the majority of organizations that have participated in this study. A shocking fact indicated that these organizations -some of which are very important- have no SISP approach or process to guide IS&T investment. Rather, a culture of ad-hoc decisions based on hunch and/ or copying others prevails without much consideration of the suitability of the acquired technologies. The lack of any competition and the subsidies covering faults, problems and failures provides the survival mechanism for such organizations. It is not uncommon to observe many disparate software and other IT systems acquired by different functions with pre-mature decommissioning. Costly IT artifacts are under-utilized or unutilized. Generally, these organizations suffer from lack of users' acceptance and the links between IS&T technologies and business are disconnect.

RESULTS AND DISCUSSION

This investigation was intended to give some insight on how UAE organizations handle their Strategic Information System Planning. This study indicates that although 35% of the participant organizations adopt formal SISP approaches, only 25% of those are evidently effective. This is alarming as it indicates that 75% of IS&T investments are not based on solid foundations that could be ensured if organizations were to commit to adopting formal and effective SISP processes.

The formal-effective cases represent only 18% of the participant organizations and it is clear that these organizations enjoy a much higher rate of IS&T success in terms of suitability, value added to the business and user acceptance. It is important to highlight that organizations with such profiles are more capable of coping with changes and the dynamics associated with the rapidity of technology and are more eligible to achieve their business goals and competitive advantage.

The informal-effective cases amount to 12% and seem to be driven by individual efforts rather than a robust system. Although the success record of IS&T in such organizations is fairly high, there is a looming risk of major setbacks in case of disappearance of the driving individuals. It is important that these organizations recognize these eminent risks and formalize their SISP processes in order to move into a team culture that manages tacit knowledge through proper measures.

The formal-ineffective cases represent 18% of the participant organizations and seem to be relevant to organizations that tend to have a limited view of the role of the technology function, and tend to handle IS&T investment as a senior management privilege. There is an evident disconnect between what the organization is trying to achieve as business goals and what IS&T applications are used for. The transformation of such organizations to benefit from their existing IT function requires senior management to establish strong links between the business and IS&T's role as enabling tools.

The survey indicates that 53% of the participating organizations have no SISP processes to guide their investment in technologies, and that they have an ad-hoc, hunch based, follow the leader approach. This approach has led to many failures and huge wastage which is concealed behind the excessive subsidies and support that these organizations enjoy. A major transformation

is needed in the culture of such organizations to redefine the relationship between the involved parties and the role of IS&T to serve the business goals. This can be achieved only through real leadership to affect change.

IMPLICATIONS AND FUTURE DIRECTIONS

This study has a number of implications that can impact the future outlook of SISP in UAE and Arab organizations. This study has successfully used the SISP classification model to characterize and classify the SISP approach in 17 UAE organizations. This is viewed as an essential first step for organizations prior to applying the shift needed to move from one SISP approach category (quadrant) to another. However, future research can help improve the model in terms of features and attributes, provide a clear mechanism with well defined procedure to make it possible for organizations undergo the assessment, and can also address the remedial strategies and actions that would be linked to the characterization and classification output.

Future research could also widen the scope to include more UAE organizations as well as Arab and international organizations. This will in effect increase the confidence in the model and the mechanism as a practical tool that can contribute to promoting SISP as a valid management tool, to guide IS&T investment and ensure that IS&T's role is efficiently utilized to support the business at a strategic and tactical level.

CONCLUSION

On the positive side, this study confirms that some UAE organizations do have SISP approaches which can be improved in both formalization and effectiveness. However, it is a major concern to know that the majority of IS&T investment was not based on proper strategic planning and thereby risking problems and failures.

There is also clear evidence that the selection of the SISP approach in any organization is influenced by different factors including: the organization's culture, business environment, management's view of IS&T's role, investment controls and other factors that can be subjects of future research. These factors clearly influence how seriously strategic planning is regarded, and how it can be utilized. However, in proposing that organizations should take the initiative to move towards effective and formal SISP approaches, the role of senior management cannot be underestimated.

The reported failures in terms of financial losses and negative impact on people and business should justify a major cultural transformation that results in a better understanding of IS&Ts and their role in modern business organizations.

This study suggests that important business gains and IS&T investment savings will be realized where a suitable SISP approach is implemented, and therefore encourages both more research in this important area as well as campaigns of organizational awareness in the UAE and the Arab world in general

REFERENCES

Abdallah, S., & Albadri, F. (2009). Strategic information systems planning: A case in the context of United Arab Emirates. *International Conference on Information Resources Management (Conf-IRM2009)*, Dubai, UAE.

Baker, B. (1995). The role of feedback in assessing information system planning effectiveness. *The Journal of Strategic Information Systems, 4*(1), 61–80. doi:10.1016/0963-8687(95)80015-I

Boynton, A. C., & Zmud, R. W. (1987). 'Information technology planning in the 1990: Direction for practice and research'. *Management Information Systems Quarterly, 1*(1), 59–71. doi:10.2307/248826

Brown, I. (2008). Investigating the impact of the external environment on strategic information systems planning: A qualitative inquiry. *Proceedings of the 2008 annual research conference of the South African Institute of Computer Scientists and Information Technologists, South Africa.*

Earl, M. J. (1993). Experiences in strategic information systems planning. *Management Information Systems Quarterly, 17*(1), 1–24. doi:10.2307/249507

Hartono, E., Lederer, A. L., Sethi, V., & Zhuang, Y. (2003). Key predictors of the implementation of strategic information systems plans. *The Data Base for Advances in Information Systems, 34*(3), 41–53.

IDC. (n.d.). *IDC*. Retrieved from http://www.idc.com/getdoc.jsp?containerId=IDC_P336

Lederer, A. L., & Sethi, V. (1988). The implementation of strategic information systems planning methodologies'. *Management Information Systems Quarterly, 12*(3), 444–461. doi:10.2307/249212

Mentzas, G. N. (1997). Implementing an IS Strategy - A Team Approach. *Long Range Planning, 30*(1), 84–95. doi:10.1016/S0024-6301(96)00099-4

Newkirk, H. E. 1; Lederer A.L., & Srinivasan C. (2003). Strategic information systems planning: too little or too much? *The Journal of Strategic Information Systems, 12*(3), 201–228. doi:10.1016/j.jsis.2003.09.001

Philip, G. (2007). IS strategic planning for operational efficiency. *Information Systems Management, 24*(3), 247–264. doi:10.1080/10580530701404504

Premkumar, G., & King, R. W. (1991). Assessing strategic information systems planning. *Long Range Planning, 24*(5), 41–58. doi:10.1016/0024-6301(91)90251-I

Pricewaterhouse Coopers. (2008). *Why isn't IT spending creating more value?* Retrieved March 2010, from http://www.pwc.com/en_US/us/increasing-it-effectiveness/assets/it_spending_creating_value.pdf

Synnott, W. R., & Gruber, W. H. 1981. *Information resource management*, New York.

Teo, T. S. H., & Ang, J. S. K. (2001). An examination of major IS problems. *International Journal of Information Management, 21*, 457–470. doi:10.1016/S0268-4012(01)00036-6

Chapter 7
ERP Post-Implementation Adoption Success Dynamics:
A Cultural Perspective

M. S. Akabawi
The American University in Cairo, Egypt

ABSTRACT

The purpose of this research is to explore the barriers of adoption and adaptation of Enterprise Resources Planning (ERP) systems at the post implementation phase within Developing Countries (DCs) business culture. Human agency-related issues and IT materiality factors specific to the DCs business environment were researched within the framework of post-implementation use of cross-functional information systems in a longitudinal case study. Focus group teams, composed of business units' end-users, managers and IT specialists (employment cohorts) were mobilized to elicit the causes of failing ERP services. The scope of this research fits within the studies that target the analysis of organizational adoption and use of ERP. Feedback loops analysis method was used for the study and presentation of the complex dynamic socio-technical behavior in enterprises to gain insight of the relationships among the many non-linear variables prevalent in the post implementation stage of the ERP lifecycle. Those variables were arrived at through canvassing stakeholder groups in the case firm, using focus group approach and the Causal Loop Diagramming as the analysis tool. The introduced methodology in this research is meant to highlight a new perspective to the understanding of the success model of ERP use and adoption.

INTRODUCTION

Over the past two decades, firms invested heavily in the acquisition and implementation of enterprise resource planning (ERP) systems as the preferred platform for managing their information and data processing needs. Many leading companies have adopted this approach since then and the world ERP market is expected to command a little under US$ 65 billion by 2009-year end according to AMR Research report, (2005). The cost expended by any firm for the acquisition and implementation of ERP exceeds millions of dollars and may affect the overall firm's earnings and revenues (Daven-

DOI: 10.4018/978-1-60960-048-8.ch007

port, 1998) and its market value, (Chatterjee et al., 2002). With such huge capital investments, many firms are focused to assess carefully their return of investment of such an infrastructure.

ERP systems are configurable commercial cross-functional software often embedding 'best practice' of business processes and operations workflow. The intent in adopting such data management strategy by firms is to reduce cost, increase productivity, customer satisfaction and supplier relations in an effort to improve their competitiveness in the global networked market (Davenport, 1998). However, the promise of high business returns proved, to many adopting firms, an elusive moving target. ERP project implementations often experience escalating budgets and exceed planned timelines (Schneider, et al., 1999). Despite these caveats, diffusion of ERP systems in organizations has, and still is, a phenomenon that drew the attention of many researchers as well as practitioners from both Information systems (IS) and Organizational Studies (OS) disciplines. The broad intention of these researchers is to establish methodologies for assessing the post implementation outcomes, accrued business benefits and the extent of adoption of those enterprise-wide systems.

To that end, most of the research work in this area is focused on identifying critical success factors that facilitate the implementation and "going live" of the ERP systems. However, there is noticeable dearth of research in the area addressing factors affecting the use of ERP systems beyond the implementation and "go live" phases, particularly within the emerging markets economies. Moon, (2007) published a comprehensive review of literature on the ERP systems which attest this assertion. In this review, he identified the major themes of the surveyed research in the area according to the following categories: (1) implementation, (2) using ERP, (3) extension, (4) value, (5) trends, and (6) education. In the first theme, given the formidable challenges posed by ERP system installation and implementa-

tion, many studies have identified factors, which positively or negatively influence the success of installing ERP systems. The theme relates mostly to the technical implementation success of such systems. The second theme, which is the focus of this research, investigates factors of success of adopting the system post the implementation that lead to improved operational performance and enhance organizational effectiveness through the use of intended information systems (IS) services within the fabrics of business processes and management decision models. Many researchers examined the impacts of those factors on the accrued value of ERP systems use to the adopting firms and the extent of success in adopting those systems at multiple organizational levels within the more advanced developed countries. In this chapter, we will address the research question: "how relevant and pertinent those success factors are for the adoption of ERP services in the evolving business organizations within developing countries (DC) business cultures"?

The focus in this research is to explore barriers of successful adoption during the post-implementation stage of ERP system operations in the context of a longitudinal case study setting within the Egyptian business culture, being one of the emerging market economies in the Arab world. The researched case firm has gone through two successive EPR implementations between 1997 and 2007 as detailed in next sections. People-related issues and factors (use and acceptance of IT artifacts) specific to the Egyptian business environment and human resources culture created barriers to the success of those two ERP implementations. Noticeably, human agencies using the first ERP were required to change attitudes, perceptions and use modality in the second one. Efforts to salvage the second ERP implementation and its adoption were sought after by the firm's management. Focus group teams, composed of business units' end-users, managers and IT specialist, were used to elicit causes impeding adoption of the new

ERP services and the apparent user community resistance to change.

To analyze the post-implementation complex socio-technical behavior in this case, we used feedback loops to understand the relationships among the many non-linear variables prevalent in the system. The use of the causal loop diagrams (CLD) tool, (Toole, 2005), during the brainstorming sessions was helpful in elucidating the interaction among the end-users and the IT specialists on the one hand, and the various levels of management on the other hand. The barriers and impediments observed were then translated into variables using feedback loops. The developed CLD graphs were deployed and used as a platform for communicating the issues and gaining consensus on the factors impairing the adoption of the ERP services within the case firm's business operations and management culture.

BACKGROUND

We address in this research the deployment, adoption and adaptation experiences with enterprise-wide information systems (IS) during the post-implementation phase in family-owned large trading and car assembly Egyptian manufacturing enterprise. The intent in using such cross-functional systems is to gain competitive advantage in the dynamic market of the automotive trading business in Egypt. At the beginning of the study, the case firm was composed of eleven affiliates which were subsequently merged into two by the start of year 2009. The firm was enlisted in the Cairo and Alexandria Stock Exchange (CASE) and the Capital Market Authority (CMA) following its public offering (IPO) in mid 2007. This resulted in the formation of Board of Directors and management restructuring. Several central corporate functional units served the activities of the two merged companies which included: IT, finance and accounting, legal, logistics and supply chain, human resources and administration, internal audit and corporate projects. The lines of business (LOB) for those two affiliates spanned the auto industry, from trading in Completely Build Units (CBU) to assembled Completely Knocked Down Units (CKD) for both passenger cars and commercial vehicles in addition to trading in auto spare parts, after sale services and other related commodities. The company has over 40 geographically dispersed sites spanning the entire Egyptian map, with the major clusters within the Greater Cairo area. Those sites range from large installations, such as assembly line factories, warehouses, showrooms, service centers, spare parts and other auto goods' outlets to small service workshops. Most sites are connected via WAN links using the Internet and/or dedicated tie links, with LAN connectivity within each site. Computing needs of the group are served through centralized datacenter, equipped with the appropriate server farm and communications infrastructure in the group's HQ in Cairo. The group employs over 6000 employees with one third of the workforce belonging to white-collar users who use IT gear in their daily work. The total revenues of the group during 2008 were well over 1 billion US$.

Use of information systems (IS) in the firm dated back to the early 90's, where some legacy and functional systems were deployed separately by the various units and departments. By 1997, the company started the implementation of its first ERP system by acquiring licenses for running BaaN ERP modules for almost all its lines of business. We observed that the implementations of those modules in the firm's operating processes were individually configured as separate instances with each unit's (company) separate master data. Many customizations were performed in the implementation phase in order to fit in the non-standard business processes of each individual affiliate company. These features encouraged non-integration between the subsidiaries and led to multiple fundamental operational difficulties. Moreover, the BaaN system vendor's weak regional presence and low market share at the time led to lack of support, e.g. internet connectivity was not used and remote support services mode was not deployed. In addition, the firm's relationship with the vendor's regional representatives was extremely strained. The failure of the BaaN ERP system to provide adequate level of information services encouraged the firm take the decision by 2004 to implement the Oracle EBS suite as the platform for operating its information management infrastructure and signed a contract with Oracle Egypt for licensing of its full EBS software suite.

The rationale for this decision was based on several key factors. There is strong local support base for Oracle Company in Egypt, in addition to the internet remote support facilities through Oracle firm's technical assistance request (TAR) model. There were evidences of sustainability and commitment of Oracle firm to the Egyptian market. The business outlook of Oracle at the time was clearly on the rise manifested by the firm's increased acquisitions and increased world-wide market share. EBS suite functionalities cover almost every line of business in the group. Also, there is abundance of local knowledge workers in the Egyptian market, trained on the full spectrum of the needed professional knowhow to operate and support the project.

It is important here to highlight that the use of longitudinal case approach in this research is chosen for several reasons. First, comprehensive and precise information on the rate of penetration of ERP system use in the Egyptian business market proved hard to compile because of the fluid stage the Egyptian business environment experienced during the last decade. Second, ERP vendors' statistics on their respective ERP product deployment in the Egyptian firms were extremely misleading (e.g. there were no differentiation between the use of few number of modules versus use of the full ERP modules, which defies the objectives of analyzing success factors for the category of Enterprise-wide IS). Third, the extremely low survey responses, commonly seen in IS studies in more advanced and developed

countries, discouraged conducting survey data collection within the less matured DCs culture. Fourth, the low number of adopting firms makes it difficult to select appropriate research sample size. Instead, use of focus group approach in a longitudinal case setting was chosen to enable inclusion of all stakeholder representations in the study. The rationale for using this method being that, people's perspectives are not fixed and are liable to change for a multitude of reasons. Consequently, research methods with potential to capture this fluidity may be more illuminating than other approaches such as surveys. The circumstances of the case firm's cultural setting thus, create a unique role for longitudinal qualitative research which can provide rich information on people's perspectives and how and why these are perceived to have changed over time.

REVIEW OF LITERATURE

The ERP research theme that addresses the analysis and measurement of success (or failure) of adopting the system within the operational and management layers in firms will be examined next, with emphasis on identifying variables that impact such endeavor within firms in different cultural settings. Assessment of the success of ERP software post its implementation is usually viewed through DeLone and McLean's (D&M) IS success model (1992, 2003) and the extensions made by many researchers such as Fryling, (2005), Gable et al., (2002), Sedera, (2002, 2004), Seddon, (1997) and Ifinedo, (2006a). D&M's model identified six independent dimensions as the principal success measurement constructs for assessing IS adoption. Gable et al., (2003), projecting those measurement constructs on ERP category, argued for limiting the dimensions to four: a) Information Quality (IQ), b) System Quality (SQ), c) Individual Impact (II), and d) Organizational Impact (OI); eliminating the User Satisfaction and Use dimensions in the original D&M model. Fryling,

(2005) in an attempt to accommodate the special characteristics of ERP project implementation, advocated *"user acceptance and IS success are highly influenced by the user community for which the ERP system is intended. The earlier a user is involved in the process the more likely they will ultimately be satisfied with the ERP and the more likely they will actually use the system"*. This assertion articulated using the original D&M IS success model during the implementation project execution phase. However, Ifinedo and Nahar, (2007), on the other hand, emphasizing the need to increase the understanding of the relationships among the relevant dimensions of ERP success in post-implementation phase, drawing on D&M model and Gable et al.'s, (2003) modification, developed ERP system success measurements model that encompassed 5 constructs: the four stipulated by Gable and colleagues, in addition to the workgroup impact construct (WI). He later updated this 5 constructs model by adding the vendor/consultant quality factor (VQ) as the sixth dimension. The novelty of those extensions for ERP systems success assessment lie in the recognition that ERP implementation and adoption is a multi faceted endeavor that encompass technological, operational, managerial, strategic, and organizational related components. In particular, the introduction of the workgroup factor as additional dimension is striking since successful implementations in large firms depends largely on the presence of reliable communication infrastructure to empower culturally-sensitive collaborative work among the user community. However, we believe the introduction of the sixth construct, the VQ factor, in the final version of his model is more pertinent for assessing success of the ERP system adoption and use within the emerging markets in DCs cultures as this research attested.

Among the critical factors that are cited in the original D&M model and its extensions in relation to ERP post-implementation is the information quality dimension. Many researchers have studied this factor. Xu et al., (2002) and Madapusi et al.,

(2007) asserted that variables crucial to achieve the desired quality information included: top management support, training, communication, employee relations, project team, quality control, and change management. As one can note, business culture, management beliefs and practices and employees' skill levels will directly influence such success factor. This article will be drawing from the above conclusions in identifying the appropriate variables and their impact on the ranking of success factors within the emerging market economies' cultures.

The impact of ERP system configuration and/ or setup revisions and enhancements on the ERP operation/use, within the larger context of system quality success factor, was also studied by score of researchers (Fryling, 2005; Nicolaou, 2007; Light, 2001; Mensching, 2004; Nah, 2001a). Many asserted that both the nature and timing of system changes are significant factors to be considered in this respect. Although these factors have not received much attention in relation to the factors impacting the adoption of ERPs' post-implementation, to most ERP practitioners, they have profound impacts during the operational phase and indirectly assert the importance of the Vendor/Consultant quality factor introduced by Ifinedo, (2007). ERP products are continuously subject to software maintenance in the form of patch fixes, version upgrades, add-ons and licensing policies changes by the ERP vendors. The likely disruptions in the operations in such mission-critical applications due to such incidents will undoubtedly send alarming and mixed signals to all stakeholders involved. Management will be obliged to consider what measures should be taken to secure business continuity, e.g. through investing in operational redundancy to achieve 24x7x365 operation mode. End-users will have to be alerted and retrained as a result of the implications of the fixes and/or upgrades on the look and operational changes upon applying them. IT organizations will have to alternate between working in project operation mode that these upgrades require, constrained by money, people and time, and at the same time guarantee the "business as usual" systemic use mode of the ERP by the organization. Moreover, these factors induce tight-coupling with the vendor/consultant to effect all these actions in safe and seamless manner, furthering the validity of Ifinendo's inclusion of vendor quality success construct in the IS success model.

Research on ERP systems adaptation and the ensued changes in business processes and organizational development cover wide spectrum. Factors such as those influencing successful implementation, from the operational (rather than technical) point of view were addressed by many researchers, e.g. the business operations coverage of the package and the number of licensed users (Francalanci, 2001; Kumar, et al., 2001; Markus, et al., 2000b; Parr and Schanks, 2000), the implementation process details (Sawyer and Southwick, 2002). The impact of system configuration and/or setup revisions and enhancements were also studied by Fryling, (2005); Nicolaou, et al., (2007); Light, (2001); Mensching, et al., (2004); Nah, (2001a), impacts of organizational and national cultures (Krumbholtz et al., 2000; Soh, et al. 2000), users and organizational learning (Fleck, 1994; Parr and Schanks, 2000), knowledge transfer from suppliers to the host organization (Lee and Lee, 2000), the configuration and setup of the package's parameter tables and their overall impact on the architecture and flexibility of the ERP packages for rapid adaptation (Fan et al., 2000; Spott, 2000). Some articles also addressed the impacts of ERP process standardization and the restructuring of organization tasks or business processes reengineering (Kumar and Van Hillgersberg, 2000). The entanglement of the ERP technology with other technological artifacts, such as networks, servers, workstations, etc., as well as commercial interests and external social practices have been addressed by Ciborra et al., (2000), Ciborra and Hanseth, (1998), Fleck, (1994), and how the interdependencies of those components impact the implementation and ad-

aptation in organizations. Many of those factors cited above are dealt with differently in various business cultures. For example, in DCs environment, operational misfits may require excessive use of user licenses – and with the management culture of reducing operating expenses (OPEX), manual and/or non-standard workarounds may impede the successful adoption of the system.

Most of the issues raised in the preceding review provide the ground work for the assessment of the success (or failure) of the second ERP system in the researched case during the post implementation stage. However, we deviate from the works of Sedera et al., (2002, 2004) and Ifinedo and Nahar, (2007) in two respects: a) cultural context, and b) methodology and analysis tool used. On the one hand, the work of Sedera, et al., (2002) focused on assessing ERP post-implementation success factors across the organizations' stakeholders (employment cohorts) within Australian Public sector Organizations, and used "a dual survey approach". In his research, he used the Delphi method to associate attributes such as data accuracy, database contents, etc. to the success factors developed by Gable et al., (2002). The statistical sample used comprised of 27 Queensland Government agencies, running live SAP systems. While Ifinedo and Nahar, (2007) addressed, more or less, the same research domain within the Baltic-Nordic region in Europe, and their sample focused on private sector organizations, again using survey measurement instrument which targeted around 500 firms in the Baltic-Nordic region. The approach used by them in associating attributes (45 in total) to the six success constructs developed in their model is based on the literature review and covered almost all the attributes developed by Sedera et al., (2002) mentioned earlier. Conversely, this research is conducted within the Egyptian business culture, targeting, more or less, the same research questions. The study addressed and considered the impact extent and ranking of the IS success factors within varied organizational and national cultures. We also used the longitudinal case study approach, as opposed to survey methodology using the Causal Loop diagramming (Toole, 2005) as the analysis tool for the study of the complex system dynamics of the human-technology interaction during the ERP post-implementation phase. Focus group deliberations during the CLD development sessions provided Delphi-like forums for the development of the attributes associated with the success factors in this research.

EXAMINING THE INFORMATION SYSTEMS SERVICES AND IT DEPLOYMENT IN THE CASE FIRM

By examining the BaaN implementation in the case firm, as detailed in previous sections, with D&M's IS model (2003) and Ifinedo's extended model (2007), we established that evidences of failure outweighed the success potential. Several reasons led to such conclusion. First, system quality is questioned, the implementation flaws mentioned resulted in less than satisfactory system functionalities. Customization actions to overcome the misfits between the functionalities offered in the package and that required by the firm, spread as a result of the workgroup-specific, country-specific, and business-practice specific business imperatives and obligations. These customizations led to proliferated uncontrolled and non-procedural updates of the databases and compromised their integrity. In addition, other classical failure factors such as lack of top management support, meagre project team competence, lack of project ownership, lack of standardization - both on the levels of data and processes, weak vendor partnership and faulty architectural choices (separate instances as opposed to group architecture implementation) contributed to the discarding of the system and management lost confidence in the system's information quality. End-users and management grew accustomed to requesting downloading data from the "open" databases to Excel sheets and reverted to end-user computing culture, with no consideration to data integrity or audit requirements, furthering the deterioration of information quality perception among the stakeholders.

Upon contracting Oracle Egypt for acquiring licences for the EBS ERP, a project for implementing the EBS suite was initiated and took around 30 months to realize against the planned 18 months timeline. The "go live" milestone was delayed due to several reasons, some technical such as the choice of running the system on MS Windows 2003 Server platform with the inherent 32 bit architecture limitations, others due to business imperatives such as the need to close financial year for auditing purposes and therefore start the new system operation by year-end. Cultural impacts were also manifested in the noticeable reluctance of end-users to switch from the old ERP practices to the new one, particularly in relation to the differences between the use of the two systems and the users' perceived usefulness to the individuals' work habits. Finally, the Oracle system "go live" signal was realized by January 1st, 2007.

To hold back the operating expenses (OPEX) of the Oracle ERP system operation, management decided to limit the used number of Oracle user licenses, implying that some transactions entry to the system will have to be batched in selected sites. The effect of such operation mode resulted in delayed transactions entry and the effectiveness and timeliness of the IS services were consequently influenced by the efficiency of human interaction, the physical movement of the source documents and the manual workflow followed thereafter. Most of these operational constraints resulted in several hindrances and slack in the effectiveness of the Oracle ERP system on both the transaction processing and management reporting levels in the case study firm.

Within a year of the operation of the Oracle ERP, it was evident that system adoption is experiencing major impediments. Manifestations of this were duly documented by the firm's auditors such as:

Figure 1. Demographics of the core focus group

Functional unit Member Role	Finance& Accounting	Supply Chain	Sales and Ordering	Manufacturing & supply Chain
Business Process specialist/ Subject Matter Reference	Director, Control Department	Director, Local Purchase	PC Sales Manager	Director, Manufacturing Warehouse and Logistics
	Director, finance and Accounting	Director, Warehouse and Logistics	CV Sales Manager	Manufacturing Costing Manager
Core Business	Bank Reconciliation Officer	W/H Officer	PC Sales Officer	Manufacturing Casting Officer (PC)
	Accounts Receivables Officer	Logistics Officer	CV Sales Officer	Manufacturing Costing Officer (CV)
Area	Accounts Payable Officer	Local Purchase Officer	PC (CBU) Ordering Officer	W/H Officer (PC) CKD
	Taxes and External Reporting Officer	Service Station Officer	PC Service Station Officer	W/H Officer (CV) CKD
Coordinators	Services Accounting Officer		CV Service Station Officer	Logistics Officer
Oracle IT Specialists	Finance Specialist	Inventory Management Specialist	Order Entry specialist	Manufacturing Specialist

- Balances were not available at the appropriate cost level,
- The inability to cut checks and pay vendors in timely manner,
- Many receivables issues required manual intervention,
- Incomplete reconciliations,
- Lack of key financial reports.

Moreover, as a result of the transparency and periodic audit reporting needs following the firm's Public Offering (IPO), pressures mounted from all stakeholders (operations staff, end-users, professionals and management) for more system agility and responsiveness to their real-time information needs. There was shared and general consensus among the stakeholders that IS support and services are less than adequate. To diagnose the root cause of those in-adequacies and establish the reasons of such shortcomings, management established an independent organizational unit mandated, as stated in its charter, *"to thoroughly analyze the shortages of the ERP services and advice on actions necessary to improve the system deliverables to the end users"*. The unit composed of members drawn from the IT organization and the core business units to fulfill this mandate. Figure 1 provides the demographics of this unit's composition.

Four business areas were identified as constituting the "hot spots" that badly need process streamlining, re-engineering and/or Oracle operational adjustments to bridge the functional misfit and gaps requirements. The four areas participated in those activities are:

1. The finance and accounting;
2. Supply chain for both automotive and non-automotive supplies;

3. Sales and Ordering for both Completely Built Units (CBU) and Completely Knocked Down (CKD) for both Passenger Cars (PC) and Commercial Vehicles (CV);
4. Manufacturing costing and supply chain.

The Analysis of Dynamics of Adopting and Use of ERP

The implementation and use of the new ERP system in the case firm represented radical change from the operational modes of previously discarded legacy and/or ERP systems. As noted by Davenport, (1998) and Brazel and Dang, (2008), ERP workflow crosses functional boundaries that exist in large organizations. The use of ERP systems in those organizations will entail changes in the workflow of business processes, either as a result of redesign of the old processes to fit the ERP's embedded "best practice" or re-alignment of existing process workflows and will also change the way interaction among various stakeholders in the organization is achieved. The changes Ifinedo, (2007) proposed to add the workgroup parameter to D & M IS success model is recalled here to manifest this assertion. Moreover, the ERP system cuts across all levels of management, it will be used as the transaction processing workhorse, populating the ERP database with the business transactions (financial and accounting data, planning and production data, sales data, order entry, …etc) initiated from the various business units with the intent of having one unified enterprise view of the organization's business conditions. To accomplish this, timeliness and integrity of the resulting ERP database must be fulfilled. For the operational management level, the ERP system will provide management reporting information to middle management for control and directing purposes. A management reporting system will thus be an important deliverable from the ERP operation. Executive and decision support information will also be a third category of information deliverables through the use of business intelligence (BI) module to manage the overall business and provide the top management with the needed integrated insights to achieve the firm's strategic goals and objectives.

To that end, synergy among the operational units in the case firm has to be provisioned to ERP user groups through a real-time platform in which some members act as creators/initiators of data items, others as consumers, with each alternating the role (initiator or consumer) according to the business model ensued as per the information manufacturing model (Ballon et al., 1998). Such a network of role playing vis-à-vis the ERP usage put the system operation at the center of the organization's information management model. The increasing reliance on seamless services of the ERP in real-time, concurrent transaction processing, also put the IT organization under pressure to provide reliable infrastructure operation, ranging from access the system through the data network, support the end user at all levels of usage and ensuring business continuity.

One can easily detect the evolving interdependencies and feedback loops inherent in the above described behavior. As the ERP operation moves through time, the feedback loops, which engulf all the variables present in this dynamic organizational behavior, will determine the ultimate fate of the adoption of the system. For example, if the creators/initiators of the data in business transactions delay posting them into the system, other consumers – such as supply chain manager or treasury officer will not have the unified picture upon which a decision is to be taken. Alternately, if the support engineers of the ERP system platform do not ensure that the link between the end-user location and the central data center location is not up and healthy, data may be lost and thus compromising the database integrity. Observing the operation of the ERP system in real-life will attest that the use of system is cast within complex social behavior prevalent with feedback loops. It is for this reason we concur with researchers such as Binbasioglu et al., (2002) in choosing "representation of the conceptual model that reflects complex social interaction". The cause-effect feedback loops model fits such a choice, (Toole, 2005). The use of this qualitative analysis tool is explained briefly in appendix (A). As will be shown later, the interpretation of the results of application of the tool led to the identification of the set of attributes necessary for the measurement of success factors that mostly influence the adoption within the cultural setting of this case.

Method

In undertaking the task of analyzing the Oracle ERP adoption success (or failure) in the case firm, focus groups were formed to gain understanding and insights of the post-implementation dynamics. Thirty focus group sessions for each area were conducted; each session lasted 3 to 4 hours with average participation of 10-15 persons comprising representatives from business unit management and Oracle end-users. Business area core members were given the responsibility of selecting appropriate candidates for the focus group meetings from their respective business areas. The demographics of the participating focus group members are given in Figure 2. Session mediation among the selected business area specialists (subject-matter process references) was provided by the researcher. The respective business area's Oracle module specialists from the IT organization (power users) were also included in those sessions. The teams aimed at eliciting information from the combined perspectives of the various stakeholders within the complex socio-technical dynamics among the Oracle ERP operation workgroups. According to Sedera, et al., (2004), those stakeholders represent the multiple perspectives for determining the perceived attributes associated with the success constructs within the case context. The intent was to qualitatively assess the extent of influence of

Figure 2. Demographics of the focus group members

Business Functional Area / Agent Role	Clerks	Unit Supervisor	Unit Manager	Functional Manager	BU Head	Total
F&A	18	8	6	3	3	38
- Controller	3	1	1	1		
- Bank A/Cing	2	1	1			
- AP	3	1	1	1		
- AR	3	2	1	1		
- Inv.	4	2	1			
- Cash Collection	3	1	1			
Supply Chain	31	12	7	3	2	55
- Inbound logistics	7	3	2	1		
- outbound logistics	8	3	2	1		
- Non-auto Purch.	7	3	1			
- Warehousing	9	3	2	1		
Sales and Ordering	22	7	5	3	2	39
- PC auto sales	6	2	1	1		
- CV auto sales	5	2	2	1		
- Spare sales	6	2	1			
- Service Centers	5	1	1	1		
Manufacturing	20	8	6	4	2	40
- CKD ordering	3	1	1			
- Assembly MRP	2	1	1	1		
- BOM	3	2	1	1		
- QA and Control	4	1	1	1		
- CKD warehouse	5	2	1	1		
- Dispatching	3	1	1			

the success constructs in the reference IS success models.

A total of 167 issues and problems were identified and mapped to the IS success model constructs, from the perspectives of usability, acceptance of the software deliverables, perceived usefulness and system quality. The focus groups referenced each of those 167 issues to the appropriate construct(s) (one or more of the constructs may associate with each issue) that mostly relate to D&M, Sedera, (2004) and Ifinedo's, (2006a, 2007) extended IS success factors. Deliberate omission of the user satisfaction construct was practiced due to the discarded earlier ERP implementation experience, which was always considered the baseline for most of the participants and represented a catalyst of resistance to the new ERP implementation. This particular substitution was necessary to off-set the cultural factors foreshadowing the Egyptian business human resources culture, e.g. retention periods for the Egyptian workforce are usually large compared to Western experiences, therefore most of the individuals involved in the new Oracle EPR were already experienced users and satisfied due to longevity of use of the ill-fated earlier one.

For each issue and/or problem from the 167 set of issues identified by the user groups, the team, together with the respective focus group members analyzed the causality pattern of the ensued issues. For example, the problem "delays in payment by Treasury due to cash flow issues as well as internal practices – checks are not issued unless the requesting party (unit, customer or supplier) follows up"; when brainstormed in the focus group forum revealed a series of cause-effect variables chain-reaction such as: flawed interaction between business processes among the involved units, reduced effectiveness of Oracle controls during Oracle transaction posting, caused delayed management decision due to ineffective Oracle reports, implicating flawed information quality which in turn pointed to the integrity of the Oracle database that is function in the volume of the posted transactions which looped back to the initial workflow issue. These series of cause-effect sequences were coded into 17 variables for all the 167 scripts that described the issues and problems surfaced from the gathered data according to each stakeholder group's perspective. These variables represented the collective attributes of the IS success factors as perceived by the stakeholders. They mostly matched the attributes established by Sedera, et al., (2002), Ainen and Hisham, (2008), and Skok, et al., (2001) within the cultural settings in their cited respective research work. Using the graphical CLD notation, the teams communicated those findings and ascertained from the participating members the sufficiency of the developed variable categories. Those are presented as the 17 top entries in Figure 3. Moreover, during a change process, resistance to change by agencies in response to the disruptive nature of the new ERP process scripts is observed. Six additional cause variables were introduced at the appropriate entry in the cause-effect loops to include such change management initiatives and assess their effect on the CLD model behavior. Those additional variables are numbered 18 to 23 in Figure 3. This set of coded variables was then used in the CLD model formulation as detailed in appendix (B).

The variables listed in Figure 3 are clustered into four main clusters: (a) Business practice induced variables cluster. This cluster addressed the following constructs given in the reference model: OI, II, WI and SQ constructs from the management group perspectives and views; (b) The Oracle system operation variables cluster addressed all six constructs in the success reference model, from the end-users (operations) group perspectives and views; (c) IT organization variable cluster addressed constructs associated with supporting the operational aspects of the ERP and relate to VQ, II and WI constructs from the IT professionals' perspectives and views. The fourth cluster addressed the exogenous directives, induced on the appropriate CLD variables, as agents

Figure 3. Case study cause-effect variables

Variable	Desired Change	Description	Domain
1	Decrease	Variance in management decision time	business
2	Increase	Degree of goodness of relations with suppliers/customers	business
3	Decrease	Variance in time of procurement/services and/or supplies delivery	business
4	Increase	Degree of interaction among business processes	business
5	Decrease	Variance in time of posting business transactions	business
6	Increase	Management reports effectiveness	IT System
7	Increase	Integrity of Oracle Database index	IT System
8	Increase	Volume of posted transactions into Oracle	IT System
9	Increase	Degree of operational ease of interface with forms/navigation	IT System
10	Increase	Effectiveness of operational controls in the interface of Oracle	IT System
11	Increase	Effectiveness/flexibility of Oracle set up and configuration	IT System
12	Decrease	Variance between implemented and actual business workflow	IT System
13	Increase	Usability of Oracle implementation and operation documentations	IT System
14	Increase	IT team competency	IT Org.
15	Increase	IT team support quality/effectiveness	IT Org.
16	Increase	IT team business workflow understanding	IT Org.
17	Increase	IT team-end user Team-spirit	IT Org.
18	Increase	Quality of managers/training/hiring	External
19	Increase	Quality of end-user training	External
20	Increase	suppliers/Customer satisfaction index	External
21	Increase	IT specialists training	External
22	Increase	Infrastructure to increase access/ process re-engineering	External
23	Increase	ERP vendor/SI Support extent	External

of change to enhance the ERP adoption success. In arriving at those exogenous variables, the researcher draws on the work of Sabherwal, et al., (2006) in which they asserted that *"as a long term strategy for improving system quality and individual impact constructs, efforts should be expended towards increased user training, attitude change and participation"* and the additional 6 variables were thus included. Those six variables, numbered 18 to 23 in Figure 3, represent change management control actions such as decisions to improve the skills of the IT team, new recruits and/or upgrade in management skill levels, end-user training, infrastructure upgrade, wider and more robust system accessibility and improved vendor support.

The 17 variables were mapped to the six success factors considered in Ifenido's model as shown in Figure 4. To identify the contribution of variables (attributes) assigned by each group to the respective IS factor, the weight for each factor is calculated by dividing the total number of the variables addressed in the groups' cluster for the factor by the total combined number of variables for all factors in the cluster. For example, in business cause cluster (management group), the number of references of OI, WI, II, and SQ factors are four, two, four and two respectively, with no references for IQ and VQ factors. Therefore, the respective weights for each success factor in the business cluster will be: 0.33, 0.17, 0.33, 0.17, 0 and 0. The calculated weights represent the perceived importance of factors by the cluster's cohorts group by the end of the deliberations of the CLD sessions within the case study context.

The causality interaction between the variables in the CLD model, as detailed in appendices (A) and (B), is based on how a change in the value of one variable will change the corresponding value in the dependent variable. Business cluster vari-

Figure 4. Variables-success factor mapping structure

Variable	1	2	3	4	5	6	7	8	9	10	11	12	13	14	15	16	17
Success factor																	
Org. Impact	X	X	X	X							X	X					
Workgroup impact				X	X		X			X		X					
Individual Impact	X		X	X	X	X		X	X				X	X	X	X	X
Information quality						X				X							
System quality		X	X	X		X	X	X	X	X	X	X	X				
Vendor quality							X		X		X	X	X			X	

ables, which represented the management perspectives, reference constructs acquiescent to the system quality influence on the end-user on the three levels of IS services: transaction processing (TPS), management reporting (MIS) and the executive and strategic support systems (EIS/SSS). Organizational, workgroup and individual culture will determine how the system use and usefulness are perceived by the user community. The second cluster, which represented the end-user perspectives, encompassed variables related to technical issues such as the designed database integrity, the set-up and configuration effectiveness, ease of use of end-user interfaces, the variance between actual process operation and the corresponding implemented process workflow inside the ERP system, the effectiveness of the implementation documentation and help screens and the effectiveness of the reporting system. Collectively, those variables reference the quality success constructs (SQ, VQ and IQ) in addition to the culturally sensitive impact constructs – OI, WI and II. The third cluster included variables pertinent to the IT organization capabilities and competence from various aspects (systems, business workflow understanding, attitudes, etc) which reference impacts of VQ, II and WI success constructs, from the perspectives of the IT professionals.

The researcher postulates that use of the causality loops to describe the dynamics of interdependency between the various variable (attributes) associated with the IS success factors, will provide more robust estimation of the perceptions of the stakeholders as opposed to the use of Likert-like scale values. The number of references to variables, in the set of CLD loops developed to describe the system dynamics, will provide more meaningful measure for the perceived impacts extent and weights for the variables. A cause variable appearing in a particular CLD loop will contribute to a change in the affected variable, thus impacting the dynamics of change in the system behavior. The frequency of a particular "cause" variable in loops implies its extent of influence on the overall system dynamics and the change process. To that end, we constructed the cause-effect matrix given in Figure 5 for the two cases of re-enforced enhancing and balanced swinging (RE/BW) loops polarities and re-enforcing excluding (RX) polarity (there were no loops with balancing stabilizing (BS) polarity and only one BW loop was found in the case's CLDs – refer to appendices A and B for more explanation of the BS, BW, RE and RX polarity states). In this matrix, cell (i,j) holds the frequencies of cause variable (i) and effect variable (j) in the total of the 40 constructed loops for polarity RE/BW and RX polarity (values are separated by a comma in each cell). For example, variable 1 in row 1 of Figure 5 is the cause variable for variable 3 (column 3 in the table) for 2 times in all loops that have RE or BW polarities and 3 times for RX polarity. The matrix is then divided logically according to the clusters of cause (rows in the matrix) variables. For each affected variable, we then calculated, cluster-wise, the percent of the cluster's effect relative to the total clusters' combined effects. The percentages of the "impact" of the cluster relative to the total clusters effect is computed for the following 8 cases for both the RX and RE/BW polarities.

Case 1: None of the external variables 18 to 23 frequencies to be considered in the calculation;

Cases 2-7: Calculation is repeated for variables 18 to 23 frequencies, one at a time, row cells sums increased by the corresponding variable's frequencies.

Case 8: All variables 18 to 23 frequencies are considered in the calculation and the summation of row cells are increased by all variables' frequencies.

The result of the calculations are presented in Figure 6 which provide the relative impact (in percentages) of clusters on each individual variable for the eight cases identified above for both BW/RE and RX polarities.

As stipulated in appendix (A), the objective of the study of the CLD model behavior is to identify variables detrimental to stabilization and hence contribute to the adoption problems. Variables, in RX loops are usually considered destabilizing since they aggravate the instability of the system. In general, lower frequency percent of a variable in RX loops (from the total frequencies of variables), (Figure 6), implies higher influence of this variable on system stability and hence improving system adoption potentials. In order to map the impact extent of those "affected" variables to the corresponding success factors, we calculated the weighted average value of the effect (impact extent) for each construct using the factor weights described earlier, and substituting in the following formula for each construct over all clusters:

Success factor percent of impact $= \sum$ weight of factor * average impact extent of variable in cluster

The calculation is limited to the RX state only — being the controlling state of the success of adoption. For brevity, we assumed only two cases out of the 8 combinations mentioned earlier: case-1 and case-8. The results are presented in Figure 7.

Analysis of the above results shows that the set of factors detrimental to the adoption success as perceived by the different stakeholder groups

Figure 5. Cause-Effect Variables Matrix or BW/RE and RX loops

Variable #	1	2	3	4	5	6	7	8	9	10	11	12	13	14	15	16	17
1			2,3														
2					1,0		1,1										
3		2,3			0,2							0,1					
4	0,1						5,0	1,0		2,1	0,1				1,0	3,0	
5							0,5	0,2	0,2		1,0						
6	1,1			2,2						1,0					1,0	1,0	
7			0,2	3,1		5,3			0,1	0,3	2,2	0,4		1,0	1,0		
8					0,1		1,8					0,2		0,1			
9							1,0	5,0						2,0			
10	1,1		0,1		0,4		1,1	0,1	0,1		1,0	0,1					
11			7,0		1,0		1,2					0,3					
12					0,2			0,3	0,6							0,1	
13																3,0	3,0
14									3,0			0,1	1,0		1,0	1,0	
15									0,1		1,0		1,0				
16							1,0	1,0			1,0		1,0	1,0	2,0		1,0
17														4,0			
18	2,3			12,4		6,3											
19								2,11	3,5								
20		2,2															
21														6,1			
22				12,3	1,9		11,17	2,11	3,5		0,1		6,0				
23				12,3			11,17			3,10	9,3	0,12	6,0	6,1	5,1	5,0	

Figure 6. Percentages of cluster impact extent on the "effect" variables for the 8 cases

BUSINESS   CLUSTER

Variable #	B None	w 18	/ 19	R 20	E 21	22	23	All	R None	X 18	19	20	21	22	23	All
1	0.0	0.0	0.0	0.0	0.0	0.0	0.0	0.0	33.3	16.7	33.3	33.3	33.3	33.3	33.3	16.7
2	100.0	100.0	100.0	50.0	100.0	100.0	100.0	50.0	100.0	100.0	100.0	60.0	100.0	100.0	100.0	60.0
3	100.0	100.0	100.0	100.0	100.0	100.0	100.0	100.0	50.0	50.0	50.0	50.0	50.0	50.0	50.0	50.0
4	0.0	0.0	0.0	0.0	0.0	0.0	0.0	0.0	0.0	0.0	0.0	0.0	0.0	0.0	0.0	0.0
5	100.0	100.0	100.0	100.0	100.0	50.0	100.0	50.0	22.2	22.2	22.2	22.2	22.2	11.1	22.2	11.1
6	0.0	0.0	0.0	0.0	0.0	0.0	0.0	0.0	0.0	0.0	0.0	0.0	0.0	0.0	0.0	0.0
7	54.5	54.5	54.5	54.5	54.5	27.3	27.3	18.2	35.3	35.3	35.3	35.3	35.3	17.6	17.6	11.8
8	14.3	14.3	11.1	14.3	14.3	11.1	14.3	9.1	33.3	33.3	11.8	33.3	33.3	11.8	33.3	7.1
9	0.0	0.0	0.0	0.0	0.0	0.0	0.0	0.0	40.0	40.0	20.0	40.0	40.0	20.0	40.0	13.3
10	66.7	66.7	66.7	66.7	66.7	66.7	33.3	33.3	10.0	10.0	10.0	10.0	10.0	10.0	5.0	5.0
11	11.1	11.1	11.1	11.1	11.1	11.1	5.6	5.6	60.0	60.0	60.0	60.0	60.0	50.0	37.5	33.3
12	0.0	0.0	0.0	0.0	0.0	0.0	0.0	0.0	8.3	8.3	8.3	8.3	8.3	8.3	4.2	4.2
13	0.0	0.0	0.0	0.0	0.0	0.0	0.0	0.0	0.0	0.0	0.0	0.0	0.0	0.0	0.0	0.0
14	0.0	0.0	0.0	0.0	0.0	0.0	0.0	0.0	0.0	0.0	0.0	0.0	0.0	0.0	0.0	0.0
15	12.5	12.5	12.5	12.5	12.5	12.5	6.3	6.3	0.0	0.0	0.0	0.0	0.0	0.0	0.0	0.0
16	37.5	37.5	37.5	37.5	37.5	37.5	18.8	18.8	0.0	0.0	0.0	0.0	0.0	0.0	0.0	0.0
17	0.0	0.0	0.0	0.0	0.0	0.0	0.0	0.0	0.0	0.0	0.0	0.0	0.0	0.0	0.0	0.0

SYSTEM   CLUSTER

Variable #	B None	w 18	/ 19	R 20	E 21	22	23	All	R None	X 18	19	20	21	22	23	All
1	100.0	50.0	100.0	100.0	100.0	100.0	100.0	50.0	66.7	33.3	66.7	66.7	66.7	66.7	66.7	33.3
2	0.0	0.0	0.0	0.0	0.0	0.0	0.0	0.0	0.0	0.0	0.0	0.0	0.0	0.0	0.0	0.0
3	0.0	0.0	0.0	0.0	0.0	0.0	0.0	0.0	50.0	50.0	50.0	50.0	50.0	50.0	50.0	50.0
4	100.0	50.0	100.0	100.0	100.0	50.0	50.0	25.0	100.0	42.9	100.0	100.0	100.0	50.0	50.0	23.1
5	0.0	0.0	0.0	0.0	0.0	0.0	0.0	0.0	77.8	77.8	77.8	77.8	77.8	38.9	77.8	38.9
6	100.0	50.0	100.0	100.0	100.0	100.0	100.0	50.0	100.0	50.0	100.0	100.0	100.0	100.0	100.0	50.0
7	36.4	36.4	36.4	36.4	36.4	18.2	18.2	12.1	64.7	64.7	64.7	64.7	64.7	32.4	32.4	21.6
8	71.4	71.4	55.6	71.4	71.4	55.6	71.4	45.5	66.7	66.7	23.5	66.7	66.7	23.5	66.7	14.3
9	0.0	0.0	0.0	0.0	0.0	0.0	0.0	0.0	40.0	40.0	20.0	40.0	40.0	20.0	40.0	13.3
10	33.3	33.3	33.3	33.3	33.3	33.3	16.7	16.7	90.0	90.0	90.0	90.0	90.0	90.0	45.0	45.0
11	33.3	33.3	33.3	33.3	33.3	33.3	16.7	16.7	40.0	40.0	40.0	40.0	40.0	33.3	25.0	22.2
12	0.0	0.0	0.0	0.0	0.0	0.0	0.0	0.0	83.3	83.3	83.3	83.3	83.3	83.3	41.7	41.7
13	0.0	0.0	0.0	0.0	0.0	0.0	0.0	0.0	0.0	0.0	0.0	0.0	0.0	0.0	0.0	0.0
14	16.7	16.7	16.7	16.7	8.3	16.7	8.3	5.6	100.0	100.0	100.0	100.0	50.0	100.0	50.0	33.3
15	50.0	50.0	50.0	50.0	50.0	50.0	25.0	25.0	100.0	100.0	100.0	100.0	100.0	100.0	50.0	50.0
16	50.0	50.0	50.0	50.0	50.0	50.0	25.0	25.0	0.0	0.0	0.0	0.0	0.0	0.0	0.0	0.0
17	75.0	75.0	75.0	75.0	75.0	75.0	75.0	75.0	0.0	0.0	0.0	0.0	0.0	0.0	0.0	0.0

IT ORG   CLUSTER

Variable #	B None	w 18	/ 19	R 20	E 21	22	23	All	R None	X 18	19	20	21	22	23	All
1	0.0	0.0	0.0	0.0	0.0	0.0	0.0	0.0	0.0	0.0	0.0	0.0	0.0	0.0	0.0	0.0
2	0.0	0.0	0.0	0.0	0.0	0.0	0.0	0.0	0.0	0.0	0.0	0.0	0.0	0.0	0.0	0.0
3	0.0	0.0	0.0	0.0	0.0	0.0	0.0	0.0	0.0	0.0	0.0	0.0	0.0	0.0	0.0	0.0
4	0.0	0.0	0.0	0.0	0.0	0.0	0.0	0.0	0.0	0.0	0.0	0.0	0.0	0.0	0.0	0.0
5	0.0	0.0	0.0	0.0	0.0	0.0	0.0	0.0	0.0	0.0	0.0	0.0	0.0	0.0	0.0	0.0
6	0.0	0.0	0.0	0.0	0.0	0.0	0.0	0.0	0.0	0.0	0.0	0.0	0.0	0.0	0.0	0.0
7	9.1	9.1	9.1	9.1	9.1	4.5	4.5	3.0	0.0	0.0	0.0	0.0	0.0	0.0	0.0	0.0
8	14.3	14.3	11.1	14.3	14.3	11.1	14.3	9.1	0.0	0.0	0.0	0.0	0.0	0.0	0.0	0.0
9	100.0	100.0	50.0	100.0	100.0	50.0	100.0	33.3	20.0	20.0	10.0	20.0	20.0	10.0	20.0	6.7
10	0.0	0.0	0.0	0.0	0.0	0.0	0.0	0.0	0.0	0.0	0.0	0.0	0.0	0.0	0.0	0.0
11	55.6	55.6	55.6	55.6	55.6	55.6	27.8	27.8	0.0	0.0	0.0	0.0	0.0	0.0	0.0	0.0
12	0.0	0.0	0.0	0.0	0.0	0.0	0.0	0.0	8.3	8.3	8.3	8.3	8.3	8.3	4.2	4.2
13	100.0	100.0	100.0	100.0	100.0	50.0	50.0	33.3	0.0	0.0	0.0	0.0	0.0	0.0	0.0	0.0
14	83.3	83.3	83.3	83.3	41.7	83.3	41.7	27.8	0.0	0.0	0.0	0.0	0.0	0.0	0.0	0.0
15	37.5	37.5	37.5	37.5	37.5	37.5	18.8	18.8	0.0	0.0	0.0	0.0	0.0	0.0	0.0	0.0
16	12.5	12.5	12.5	12.5	12.5	12.5	6.3	6.3	0.0	0.0	0.0	0.0	0.0	0.0	0.0	0.0
17	25.0	25.0	25.0	25.0	25.0	25.0	25.0	25.0	0.0	0.0	0.0	0.0	0.0	0.0	0.0	0.0

resulted in the following ranking of the factors of success in descending order (the higher the rank, the higher effect on adoption success):

In case of not implementing change management initiative:

Management group: IQ → VQ → WI → OI → II → SQ

End-user group: IQ → WI → OI → VQ → II → SQ

IT professionals group: IQ → OI → WI → II → VQ → SQ

In case of implementing change management initiative

Management group: WI → IQ → II → SQ → OI → VQ

End-user group: OI → II → SQ → IQ → VQ → WI

Figure 7. Stakeholders perceived effects on success factors

RX	Management		End-users		IT Professionals	
Factor	M-before	M-after	EU-before	EU-after	ITP-before	ITP-after
VQ	12.27	14.28	13.21	8.36	10.42	6.25
OI	18.61	14.03	12.84	2.17	4.17	16.67
SQ	26.48	11.61	24.09	5.28	16.67	4.17
II	23.52	8.23	20.24	5.03	8.33	8.33
IQ	4.91	6.05	5.05	5.65	4.17	10.42
WI	14.21	2.42	12.06	10.23	6.25	4.17

IT professionals group: WI → SQ → VQ → II → IQ → OI

Within the case study context, IQ factor is unanimously perceived by all groups as the most impacting factor for success. The three groups held different views on the second most important success factor. While management group perceived VQ as the second, the end-users group foresaw the WI in second rank, while IT professionals group perceived OI in second rank. Management and IT professionals evaluated the WI with lesser rank. The II factor was also perceived by management and end-users groups less influential, while IT professionals placed it in the midrange rank. This behavior may be attributed to the belief by IT professionals, in that - trained and qualified users (operations and management) will make a difference in intelligently using the system and therefore their impact will be a bit higher than the perception of the other two groups.

SQ ranking was somewhat unexpected, all groups perceived the factor with least impact rank for system adoption success. One offered explanation to this could be attributed to the hype: "well reputed system's quality is unquestionable". The unwarranted confidence in foreign products, perceived by many in the DCs, may have also bearing on such ranking. Further, OI factor effect on success was moderately perceived by the end-users and management groups, with the former placing it a step higher than the latter. This reflects the tendency of end-users in blaming the organization culture whenever confronted with defective operation, rather than admitting incompetence or inefficiencies.

With adoption of the change management initiative, considerable changes in perceptions were observed. The WI and OI ranks have seen change to higher levels of importance for success by both management and IT professionals groups, while WI assumed the lowest rank by the end-users group, placing OI factor highest instead. Such changes in the perceptions of those two factors reflected, in our view, the varying emphasis of each group vis-à-vis the outcomes of the change management process. Both management and IT professionals came to the realization of the importance of synergy among involved workforce in order to streamline business process workflows, which is dependent on workgroup synergy. Pressures and accusations in not fulfilling this streamlining would therefore be directed to the operations personnel – the end-users. As a defensive attitude, this group will revert to denial attitude and hence deliberate lowering in the factor importance was observed. The novelty of the use of CLD tool is manifested here in this result when the causality process captured these differences in perceptions by each participating group and embedding their prejudices in the loops structures. By the same token, one can rationalize the change observed in the OI factor ranking by the other two groups.

IQ factor, nonetheless, has seen retreat from being 'the' most critical success factor by all groups prior change management, reduced by the end-users and the IT professionals, while assumed high rank (second) by the management. This is quite explicable behavior, stemming from the varying views of the three groups towards the role of the IQ in the system operation, with the two groups: end-users and IT professionals, placing WI as top critical success factor (CSF), thus perceiving improved IQ. The IQ factor, per se, from their point of view is not the target success factor, rather the consequence. Invariably, IQ factor, being the factor management relies on for quality decisions, will be consolidating their point in putting it in a higher rank. Perceptions on the SQ have also seen more drastic changes from being the least CSF to become the second highest by the IT professionals, moderately high by the end-users group and moderately low by the management group. These variations in the SQ ranking point to the success of the change management programs in helping eliminate the myth of the cultural hype "trusted systems usually come from foreign reputed firms".

The increased awareness of individuals' roles for management and end-users groups, following the implementation of the change management, is evidenced in the increase in II rank as a CSF for both. However, IT professionals group's perception for this factor has witnessed no change. Also, the ranking of VQ factor has shown mixed changes. Management group reduced its rank to a minimal impact implying, while success in the early stages of the implementation of the system would be highly dependent on VQ and support, such dependency will diminish by time during the post-implementation phase. The reverse of this perception is seen in the IT professionals' group views. The increased ranking of this factor by the IT professionals, after the change management, imply their need for vendor support against the

other groups' contempt during the post implementation stage through securing tighter coupling with the vendor.

In considering the effects of the change management attributes (variables in the CLD model), as noted in the above insights and from Figure 6, appreciable reduction in the percentages of all variables in the RX polarity loops were observed upon their introduction, implying increased potentials for system adoption. Interpreting this assertion will imply the following remedial actions:

- For management decisions timeliness, the maximum impact is gained by adopting directives for raising the quality of managers, either through intensive professional development programs for the existing management staff or recruiting managers with the appropriate skills;

- Improving relations with suppliers and customers is directly linked to adopting new programs for Customer Relationship Management (CRM) and Suppliers Relationship Management (SRM);

- Fostering the integration among the business processes requires adopting combination of directives calling for increasing the managers' decision making quality, improving the communication among the business units by advancing collaborative infrastructure, and working towards stronger partnership/ increased support of the ERP system vendor;

- Containing delays observed in posting the business transactions into the system, improvement of the technology infrastructure;

- To increase the effectiveness and usability of Oracle reporting system requires highly qualified and trained managers;

- Increasing the integrity of the Oracle database is proportionally related to adopting all proposed directives with the actions related to infrastructure improvement and increased vendor support being the most critical;

- Increasing use of Oracle transaction processing capabilities (leading to increased volume of transactions posting) is highly related to the adoption of the two directives associated with improving the end-user training and enhancing the technology infrastructure in the firm. The other directives have also relatively lesser impact;
- Improving ease of use of Oracle user interface for system navigation and data-entry depends on adopting the above two mentioned directives, namely: improving the end-user training and enhancing the technology infrastructure;
- To increase the effectiveness of Oracle controls in the user-interface (during transaction posting for avoiding entry of wrong values and the like), the directive calling for the increase in vendor support is critical. All the remaining initiatives cause lesser impact extent;
- Increasing effectiveness and flexibility of the Oracle ERP set-up and configuration also shows the same dependency as the preceding point;
- Eliminating variances observed between the Oracle implemented workflow and the actual business workflow showed no dependency on any of the proposed directives implying that such need will have to be approached with other more fundamental initiatives such as Business process redesign and/or re-configuration of the Oracle system (in essence undertaking a re-implementation project);
- Improving usability of Oracle's system and user documentations is achieved through implementing the two directives calling for enhancing the technology infrastructure in the firm and increased vendor partnership and support;

- The IT team competence improvement is linked to the adoption of combination of increased IT team professional development, training and hiring new recruitments in addition to increased vendor partnership and support;
- The IT team support quality and effectiveness is achieved through adopting all directives with emphasis on increased vendor partnership and support;
- Improving IT team business workflow understanding shows the same pattern of dependency as the preceding point;
- Finally, improvement of the IT/end-user collaboration and partnership spirits (attitude change) will in general be dependent on adopting all proposed initiatives.

By noting the impact extent of the change management variables proposed in Figure 6 and their remedial implications listed above, we ranked (in descending order of priority) the strategies for improving success potentials of adoption within the cultural setting of the case:

1. Establishment of an on-going professional development programs for existing management staff (all levels) and contemplate new recruits for management posts that need visionary and agile management;
2. Establishment of a comprehensive and perpetual end-user technology and systems training and/or professional development programs;
3. Establishment of comprehensive and perpetual IT team technology and systems professional development programs, particularly in soft skills;
4. Increased partnership with the ERP vendor for committed support; and improvement of the system quality
5. Commitment to perpetual improvement of the infrastructure on all levels: networking (WAN and LANs), server farm, storage

network, end-user stations, increase Oracle user licenses, security infrastructure etc;

6. Implementing state-of-the-art CRM and SRM to improve relations with customers and suppliers and streamline their interfacing processes (second wave implementation);

7. Finally contemplating undertaking business process redesign initiative.

CONCLUSION AND FUTURE RESEARCH

This study contributed to the understanding of factors influencing success of ERP systems adoption within DCs business culture. We canvassed the stakeholders' perceptions and attitudes and their assessment of the success (or failure) dimensions of an operational ERP system in a longitudinal case study firm. Focus group sessions and feedback loops were used as the interpretive tool to gain understanding of the socio-technical dynamics among the stakeholders. The stakeholder groups involved in this study included managers (both strategic and operational), end-users and IT professionals. Variations in perceptions of success (or failure) by each group were collected during the causal feedback loops elicitation sessions. Causal-Loop diagrams (CLD) were used to capture these variations by way of abstracting the perceived success or impediments into variables. The variables set arrived at were generated by the participants in similar way as the attributes of the success factors enlisted by Sedera, et al., (2002) using Delphi method. Changes in variable values represented the changes required for the remedy of the noted post-implementation shortcomings.

The study highlighted the variations in the success constructs ranks as postulated in the IS success reference models (M&D, 2003; Gable et al., 2002; Sedera, et al., 2002, 2004; Ifinedo and Nahar, 2007) when applied within DCs cultural and business settings. To highlight the cultural impacts on the rankings of ERP adoption success factors by the various stakeholders groups, the results in the case study are compared next with rankings in studies undertaken by Sedera et al., (2002) and Ifinedo et al., (2007) applied within different cultures than that of the studied case's culture.

Variations in Management's Perceptions of Success Factors Ranking

The IQ factor's importance in the case study was placed at the highest rank before, and second after, the change management program adoption. This result conforms to both Ifinedo and Sedera's rankings with minor differences (e.g. the former placed the IQ factor rank at the top, while the latter placed it at the second to the top). This suggests that, regardless of cultural differences, management cohorts view the IQ as the determining factor for the successful adoption of ERP systems. While Sedera's study did not include the VQ factor, the comparison between the case's results and Ifinedo's suggests (for the VQ factor) that, without change management programs adoption, ranking of it shows no difference. However, with change management in place, the case's result revealed reversed rankings. This manifests the cultural differences in perceptions regarding the role of vendors/consultants from the two cultural perspectives. In the Egyptian culture, we noticed the eagerness (giving VQ higher rank) by the IT professionals to leverage their power by tightening the relationship with vendors/consultants.

Improved professional and technical awareness of the benefits of using integrated cross-functional systems by management group, after the change management program, imply putting more emphasis on WI and II factors vis-à-vis the system adoption. In more developed business cultures, employment laws permit recruiting the appropriate workforce for the operations of the ERP systems, whereas, in the Egyptian business culture, the labor laws rule out such freedom.

Therefore, management will be more concerned with existing personnel's skill set, interpersonal communications, work group spirit, etc. on the success potentials, hence emphasizing the critical roles of II and WI factors.

The SQ factor seems to have no major differences in ranks among the three studies with moderate success impact, with the exception of the case study's rank before the application of the change management program. This variation suggests that the group's knowledge about the ERP systems' misfits and loss of the vendor's credibility hype help in putting system quality in its proper perspective and rank.

Variations in End-Users' Perceptions to the Success Factors Ranking

The comparative analysis in this group is limited to the results of this research and those of Sedera's work (2002), since Ifinedo's (2007) did not include those cohorts category in his cited work. Moreover, Sedera's work included only four success factors SQ, II, IQ and OI, whereas the case's results cover the full six constructs.

The factors ranking in the case (after the adoption of the change management program) and Sedera's rankings for the OI and SQ factors are in agreement, however, without the adoption of the change management, the rankings were lowered for both factors, with the same order among the two factors. This suggests that impacts of cultural differences have no bearings on the other two factors, and more education and training will imply conformance to the pattern seen in the more developed business cultures.

The IQ and II factors also showed reversed ordering of ranks relative to each other in both studies. However, the results of the case study, with no change management adoption, showed conformance with Sedera's ranking pattern with higher ranking order (1 and 3 as opposed to 2 and 4). These observations suggest that cultural differences do exist as related to the individual's

emphasis on their roles in the success rather than the information quality, with WI factor playing a decisive role in balancing the reduction of the IQ importance.

Variations in IT Professionals' Perceptions to the Success Factors Ranking

The differences noted among the four set of rankings of success factors (two for the case study, Sedera's and Ifinedo's) for the adoption of the ERP suggest the significance of cultural variations between IT professionals in DCs and developed countries' contexts. Although Sedera and Ifinedo's results show consistent ranking among the four common factors with the exception of exchanged ranks of the SQ and IQ factors, the figures related to the case study (both before and after the adoption of the change management) show no consistent pattern among the compared studies. We attribute this to factors such as professional conduct, competency, longevity in working with such a complex system operations and training levels. However, it is our belief that this category of cohorts posed challenge to the understanding of the ensued cultural factors which have resulted in such varied perceptions among the three works.

In conclusion, the approach introduced in this research is meant to highlight a new perspective to the understanding of the success model of ERP in post-implementation phase. Further research related to this paper scope and the model it purports is to move beyond the CLD model's qualitative interpretive reasoning and combine it with the Stock and Flow diagrams (SFD) as contended by Haraldsson et al., (2006) to enable quantitative assessment of the impacts of ERP success attributes during the post-implementation phase and the magnitude of changes sought when applying change management programs.

REFERENCES

Ainin, S. & Hisham, N. (2008). Applying Importance –performance Analysis to Information Systems: An Exploratory Case Study. *Journal of Information, Information Technology and Organization, 3.*

AMR. (n.d.) *AMR Research (currently part of Gartner Inc.).* Retrieved from http://www.amrresearch.com/Content/View.asp

Bagozzi, R. (2007). The legacy of the technology acceptance model and a proposal for a paradigm shift. *Journal of the Association for Information Systems*, 8(4), 244–254.

Ballon, D. (1998). Modeling Information Manufacturing Systems to Determine Information Product Quality. *Management Science, 44*(4), 462–484. doi:10.1287/mnsc.44.4.462

Binbasioglu, M., & Winston, E. (2002). *Are you solving the right Problem? A Study of Packaged Software Implementation, International Association for Computer Information Systems*. Fort Lauderdale, FL: IACIS.

Brazel J. F., & Dang, Li. (2008). The Effect of ERP System Implementations on the Management of Earnings and Earnings Release Dates. *Journal of Information Systems 22*(2) fall 2008, 1–21.

Chatterjee, D. (2002). Shaping up for E-commerce: Institution Enablers of the organizational assimilation of Web technologies. *Management Information Systems Quarterly*, 26(2), 65. doi:10.2307/4132321

Ciborra, C. (2000). *From control to Drift*. Oxford, UK: Oxford University Press.

Ciborra, C., & Hanseth, O. (1998). From tool to Gestell. Agendas for managing information infrastructures. *Information Technology & People*, 11(4), 305–327. doi:10.1108/09593849810246129

Davenport, T. (1998). Putting the enterprise into the Enterprise System. *Harvard Business Review*, 76(4), 121–131.

DeLone, W., & McLean, E. (1992). Information systems success: the quest for the dependent variable. *Information Systems Research, 3*(1), 60–95. doi:10.1287/isre.3.1.60

DeLone, W. and McLean, E. (2003). The DeLone and McLean Model of Information Systems Success: A Ten-Year Update. *Journal of Management Information Systems, spring 2003, 19*(4), 9-30.

Fan, M. (2000). The adoption and design methodologies of component-based enterprise systems. *European Journal of Information Systems, 9,* 25–35. doi:10.1057/palgrave.ejis.3000343

Fleck, J. (1994). Learning by trying: the implementation of configurational technology. *Research Policy, 23,* 637–652. doi:10.1016/0048-7333(94)90014-0

Forrester, J. W. (1968). *Principles of Systems.* Cambridge, MA: Wright Allen.

Francalanci, C. (2001). Predicting the implementation effort of ERP projects: empirical evidences on SAP R/3. *Journal of Information Technology, 16*(1), 33–48. doi:10.1080/02683960010035943

Fryling, M. (2005). ERP Implementation Dynamics. *Information Science and Policy University at Albany. State University of New York, Oct.2005*

Gable, G. (2001). Large packaged application software maintenance: a research framework. *Journal of Software Maintenance and Evolution: Research and Practice, 13*(6), 351–371. doi:10.1002/smr.237

Gable, G., et al. (2002,). Enterprise Resources Planning Systems Impacts: A Delphi study of Australian public sector. In *Proceedings of the sixth Australasian Conference of Information Systems (PACIS 2002), 2-4 September 2002, Tokyo, Japan.*

Gable, G., et al. (2003, December). *Enterprise systems success: a measurement model.* Paper presented at twenty-fourth International Conference on Information Systems, 14-17, Seattle, WA.

Haraldsson, H., Belyazid, S., & Sverdrup, H. (2006, June). *Causal Loop Diagrams–promoting deep learning of complex systems in engineering education.* Paper presented at the 4th Pedagogical Inspiration Conference 1. Lund University, Sweden.

Ifinedo, P. (2006a), Extending the Gable et al. enterprise systems success measurement model: a preliminary study. *Journal of Information Technology Management, 17*(1), 14-33.

Ifinedo, P. (2007). Investigating the relationships among ERP systems success dimensions: a structural equation model. *Issues in Information Systems, 8*(2).

Ifinedo, P., & Nahar, N. (2007). ERP system success: an empirical analysis of how two organizational stakeholder groups prioritize and evaluate relevant measures. *Enterprise Information Systems, 1*(February), 25–48. doi:10.1080/17517570601088539

Kirkwood, C. W. (1998). *System Dynamics Methods: A Quick Introduction.* Ventana Systems, Inc.

Krumbholz, M. (2000). Implementing enterprise resource planning packages in different corporate and national cultures. *Journal of Information Technology, 15*(4), 267–280. doi:10.1080/02683960010008962

Kumar, K., & Van Hillegersberg. (2000). ERP: Experience and evolution. *Communications of the ACM, 43*(April), 23–26.

Kumar, V. (2001). *An investigation of critical management issues in ERP implementation: empirical evidences from Canadian organizations.* Technovation.

Lapp, C., & Ossimitz, G. (2008, June). Proposing a classification of feedback loops in four types. *Scientific Inquiry, 9*(1), 29–36.

Lee, Z., & Lee, J. (2000). An ERP implementation study for a knowledge transfer perspective. *Journal of Information Technology, 15*(4), 281–288. doi:10.1080/02683960010009060

Light, B. (2001). The maintenance implications of the customization of ERP software. *Journal of software maintenance and Evolution. Research and Practice, 13*(6), 415–429.

Loudon, K. (1997). *Compiler Construction Principles and Practice.* New York: PWS publication.

Madapusi, A., & Kuo, C. (2007). Assessing Data and Information Quality in ERP Systems. In *Proceedings of the Decision Sciences Institute Annual Meeting, Arizona.*

Markus, M., & Tanis, C. (2001b). The Enterprise System Experience – From adoption to success. In Zmud, R. (Ed.), *Framing the Domains of IT management: Projecting the Future through the Past* (pp. 173–207). OH: Pinnflex Educational Resources Cincinnati.

Mensching, J., & Corbitt, G. (2004). EPR data archiving- a critical analysis. *Journal of Enterprise Information Management, 17*(2), 131–141. doi:10.1108/17410390410518772

Moon, Y. B. (2007). Enterprise Resource Planning (ERP): a review of the literature. *International Journal of Management and Enterprise Development, 4*(3), 235–264. doi:10.1504/IJMED.2007.012679

Nah, F. H. (2001a). Characteristics of ERP software maintenance: a multi-cause study. *Journal of software maintenance and Evolution. Research and Practice, 13*(6), 339–414.

Nicolaou, I., Bhattacharya, S. (2007). Organizational performance effects of ERP systems usage: The impact of post-implementation changes. *International Journal of Accounting Information Systems. On-line edition, 7*(1), 18-35.

Parr, A., & Schanks, G. (2000). A model of ERP project implementation. *Journal of Information Technology, 15*(4), 289–304. doi:10.1080/02683960010009051

Sabherwal, R., et al (2006). Information system success: individual and organizational determinants. *Management Science* DEC-06.

Sawyer, S., & Southwick, R. (2002). Temporal issues in information and communication technology enabled organizational change. *The Information Society, 18*, 263–280. doi:10.1080/01972240290075110

Schneider, P. et al. (1999). ERPeople skills. *CIO magazine, 12*(10), 30-37.

Seddon, P. B. (1997). A re-specification and extension of the DeLone and McLean model of IS success. *Information Systems Research, 18*(3), 240–253. doi:10.1287/isre.8.3.240

Sedera, D., et al. (2002). Enterprise resource planning systems impacts: a Delphi study of Australian public sector organizations. In *Proceedings of the Pacific Asia Conference on Information Systems (PACIS)*, 584-600, Tokyo, Japan.

Sedera, D., et al. (2004). Measuring enterprise systems success: the importance of a multiple stakeholder perspective. In *Proceedings of the 12th European Conference on Information Systems* (pp. 1-13). Turku, Finland.

Skok, W. (2001). *Potential Impact of Cultural Differences on Enterprise Resource Planning (ERP) Projects*. The Electronic Journal on Information Systems in Developing Countries.

Soh, C. (2000). Cultural Fits and Misfits: Is ERP a Universal Solution? *Communications of the ACM, 43*(4), 47–51. doi:10.1145/332051.332070

Spott, D. (2000). Componentizing the enterprise applications packages. *Communications of the ACM, 43*(4), 63–90. doi:10.1145/332051.332074

Toole, T. (2005). *A Project Management Causal Loop Diagram*. Paper presented at ARCOM Conference, London UK., Sept. 5-7.

Xu, H. (2002). Data Quality Issues in Implementing an ERP. *Industrial Management & Data Systems, 102*(1), 47–58. doi:10.1108/02635570210414668

APPENDIX (A): QUALITATIVE ANALYSIS OF COMPLEX SYSTEMS BEHAVIOR USING CAUSAL LOOP DIAGRAMS

To characterize ERP system structures and study the dynamic behavior of those systems for the purpose of eliminating problems or unwanted behavior, one first needs to conceptualize from the specific events (variables) associated with those problems. Usually this requires investigating how one or more variable of interest change over time. In a business setting, variables of interest might be such things as cost, sales, revenue, profit, market share, and so forth. The intent here is to study patterns of behavior of those variables. The systems approach gains much of its power as a problem solving method from the fact that similar patterns of behavior show up in a variety of different situations, and the underlying system structures that cause these characteristic patterns are known, (Kirkwood, 1998). Thus, once we have identified a pattern of behavior that is problematic, we can look for the system structure that is known to cause that pattern. By finding and modifying this system structure, there is a likelihood of permanently eliminating the problem pattern of behavior. To understand the system structures, which cause the patterns of behavior, a graphical notation for representing system structure was developed by Forrester, (1968) known as the Causal-Loop diagramming (CLD) as depicted in Figure 8.

The Causal Loop Diagram (CLD) is a qualitative analysis tool for systematically identifying and communicating feedback loop structures (Toole, 2005). It enables communication of complex system behavior into simplified circular loop feedback structure. CLD promotes 'continuous' thinking, i.e. "*a story of a problem is read through the diagram and its development projected on a 'time scale graph' in order to understand the interaction of the feedback loop structure in the diagram*" Haraldsson et al., (2006). It is based on cause-and-effect relations between variables and shows their systemic behavior through the combination of several such relations in a network of connecting arcs between cause and effect variables. The Cause and effect variables either change in the same direction (indicated with a "plus" on the tip of the arc connecting them) or change in opposite direction (indicated with a "minus" on the tip of the arc connecting them).

Figure 8. Cause-Loop diagram (numbers refer to variables in Figure 3)

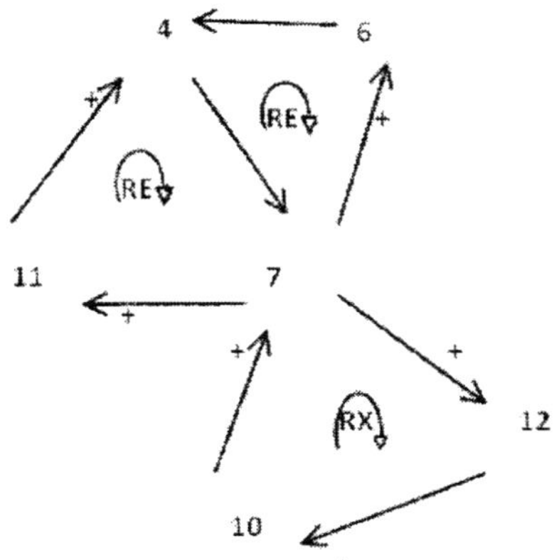

Processes that feedback in the same direction are called reinforced processes (indicated with R) since they amplify the condition. Similarly, the processes that feedback to give a change in opposite direction (indicated with B) balance (dampen) out a condition (Figure 8). Lapp and Ossimitz, (2008), have further extended those two feedback loops notations to encompass variety of behavior within the two. Reinforcing loops with zero negative links will depict enhancing behavior, however if the number of negative links is even and greater than zero it will depict an excluding loops. Likewise, balancing loops may depict either a stabilizing or swinging behavior depending whether time delay have (the latter case) or have no (the former case) impact on the cause-effect process. In this article, we will denote to the four alternate conditions proposed by Lapp and Ossimitz as BS, BW, RX and RE for *balanced stabilizing, balanced swinging, reinforcing excluding and reinforcing enhancing* respectively.

Projecting this system characterization in the context of ERP operation, one can equate the balanced stabilizing behavior of the feed-back loops of variables as a healthy behavior in which smooth adoption of the new operation order will ensue. While for both the balanced swinging and re-enforced enhancing states, the resulting dynamics of the variables' behavior will ultimately stabilize over a periods of time in the former state (BW), thus contributing to smooth adoption of the new operation order of the system; and in the latter state (RE), regenerates a new enhanced interaction which will ultimately result in behavioral adaptation of the new operation order of the system without much resistance or user refusal of ERP adoption. However, the RX system state implies that resistance to the adoption of ERP will ensue and therefore measures for reducing the number of those causalities that generate such condition should be sought after.

The intent from modeling the system in the CLD form is thus two folds: a) fortifying the stabilizing consequences of the preceding behavior (BS, BW and RE); b) identifying the underlying system structures that work against the stabilization and identify the measures and modifications necessary to eliminate those patterns of behavior, by noting those loops that contribute to this destabilization (RXs) and introduce measures and solutions to convert or eliminate them to stabilizing states.

CLDs are constructed to convey logically sound system states accepted by stakeholders and others. Modelers build logically sound CLD's and at some point, will also present their CLD's to others to communicate and elicit action. To ensure validity of developed CLD models some conditions are examined to assert the logical constructs developed. These are: (a) Clarity; (b) Quantity existence; (c) Causality existence; (d) Cause insufficiency; (e) Additional cause; (f) Cause-effect reversal; and (g) Predicted effect existence. These modeling guidelines along with the above cited verification conditions will be applied on the case studied to examine the ERP adoption and post-implementation dynamics.

APPENDIX (B)

The CLD model for the 23 variables identifies as per Figure 3, was formed by all participating stakeholders using the graphical CLD diagrams (refer to example given in Figure 8). Because of the large number of the interacting variables, which yield a cumbersome mesh of interweaving graphical diagrams, we opted for devising expression representation of the CLD graphs, drawing from the Backus-Naur Form (BNF) (Loudon, 1997) representation for computer language syntactic constructs. For this Case's cause-effect loops, one can represent any loop using the following abstract notations:

```
Loop:: = {P V_i {B E_k {S E_l}* B}* S V_j {B E_k {S E_l}* B}* }{S {P V_i {B E_k {S E_l}*
B}* S V_j {B E_k {S E_l}* B}*}^n
```
    where

```
P:: = +|-              denoting polarity of the arc connecting cause and effect
variables
S:: =, |;             variables/ relations separator
B:: = (|)        open/close bracket
{ }:: = expression
{ }*:: = repeated expression 0 or more times,
{ }^n:: = repeated expression n times           for n > 0
V_i                cause/effect variable i          V i =1 to 17
E_k                exogenous variable k              V k=18 to 23
```

40 CLD loops were constructed for the 23 identified variables during the focus group deliberations. Figure 9 provides the BNF expressions for those 40 graphical diagrams using the above loop syntactic construct. The intent for using this method is to have a tabular form for reporting the results, albeit loosing the novelty and ease of communication of the CLD graphs. However, we claim that parsing those loop expressions to draw the graphs is a straightforward task.

Figure 9. CLD graphs of the Case study represented as BNF expressions

Loop #	Syntactic Construct	Loop Type
1	-3,2;+2,11(23);+11(23),7(22,23);+7(22,23),6(18);-6(18),1(18);+1(18),3	RX
2	+4(18,22,23),7(22,23);+7(22,23),6(18);+6(18),4(18,22,23)	RE
3	+7(22,23),9(19,22);+9(19,22),8(19,22);-8(19,22),5(22);-5(22),7(22,23)	RX
4	-8(19,22),12(23);-12(23),10(23);+10(23),8(19,22)	RX
5	+3,5(22);-5(22),7(22,23);+7(22,23),10(23);-10(23),3	RX
6	+10(23),7(22,23);-7(22,23),12(23);-12(23),10(23)	RX
7	-3,2(20);+2(20),11(23);+11(23),7(22,23);+7(22,23),6(18);+6(18),4(18,22,23);+4(18,22,23),10(23);- 10(23),1(18);+1(18),3	RX
8	-3,2(20);2(20),5(22);+5(22),11(23);+11(23),7(22,23);+7(22,23),6(18);+6(18),4(18,22,23);+4(18,22,23),10(23);-10(23),1(18);+1(18),3	BW
9	+6(18),16(23);+16(23),15(23);+15(23),11(23);+11(23),6(18)	RE
10	+11(23),4(18,22,23);+4(18,22,23),7(22,23);+7(22,23),6(18);+6(18),15(23);+15(23),11(23)	RE
11	- 3,2(20);+2(20),7(22,23);+7(22,23),6(18);-6(18),1(18);+1(18),3	RE
12	- 3,2(20);+2(20),7(22,23);+7(22,23),6(18);+6(18),4(18,22,23);-4(18,22,23),1(18);+1(18),3	RX
13	+3,5(22);-5(22),9(19,22);+9(19,22),8(19,22);+8(19,22),7(22,23);7(22,23),3	RX
14	+12(23),5(22);-5(22),9(19,22);+9(19,22),8(19,22);+8(19,22),7(22,23);+7(22,23),10(23);-10(23),12(23)	RX
15	-10(23),5(22);-5(22),8(19,22);+8(19,22),7(22,23);+7(22,23),10(23)	RX
16	+4(18,22,23),7(22,23);+7(22,23),11(23);+11(23),4(18,22,23)	RE
17	+4(18,22,23),10(23);+10(23),7(22,23);+7(22,23),11(23);+11(23),4(18,22,23)	RE
18	+4(18,22,23),7(22,23);+7(22,23),6(18);+6(18),10(23);+10(23),11(23);+11(23),4(18,22,23)	RE
19	+4(18,22,23),9(19,22);+9(19,22),7(22,23);+7(22,23),4(18,22,23)	RE

Chapter 8
Preparing People to Manage, Support and Use Enterprise Systems in an Arabian Gulf Context

Hamed Salim Al-Hinai
University of Sunderland, UK

Helen Edwards
University of Sunderland, UK

ABSTRACT

Preparing organizations and assessing their readiness to implement enterprise systems is a growing agenda in information systems research. This chapter examines the challenges that face the preparation of organizations' stakeholders in managing and supporting the implementation and using enterprise systems in an Arabian Gulf context. The chapter points out the major issues that have been identified by enterprise systems' professionals such as: lack of business and IT strategies, continuous top management support, insufficient training, inefficient communication procedures, inefficient change management and cultural barriers. To overcome these challenges, the chapter suggests that the preparation should not be limited to end users since the preparation of other stakeholders, such as top managers, project manager and project teams, is essential to successfully implement and use the enterprise system. Based on insights gained by empirical research with enterprise systems professionals, the chapter recommends a number of factors that need to be considered. These have been grouped into the categories of: organizational, managerial, technical, external, and human/social factors.

INTRODUCTION

In recent years we have witnessed a growth in information systems research focused on preparing organizations and assessing their readiness to implement enterprise systems. This concentration of research could be attributed to the high demand and investment into the implementation of enterprise systems around the world, coupled with the various challenges and issues that have led to high failure rates.

DOI: 10.4018/978-1-60960-048-8.ch008

Technical factors tended to be the main focus of early research of enterprise systems. However, more recent investigations of enterprise systems' critical success factors (CSFs) have revealed the importance of the human domain in managing and supporting the implementation, and in using enterprise systems. In particular, researchers have investigated the attitudes and behavior of enterprises systems' stakeholders in accepting and using these systems, with especial interest given to end users. We may surmise from this that the human domain is a crucial domain that needs to be carefully considered in preparing to manage and support both the implementation and use of such systems.

This chapter considers this human domain, but specifically within the context of enterprise systems implementation in the Arabian Gulf. Although many of the challenges that have been identified might seem similar to those faced by organizations in other parts of the world, the severity and criticality of these factors differ from region to region and country to country.

The chapter makes explicit the range of major stakeholders' categories that are involved in enterprise systems implementation, such as: top management, project manager, project teams, key users and end users. It then also examines the factors that have been known to be relevant in the Arabian Gulf context. The list of issues include: lack of business and IT strategies, disturbed top management support, under-skilled resources, inefficient communication procedures, inefficient change management and inadequate training. The factors presented and discussed in this chapter have been identified by both a thorough review of the research literature and empirical data collected from stakeholders involved in enterprise systems within the Arabian Gulf. These factors, that influence the preparation of stakeholders, have been classified into the categories of: organizational, project management, technical, external and human/ social.

BACKGROUND

Enterprise systems, such as E-Government, enterprise resource planning (ERP), customer relationship management (CRM) and data warehouse systems are complex solutions that seek to integrate the different functions in an organization into appropriate business processes. Their aim is to provide holistic integrated, dynamic cross-departmental business processes that enable the organization to cut down the time required to complete identified transactions or processes (Aladwani, 2001; Appuswamy, 2000). Where this aim is realized the benefits to the organization are often in improved customer service (and streamlined processes); for instance, the customer no longer has to wait in long queues or visit different offices in order to complete his transaction (Appuswamy, 2000). Such integrated systems can cover functions such as finance, procurement, supply chain, maintenance, human resources, etc (Abdinnour-Helm, Lengnick-Hall, & Lengnick-Hal, 2003). The integration is not limited to functions but can also draw in the wider sociotechnical system which includes the integration of data and people within the organization. In such cases the transparency and availability of information to decision makers in the organization is tremendously improved. The promised benefits of enterprise systems have led to high investments by organizations around the world in order to introduce these systems: either to get a competitive edge in their marketplace, or, in certain cases, simply to survive. Similarly, organizations in the Arabian Gulf countries are also investing and introducing enterprise systems to improve and streamline their business processes: especially given the economic boom that they have experienced in recent years. For example, government organizations are implementing E-government solutions while private sector organizations are introducing ERP and CRM solution to support much of their business activity.

However, the implementation of enterprise systems is a non-trivial task and is surrounded by various challenges, issues and risks (Appuswamy, 2000; Nah, Lau, & Kaung, 2001; Ngai, Law, & Wat, 2008). Some of the major issues that confront implementers are: people, business processes, cost overruns, time overruns, hardware, networking and data migration, configuration and problems of integration with legacy systems (Appuswamy, 2000). To understand why these issues occur, we can consider the simple example of the packaged practices that are inbuilt into specific ERP solutions. These are considered, by the vendors to be "best practice", however, they may not fit a specific organization's requirements in terms of supporting its business functions, language and culture. Therefore the ERP implementation is likely to be hampered by these mismatches (Holsapple, Wang, & Wu, 2006). In many of the enterprise system case studies that have been reported in the literature such implementations have been classed as either complete failures, or as projects with severe cost and time overruns. (Stewart, Milford, Jewels, Hunter, & Hunter, 2000) contended that ERP implementation is "a significant intervention" to the life of the organization as it implies various organizational changes that impacts and is impacted by a number of variables including organization's culture, decision-making strategies, risk taking orientation, leadership strategies and perception of the value of information technology. It has been suggested that the failure rates for such systems could be as high as 90% (Magnusson, Nilsson, & Carlsson, 2004), and often the root causes of the failures have been traced back to challenges related to organizational, technical, economic, and people issues (Sarker & Lee, 2003).

To tackle the problem, investigators have conducted a wealth of research on the enterprise systems phenomenon trying to determine the most suitable environment to enhance the likelihood of successful implementation. Practitioners and consultants have adopted different methodologies and frameworks to achieve success. Unfortunately,

although these approaches have addressed some of the contextual elements and identified critical success factors that can be considered when implementing enterprise systems, the failure rate of these systems have not markedly reduced.

Initially, research on ERP systems concentrated on the technical issues (Stapleton & Rezak, 2004). Rengersen (2006) traced ERP implementation methods back to industrial engineering and software development and identified that this engineering focus resulted in implementation stages being design-oriented and paying "hardly any attention to human aspects". However, examinations of recent research on enterprise system failures have identified a variety of factors including human, organizational, technical and economic (Sarker & Lee, 2003). Heeks & Bhatnagar (1999) suggest that success and failure factors of information systems are based on ten categories, namely: information, technical, people, management, process, cultural, structural, strategic, political, and environmental factors. This implies that different strategic, organizational, technical and people factors are required during the implementation of enterprise or integrated systems (Aladwani, 2001; Wainwright & Waring, 2004). Indeed, a growing consensus emerged among many researchers that human factors are of critical importance and cannot be ignored (Sarker & Lee, 2003). Hence, gaining people confidence becomes essential (Appuswamy, 2000). People need to be carefully prepared to manage and support the implementation, while special considerations should be undertaken in supporting people who are going to use enterprise systems such as E-Government and ERP systems.

A number of authors have conducted literature reviews of reported studies in order to derive comprehensive lists of enterprise systems implementation critical success factors (CSFs) and classify these factors to different groups or categories. However, each author has classified these CSFs differently, which makes cross comparisons difficult. For instance, Esteves & Pastor

(2000) identified 20 ERP implementation critical success factors and divided them into four perspectives: organizational, technological, strategic and tactical. Factors were grouped as organizational or technological and then each group was sub-divided into strategic and tactical. On the other hand, Nah, et al. (2001) distributed 11 ERP implementation CSFs across the four phases of Markus & Tanis' (2000) ERP life cycle model: chartering, project, shakedown, upward and downward phases. Another research group, Al-Mashari, Al-Mudimigh, & Zairi (2003) presented a taxonomy of ERP critical success factors based 12 CSFs and grouped them into three groups: setup, deployment and evaluation. Whereas Ngai, et al. (2008) specified 18 ERP implementation CSFs, with 12 of these decomposed into sub-factors of country-related, vendor-related and organizational-related.

Other authors have attempted to examine different enterprise systems implementations in country-specific environments. These attempts have identified diverse challenges and hurdles that face the successful implementation. In addition, these investigations lead to derivation of a range of recommendations in order to carry out a successful enterprise system implementation process. For instance, El Sawah, Tharwat, & Rasmy (2008) conducted a study about the challenges facing ERP implementation in the Egyptian market. The result of their study demonstrated that most of the problems occurred due to: the lack of vendor support, limited role of consultants in the implementation process, and the lack of knowledge about the details and capabilities of the ERP. They concluded that while CSFs positively reinforce implementation success, the Egyptian organizational culture hinders its progress. Wang & Chen (2006) study in the Taiwanese market evaluated the importance of internal and external support. They concluded that consultants are critical in influencing the effectiveness of communication and conflict resolution in the consulting process. Their study also suggested that top management support impacts the ERP, while user support

neither directly nor indirectly affects the quality of the delivered system. Al-Hinai (2004) study on Oman's E-Government economic and social impact highlighted a number of challenges such as digital divide, top management support, preparing and appointing skilled staff, and resistance to change.

Al-Mashari & Al-Mudimigh (2003) investigated the reasons behind the failure of ERP implementation in a middle-eastern manufacturer (Comp Group). They identified challenges that had led to failure of the project such as scope creep, lack of top management support, lack of change management, lack of communication, lack of performance management and propensity to isolate IT from business affairs. Soh, Kien, & Tay-Yap (2000) identified three types of cultural misfits between ERP solution and a Singaporean hospital that implemented the solution: data, functional and output. These misfits happened as a result of company-specific, public sector-specific, and country-specific requirements that did not match with the capabilities provided by the ERP. They suggest that procedures in Asian organizations are different from the west as they have evolved in a different cultural, economic, and regulatory context.

Ehie & Madsen (2005) carried out a study to identify the critical issues in ERP implementation. They found that 6 of 8 factors were strongly correlated with ERP implementation. These factors were project management principles, feasibility and evaluation of ERP project in the firm, top management support, business process re-engineering, consulting services, and cost/budget issues. Human resource development and IT infrastructure were found to be non-significantly correlated to successful ERP implementation. Jang, Lin, & Pan (2009) examined, via a survey, the impact of business strategies in the adoption of ERP in Taiwan. They concluded that the size of the organization and both internal and external business strategies significantly impact the decision to adopt ERP. On the other hand, they found that the perceived advantage of IT does not have a significant direct effect on the decision to adopt ERP, which contradicts with many other studies that demonstrated the importance of perceived advantage or usefulness of ERP system is a significant variable that influence organizations' decisions to implement ERP system. Shiau, Hsu, & Wang (2009) also conducted a study in Taiwan. The aim here was to generate a model to assess the ERP adoption of small and medium size enterprises (SMEs). They determined that the characteristics of the CEO, and perceived benefits of an ERP system, have a positive impact on ERP adoption. On the other hand, cost and technology complexity have negative effects on ERP adoption. However, in practice only the benefits of the ERP systems significantly affect ERP adoption. This study highlights the importance of educating top management and CEOs about the benefits of ERP system and the importance of assessing this realization.

Wang, Shih, Jiang, & Klein (2008)investigated the fit and interaction between six ERP implementation facilitating factors. The factors examined in the study include: vendor support, consultant competence, ERP project team member competence, ERP project manager leadership, top management support, and user support. They advocated that, in order to implement ERP successfully, alignment should not be limited to internal factors (ERP project team member competence, ERP project manager leadership, top management support, and user support) but also with external factors (vendor support and consultant competence). Ramizer & Garcia (2005)empirical investigation into the success of ERP systems in Chile tested the significance of eight ERP explanatory critical success factors in four dimensions that were used to measure the consequences of the implementation (this was based on DeLone and McLean IS success model). The ERP explanatory success factors, that were examined, were: strategic planning of the information technologies, executive commitment, project management, information technologies skills, business process skills, ERP training, learning and change readiness. The success dimensions were: system quality, information quality, service quality and net benefits. They identified a significant relationship between information technologies skills, ERP training, learning and change readiness and the dimensions of the system success. Other explanatory success factors were either partially supported or not supported.

Preparing people to manage and support the implementation and use of enterprise systems is one of the major challenges that has been addressed by both researchers and practitioners. Indeed, this issue is not new since many authors have pointed out the importance of this activity. For instance, Davenport (2000) reflected this concern by categorizing the "rational approach" in implementing ERP into two parts: preparing people and preparing the technical system. He believes that preparing people involves creating the structure with the specific roles such as executive sponsor, project manager, super users, etc. In addition, preparing people must ensure support from future users of the ERP, considering their training, and familiarizing them with the new business processes.

The issue of preparing people is a world-wide phenomenon: and is discussed in literature focusing on both developed and developing countries. However, the complexity and the capacity can differentiate between developed and developing countries or even from one organization to another within the same country (depending on the context of enterprise system implementation). Because enterprise systems are organizational-wide solutions and cut across functional boundaries, their implementation often involves multiple stakeholders in scattered geographical locations. Hence, it is essential to achieve mutual trust and commitment in order to ensure shared beliefs among the systems' stakeholders (Amoako-Gyampah & Salam, 2004).

The stakeholder groups within the organization that is intending to introduce enterprise system usually consist of:

- **Senior/Top managers:** Top managers within the organization are the people who develop the organization strategy, approve the project, make decisions, align the project with strategic business goals, legitimize new goals and objectives and mediate between parties in case of conflicts (Esteves & Pastor, 2000; Nah, et al., 2001). Their role is vital in supporting the implementation of the enterprise system. They should be involved from the start of the project until post implementation (Finney & Corbett, 2007; Nah, et al., 2001).
- **Project Manager:** Managing an enterprise system project is different from a normal information system project since the enterprise system usually is bought "off the shelf". This brings with it considerable changes to the business processes, organization culture and people's mind sets. A careful scoping, planning and scheduling of the project are very important to complete the project on time, on budget and on scope. Awareness, education, and training programmers to all stakeholders need to be well planned. The project manager has to pay high attention and make sure the quality of training provided to key users and end users is appropriate as it has a significant influence in their perceptions and beliefs about the enterprise system and their attitudes to use of the system.
- **Project teams:** Project teams play a fundamental role in providing the requirements of the new enterprise system. The best staff from both business and information technology departments should be selected to be part of the project teams. Project teams are usually supported by consultants to expand the knowledge and experience base.
- **Key users:** Key users work with the consultants so that the consultants understand the current business processes "as-is". Key users also help the consultant in defining

new requirements, which will help the consultant to identify new business processes "to-be". The key users working with the consultant should be experts in their functional area. Some key users will be required to train and support end users before and after the go-live of the enterprise system.

- **End users:** End users are those who will to operate the enterprise system after the go-live date. Special care should be provided to end users in order to reduce their resistance to the new system and to accept and use the system. For instance, adequate quality training should be provided. In addition, on-the-job assistance should be provided to end users to help them operate the new system effectively.

CURRENT CHALLENGES FACING STAKEHOLDERS' PREPARATION

To investigate the challenges that face implementers in preparing stakeholders to manage, support the implementation, and use enterprise systems in Arabian Gulf countries, the authors distributed 60 survey questionnaires to professionals who had been involved in ERP and e-government implementations. The questionnaire probed the professionals about their experience with solutions implementation, challenges they had faced, and the factors that they believed had influenced their readiness to manage, support and use the enterprise system. Respondents were also given the opportunity to add extra comments. 25 responses were received that could be used within the research. The demographic data of the respondents is given in Tables 1, 2.

The responses confirmed that a number of specific organizational, external, technical and human/social factors influence the preparation of stakeholders. The analysis also indicated that addressing some of these factors can be considered as a challenge which some organizations could

Table 1. Respondents per sector

Industry	No of Respondents	Percentage
Government	7	28%
Airline	1	4%
Petrochemical	1	4%
Academic	7	28%
Oil and Gas	1	4%
Health Services	1	4%
Utilities	2	8%
General Trade	2	8%
Banking & Finance	1	4%
IT	2	8%
Total	**25**	**100%**

Table 2. Respondents' roles during the implementation

Role	No of Respondents	Percentage
Senior Manager (Business)	3	12%
Senior Manager (IT)	4	16%
Enterprise System's (E-Gov. or ERP) Project Manager	3	12%
Project Team Member	10	40%
Key User	3	12%
End User	2	8%
Total	**25**	**100%**

not overcome and, as a consequence, they have negatively impacted the preparation of stakeholders. In many cases these challenges led to cost and time overruns.

The authors' empirical data mapped onto many of the challenges identified in the literature as facing implementers in preparing organizations' stakeholders. Among these challenges are:

- **Lack of business and IT strategies**: Organizations need to set up both business and IT strategies before engaging in enterprise systems initiation stages. Markus, Tanis, & van Fenema (2000) suggested that organizations need to start planning enterprise implementations at the strategic level before proceeding to the technical (software and hardware) levels. Unfortunately, many organizations in the Arabian Gulf region start the implementation without establishing business and IT strategies which contributes to cost and time overruns, or partial success and limited use of enterprise systems.

- **Sporadic top management support**: Top Managers usually take the decision to introduce enterprise systems. They are usually enthusiastic in the beginning of the proj-

ect, i.e. in the initiation stage. However, as time goes by their enthusiasm wanes, especially if the project faces unexpected risks. For example, top management support could face obstacles when extension of time, or extra budget is required, or there are conflicts between different stakeholders and external stakeholders, or between organization's stakeholders themselves. "Giving blessings" to the inauguration of the enterprise system or supporting technical requirements are not sufficient (Chen, 2001). Unfortunately, as pointed out by some respondents, top managers in a number of organizations in the Arabian Gulf have not provided the required support; for instance, sometimes decision making activities are left to the consultants. This issue is also reported by (Al-Mashari & Al-Mudimigh, 2003). Typically, other stakeholders' performance and acceptance of enterprise systems are influenced by their top managers' presence, involvement and support, this is termed "social influence" in the literature (Venkatesh, Morris, Davis, & Davis, 2003). As a consequence, failures of top managers in performing their duties could lead to loss of trust among other stakeholders. It is also noticeable that organizations in the Gulf region tended to be built on vertical structure rather than horizontal, which contributes to the sporadic top management support.

- **Under-skilled staff**: Project managers can be under various pressures from top management, users and external resources such as consultants and implementers. Furthermore, project managers can be over optimistic due to the lack of experience in managing large information systems. As a result, such pressures or optimism can result in inappropriate implementation schedules: where insufficient resources or under-skilled staff are assigned. In addition, organizations often cannot devote the required project team resources on a full-time basis, and less qualified staff are added as project team members. This can lead to various risks such as accepting enterprise systems with inadequate quality or the inability of staff to manage and support the system.

- **Insufficient training**: This problem is a major barrier to the successful implementation of enterprise systems. Many organizations do not allocate sufficient budget to train stakeholders who are going to support, manage and use the system. The assumption that stakeholders are capable of understanding the technology, or that the technology is easy to understand, is one of the mistakes that organizations often make, especially when these organizations have not used complex systems before.

- **Inefficient communication procedures**: Where communication procedures are poor, users do not know in advance about the new enterprise system. They are not participants in setting up requirements, and the project benefits and its progress are not communicated. As a consequence, users deal with the enterprise system as a "strange object" and the implementation faces both high resistance and low acceptance by end users.

- **Inefficient change management**: Implementing an enterprise system is more than buying software as it includes the vendors' view of the best business practices (El Sawah, et al., 2008). Enterprise systems implementation involves various changes in business processes which require a comprehensive change management awareness program. The failure in properly preparing people on imminent changes will cause rejection, resistance, and chaos as a consequence of the changes created by the implementation (Umble, Haft, & Umble, 2003). Unfortunately, some organizations assume that stakeholders will be easily persuaded and will accept the new business process without difficulty.

- **Cultural barriers**: As mentioned earlier, enterprise systems contain "best business practices", which often reflect the western (American and European) cultures and practices (Soh, et al., 2000). It is noted that the embedded culture within enterprises' systems might conflict with the organizational and Arabian Gulf countries' cultures. This conflict could lead to a strong resistance from users who have not yet worked with western best business practices. Another domain of cultural conflict between enterprise systems and the Arabian Gulf context is language. Enterprise systems are typically presented in English whilst the first language in the Arabian Gulf is Arabic. This challenge is more noticeable with the older generation of stakeholders for whom English has not been used in their day-to-day activities, or for those organizations with a higher percentage of nationals than foreigners among the workforce.

SOLUTIONS AND RECOMMENDATIONS

To prepare organizations' stakeholders to manage and support the implementation and use of enterprise systems and to overcome the mentioned challenges, a number of factors need to be considered and monitored. The analysis of responses from survey and interviews has revealed that organizational, project management, technical, external and human/social considerations influence stakeholders' preparation.

- **Organizational factors:** These factors are defined and need to be set up at the top organizational or strategic level. As these factors impact on all employees in the organization, a number of factors influence the preparation and readiness of enterprise system's stakeholders to manage, support and use the system. Among these factors are:

  - **Establishment of strategies, goals and objectives:** Establishing both business and IT strategies, goals and objectives, and the alignment between business and IT strategies help stakeholders in understanding organization's vision. Umble, et al. (2003) suggest that organizations need to establish a clear and compelling vision that defines how the organization operate taking into consideration its customers satisfaction, employees empowerment and suppliers facilitation. They also assert that clear goals, objectives expectations, deliverables should be clearly defined.

  - **Organizational structure with clear roles and responsibilities:** Before embarking on enterprise systems implementation, the organization needs to create a structure that include specific roles such as an executive sponsor, project manager, process owners, super users, vision and planning and implementation teams (Davenport, 2000). The structure, roles and responsibilities should be communicated to all employees involved (Nah, et al., 2001)

  - **Empowerment and motivation:** It is necessary to empower the enterprise systems project teams to take necessary decisions (Esteves & Pastor, 2000; Finney & Corbett, 2007; Ngai, et al., 2008; Umble, et al., 2003). This can be achieved by establishing an interesting work environment and recognizing the work done by stakeholders (Finney & Corbett, 2007). Empowerment and motivation are crucial in maintaining high morale and motivation amongst stakeholders, which facilitates staff retention (Finney & Corbett, 2007). Empowering employees to participate and recommend changes enables them to feel that they are part of the project and the change process. Umble, et al. (2003) suggested that when teams reach their assigned goals, rewards should be presented in a very visible way. Indeed, in a number of Arabian Gulf countries organizations that have implemented enterprise systems delegated the decision-making to stakeholders based on their roles and responsibilities: some of these organizations acknowledged stakeholders with different types of rewards after the success of enterprise systems implementation success.

  - **Allocation of adequate training budget:** Adequate and quality training can play a very important role in enterprise system implementation success (Sarker & Lee, 2003; Ward, Hemingway, & Daniel, 2005). Organizations must provide awareness and education programs to all stakeholders including top management. Education equips stakeholders with the fundamental knowledge about the enterprise system: its benefits, challenges, issues and resolutions (Calisir, Gumussory, & Bayram, 2009). For example, top management will understand not only the anticipated benefits for their organization but will also appreciate the complexity of enterprise systems: as a consequence, they should value the efforts made by other stakeholders. The benefit of the educational process should be senior managers who are actively committed to the project and support and track the project progress. Training, on the other hand, can enhance stakeholders' capabilities, depending on stakeholder role. For example, high quality, relevant, training enables end users to understand the functionalities of the enterprise system, and thus can promote positive attitudes to it (Amoako-Gyampah & Salam, 2004). It is noticeable that some organizations in the Arabian Gulf countries, especially in the recent years, have invested in training their staff, both in the business and technical departments, to externally recognize certification standards. Such certificated training has been seen to positively impact the capabilities of staff and helped these organizations to accelerate the implementation process.

- **Managerial factors**: Organizations' management is required to fulfill the following managerial factors:

- **Project management:** Managing enterprises systems project is unlike managing normal information system project. Amoako-Gyampah & Salam (2004) argued that typical project management principles, used to manage simple IT systems, would not be sufficient in ERP implementation since the challenges are magnified in such complex environments, often resulting in failed implementations (Markus, et al., 2000). The implementation of enterprise systems is complex and covers a combination of hardware, software, and organizational issues; effective project management allows organizations to plan, coordinate, and monitor various activities in different stages of implementation (El Sawah, et al., 2008; Ngai, et al., 2008). Hence, such projects require the organization to engage in excellent project management, where clear definitions of objectives and scope are set, aggressive but achievable schedules are put in place, work and resources plans are established, and tracking and control planning is developed (Umble, et al., 2003). Finney & Corbett (2007) and Nah, et al. (2001) suggested that the project should be formally defined in terms of its milestones, which involves the definition of the critical path, the time lines for tasks and decisions. Deadlines should be met and decisions timelines should be managed. They also asserted that the project scope should be clearly defined and limited to the modules of software that will be implemented, targeted business units, and the amount of business process re-engineering. The project should be periodically monitored by project team members who are stakeholders (El Sawah, et al., 2008).

- **Effective change management:** Change management is referred to "the requirement that the implementation team should formally draw a change management program" (Finney & Corbett, 2007; Nah, et al., 2001). The aim of change management is to build acceptance and employees' positive attitude toward the system (Finney & Corbett, 2007). Ngai, et al. (2008) argued that effective implementation requires change management strategies and understanding of the organizational culture, so that forces for change can overcome the forces of resistance. In addition, Hong & Kim (2002) asserted that the implementation of an enterprise systems (especially ERP) is much more than simply a technological change since it introduces changes in the organization such as structure, measurement compensation, organizational culture, training, etc. Therefore, effective change management must enable an organization to prepare stakeholders by giving sufficient knowledge about the new business processes and planned organizational changes before the actual implementation of the system. A successful organizational change depends on the proper integration of people, technology and processes (Esteves & Pastor, 2000).
- **Effective communication:** Communication between various stakeholders, from senior management to shop-floor employees within the organization is essential (Aladwani, 2001; Appuswamy, 2000;

Finney & Corbett, 2007; Nah, et al., 2001). Ngai, et al., (2008) has seen that effective communication is necessary at all levels before and during the implementation of system solutions. It is not sufficient that communication to be top-down approach but it must be 'inward' to the project team and 'outward' to the whole organization and in regular basis (Esteves & Pastor, 2000). For instance, employees should be informed about the scope, objectives, activities, updates and changes, while users input, requirements, comments, reactions and approvals should be communicated upward (Nah, et al., 2001). Various benefits can be realized through effective communication. For instance, Wang & Chen (2006) identified that it improves relationships between different functional areas and establishes bases for the resolution of conflicts. Bueno & Salmeron (2008) point to the link between communication and user trust of the systems, and consequently user acceptance. Effective communication ensures that end users, who are often forgotten, are informed and consulted, and thus better prepared for subsequent training in use of the system. Arabian Gulf organizations have used different ways to communicate both anticipated project benefits and its progress to all stakeholders; these include: emails, news bulletins, workshops and social gatherings. Socialization is a well accepted process in the Arabian Gulf culture that helps in establishing strong bonds between stakeholders and this can be used to prepare the ground for the system solution before its implementation.

- **Technical factors**. These factors focus on the technical aspects of the enterprise system and the technical setup that should be prepared and tested during the implementation process. These factors influence stakeholders' preparation attitudes and behavior, especially users. In the Arabian Gulf context, the following technical factors have been identified:
  - **Software quality:** These issues relate to the enterprise system characteristics such as ease of learning, understandability and use, data consistency and reliability, and output quality. Software quality has a direct influence on user satisfaction (Gable, Sedera, & Chan, 2003).
  - **Hardware reliability:** Enterprise systems require the installation of high end servers with high capacities of memory and data storage devices. Using inadequate capacities leads to unexpected results, which in turn cause the users to lose trust in the enterprise system.
  - Network reliability. The data network is the nervous system of an information system. Without a reliable network with required capacity of bandwidth and reliable communication devices, data flow between the different operational sites of the enterprise system is difficult.
- **External factors**. These factors are external to the organization but impact upon the implementation success in general, and the organization's stakeholders in particular. Among the relevant issues are:
  - **Consultant support:** Many organizations in the Arabian Gulf countries, especially large organizations that have implemented multiple modules of enterprise systems, sign contracts with consultants in order to help them

with specialized skills and experience, and the "know how" that is not available (or is expensive to build) internally (Gable, et al., 2003). In this context, consultants play a major role in persuading senior managers to take particular decisions. However, as mentioned in the challenges section, senior managers must not delegate the decision-making process to consultants.
  - **Vendor and implementer support:** This represents an important factor within any software package purchase. Contracts need to specify what is required in terms of extended technical assistance, emergency maintenance, updates, and special user training (El Sawah, et al., 2008; Wang & Chen, 2006). Organizations in Arabian Gulf countries rarely have internal staff who can carry out the initial implementation of an enterprise system, especially ERP systems. Nevertheless, some of these organizations have trained their staff, with the help of vendors and implementers, to carry out updates and upgrades of their enterprise systems.
- **Human and social factors**: These focus on the stakeholders themselves and consider the personal characteristics and behaviors, as well as the social relationships between the stakeholders. These factors include:
  - **Top management support and commitment:** This describes the extent to which client executives provide the attention, resources, and authority required for enterprise system implementation (Wang & Chen, 2006). Wang and Chen stressed further that the implementations triggers a range of changes in organizational cultures and business practices,

which emphasizes the role of top management. Similarly, El Sawah, et al. (2008) contended that ERP systems, for instance, introduce radical changes in an organization's work norms and procedures which requires organizational alignment. It would be very difficult to achieve such changes without the continuous involvement of senior managers at all stages of implementation. Such commitment from senior managers can encourage end users to engage the implementation and reduce their resistance. Stakeholders from different organizations from Arabian Gulf countries recognize that the involvement, support and commitment of their senior managers has contributed in building the trust between stakeholders and highlights the importance of the enterprise system to the organization.

- **Stakeholders' skills and competencies:** It is not sufficient to identify the stakeholders without identifying the skills and characteristics of each stakeholder. For example, top management leadership (Magnusson, et al., 2004), project manager skills, and ERP project team competences are crucial. A number of organizations in Arabian Gulf countries have carried out assessments of project teams' skills and competencies in order to identify their capabilities to manage and support enterprise systems implementation. Those who lack the required skills and competencies have then either received training or been withdrawn from the project teams.
- **Social factors:** These include issues such as power, control, politics, conflict and change. It is key that the social factors in play between differ-

ent stakeholders are identified. For instance, the way in which different interest groups or stakeholders resolve their conflicts, or reconcile their different interests, has been found to be essential in enterprise system implementation success (Ward, et al., 2005). Also, it is vital that stakeholders understand the communication procedures that govern the interaction between stakeholders prior to the enterprise system implementation (Ward, et al., 2005).

- **Perceived benefits:** The benefits of enterprise system can impact the individual and the organization (Gable, et al., 2003). The individual perceived benefits influence an individual's behavior; whilst the perceived organizational benefits influence both top management support and user satisfaction and use of the enterprise systems.
- **Beliefs and Attitudes:** Beliefs and attitudes affect the behavior of a stakeholder both during and after the implementation phase (Amoako-Gyampah & Salam, 2004). Often such attitudes have a marked impact on the success or failure of the implementation (Abdinnour-Helm, et al., 2003).

FUTURE RESEARCH DIRECTIONS

This chapter has identified challenges and issues that are encountered when embedding an enterprise system in an organization. Identifying factors that influence stakeholders for enterprise systems implementation forms a component of an on-going research project to develop a model to prepare ERP stakeholders. It is apparently the significance of each factor that differs from stage to stage of and its influence on stakeholders that varies. For instance, end user training will not influence the preparation of users in the initiation stage but it will have a big influence on users' attitudes at the use (go-live) stage. Therefore, the aim of the research under consideration is to empirically investigate the significance of each factor to each stakeholder at each stage of ERP implementation. The outcome of the research is the development of a model that can be used by organizations, especially organizations in the Arabian Gulf, to assess the preparedness of their stakeholders in implementing managing, supporting and using ERP solutions.

CONCLUSION

This chapter has shed lights on the challenges, issues and success factors that have been observed by practitioners during the implementation of enterprise systems. In particular the focus has been on those factors that are of importance in such implementations in the Arabian Gulf context. Identifying the issues and success factors in such projects and dealing with the associated challenges and risks are essential in any project. However, this is magnified when we deal with large and complex projects such as the implementation of enterprise systems. To deal with the issues raised it is essential to have clear procedures that overcome the challenges, mitigate the risks and utilize the success factors. Moreover, these procedures not only need to be set in train, but they also need to be supported by a measurement program that can monitor and audit progress and success, so that current projects are more likely to succeed, and future projects can learn the lessons from the past.

This chapter has raised some of the challenges and factors that influence the preparation of stakeholders in enterprise systems implementation. However information provided is context-based and cannot be exhaustive, since challenges differ in different settings: such developed or develop-

ing country, organization size, industrial sector, complexity of the organization. Some of these differences happen due to the cultural differences between different countries and organizations. Therefore, although guidance can be provided, each enterprise system implementation team needs to be aware of their context and tailor their process to their organization, and stakeholders.

REFERENCES

Abdinnour-Helm, S., Lengnick-Hall, M. L., & Lengnick-Hal, A. (2003). Pre-Implementation Attitudes and Organisational Readiness for Implementing Enterprise Resource Planning Systems. *European Journal of Operational Research, 146*(2), 258–273. doi:10.1016/S0377-2217(02)00548-9

Al-Hinai, H. (2004). *The Sultanate of Oman's E-Government Initiative and its Economic and Social Impact. Unpublished Master disertaion.* Salford, UK: University of Salford.

Al-Mashari, M., & Al-Mudimigh, A. (2003). ERP Implementation: Lessons from a Case Study. *Information Technology & People, 16*(1), 21–33. doi:10.1108/09593840310463005

Al-Mashari, M., Al-Mudimigh, A., & Zairi, M. (2003). Enterprise Resource Planning: A Taxonomy of Critical Success Factors. *European Journal of Operational Research*, 352–364. doi:10.1016/S0377-2217(02)00554-4

Aladwani, A. M. (2001). Change Management Strategies for Successful ERP Implementation. *Business Process Management Journal, 7*(3), 266–275. doi:10.1108/14637150110392764

Amoako-Gyampah, K., & Salam, A. F. (2004). An extension of the technology acceptance model in an ERP implementation environment. *Information & Management, 41*(6), 731–745. doi:10.1016/j.im.2003.08.010

Appuswamy, R. (2000). *Implementations Issues in ERP*. Paper presented at the 1st International Conference on Systems Thinking in Management (ICSTM2000).

Bueno, S., & Salmeron, J. (2008). TAM-based success modeling in ERP. *Interacting with Computers, 20*(6), 515–523. doi:10.1016/j.intcom.2008.08.003

Calisir, F., Gumussory, C. A., & Bayram, A. (2009). Predicting the behavioral intention to use enterprise resource planning systems: An exploratory extension of the technology acceptance model. *Management Research News, 32*(7).

Chen, I. J. (2001). Planning for ERP systems: analysis and future trend. *Business Process Management Journal, 7*(5), 374–386. doi:10.1108/14637150110406768

Davenport, T. H. (2000). *Mission Critical: Realizing the Promise of Enterprise Systems*. Boston, USA: Harvard Business School Press.

Ehie, I. C., & Madsen, M. (2005). Identifying Critical Issues in Enterprise Resource Planning (ERP) Implementation. *Computers in Industry, 50*(6), 545–557. doi:10.1016/j.compind.2005.02.006

El Sawah, S., Tharwat, A. A. E., & Rasmy, M. H. (2008). A quantitative model to predict the Egyptian ERP implementation success index. *Business Process Management Journal, 14*(3), 288–306. doi:10.1108/14637150810876643

Esteves, J., & Pastor, J. (2000). *Towards the Unification of Critical Success Factors of ERP Implementation*. Paper presented at the 10 th Annual Business Information Technology (BIT) 2000.

Finney, S., & Corbett, M. (2007). ERP Implementation: A Compilation and Analysis of Critical Success Factors. *Business Process Management Journal, 13*(3), 329–347. doi:10.1108/14637150710752272

Gable, G., Sedera, D., & Chan, T. (2003). *Enterprise Systems Success: A Meaurement Model*. Paper presented at the 24th International Conference in Information Systems.

Heeks, R., & Bhatnagar, S. (1999). Understanding Success and Failure in Information Age Reform. In Heeks, R. (Ed.), *Reinventing Government in the Information Age: International Practice in IT-Enabled Public Sector Reform* (*Vol. 1*, pp. 49–74). London: Routledge.

Holsapple, C. W., Wang, Y., & Wu, J. (2006). Empirically Testing User Characteristics and Fitness Factors in Enterprise Resource Planning Success. *International Journal of Human-Computer Interaction, 9*(3), 323–342.

Hong, K., & Kim, Y. (2002). The critical success factors for ERP implementation: an organizational fit perspective. *Information & Management, 40*(1), 25–40. doi:10.1016/S0378-7206(01)00134-3

Jang, W., Lin, C., & Pan, M. (2009). Business strategies and the adoption of ERP: Evidence from Taiwan's communications industry. *Journal of Manufacturing Technology Management, 20*(8), 1084–1098. doi:10.1108/17410380910997227

Magnusson, J., Nilsson, A., & Carlsson, F. (2004). *Forecasting ERP Implementation Success – Towards A Grounded Framework*. Paper presented at the 13th European Conference on Information Systems (ECIS 2004), Turku, Finland.

Markus, M. L., & Tanis, C. (2000). the enterprise system experience - from adoption to success. In R. W. Zmud (Ed.), *Framing the Domains of IT Management: Projecting the Future Through the Past* (pp. 173-207). Cincinnatti, OH, USA: Pinnaflex Educational Resources, Inc.

Markus, M. L., Tanis, T., & van Fenema, P. C. (2000). Enterprise Resource Planning: Multisite ERP Implementations. *Communications of the ACM, 43*(4), 42–46. doi:10.1145/332051.332068

Nah, F. F., Lau, J. L., & Kaung, J. (2001). Critical factors for successful implementation of enterprise systems. *Business Process Management Journal, 7*(21), 285–296. doi:10.1108/14637150110392782

Ngai, E., Law, C., & Wat, F. (2008). Examining the Critical Success Factors in the Adoption of Enterprise Resource Planning. *Computers in Industry,* 548–564. doi:10.1016/j.compind.2007.12.001

Ramizer, P., & Garcia, R. (2005). *Success of ERP Systems in Chile: An Imperical Study.* Paper presented at the European, Mediterranean and Middle Eastern Conference in Information Systems (EMCIS2005).

Rengersen, L. (2006). Sustainable Software Implementation: Using Emergent Change to Increase Enterprise Recourse Planning (ERP) Systems' Implementation, management and Governance. Retrieved 15 March, 2009, from http://www.os.utwente.nl/tsr/files/42-47.pdf

Sarker, S., & Lee, A. S. (2003). Using a case study to test the role of three key social enablers in ERP implementation. *Information & Management, 40*(8), 813–829. doi:10.1016/S0378-7206(02)00103-9

Shiau, W., Hsu, P., & Wang, J. (2009). Development of measures to assess the ERP adoption of small and medium enterprises. *Journal of Enterprise Information Management, 22*(1), 99–118. doi:10.1108/17410390910922859

Soh, C., Kien, S. S., & Tay-Yap, J. (2000). Cultural fits and misfits: is ERP a universal solution? *Communications of the ACM, 43*(4), 47–51. doi:10.1145/332051.332070

Stapleton, G., & Rezak, C. J. (2004). Change Management Underpins a Successful ERP Implementation at Marathon Oil. *Journal of Organizational Excellence, 23*(4), 15–22. doi:10.1002/npr.20022

Stewart, G., Milford, M., Jewels, T., Hunter, T., & Hunter, B. (2000). *Organisational Readiness for ERP Implementation.* Paper presented at the Americas Conference on Information Systems (AMCIS2000), USA.

Umble, E. J., Haft, R. R., & Umble, M. M. (2003). Enterprise Resource Planning: Implementation Procedures and Critical Success Factors. *European Journal of Operational Research, 146*(2), 241–257. doi:10.1016/S0377-2217(02)00547-7

Venkatesh, V., Morris, M. G., Davis, G. B., & Davis, F. D. (2003). User Acceptance of Information Technology: Toward a Unified View. *Management Information Systems Quarterly, 27*(3), 425–478.

Wainwright, D., & Waring, T. (2004). Three domains for implementing integrated information systems: redressing the balance between technology, strategic and organisational analysis. *International Journal of Information Management, 24*(4), 329–346. doi:10.1016/j.ijinfomgt.2004.04.001

Wang, E. T. G., & Chen, J. H. F. (2006). Effects of internal support and consultant quality on the consulting process and ERP system quality. *Decision Support Systems, 42*(2), 1029–1041. doi:10.1016/j.dss.2005.08.005

Wang, E. T. G., Shih, S.-P., Jiang, J. J., & Klein, G. (2008). The consistency among facilitating factors and ERP implementation success: A holistic view of fit. *Journal of Systems and Software, 81*(9), 1609–1621. doi:10.1016/j.jss.2007.11.722

Ward, J., Hemingway, C., & Daniel, E. (2005). A framework for addressing the organisational issues of enterprise systems implementation. *The Journal of Strategic Information Systems, 14*(2), 97–119. doi:10.1016/j.jsis.2005.04.005

Chapter 9
Key Determinants of Successful Implementation of IT in International Commercial Bank

Tarek Hatem
The American University in Cairo, Egypt

Elham Metwally
Misr International University, Egypt & The American University in Cairo, Egypt

ABSTRACT

This research reports the results of a single case study that covers a successful project of IT implementation in International Commercial Bank (ICB) from the Egyptian banking industry. The case highlights leadership actions, as well as other related factors regarding effectiveness of IT implementation that are linked to strategic competitiveness and value creation. Multiple sources of data were used. Primary sources include in-depth interviews in semi-structured format with industry authorities, IT and retail banking managers, and the bank's executives in general; whereas, secondary sources of data include annual reports, website information, and financial statements. Findings show that successful implementation was influenced by the interplay of several management practices, which eventually, had an impact on strategic competitiveness through their impact on some in-house attributes; notably, a dominating constructive cultural pattern leading to higher levels of organizational commitment, and the bank's value chain.

INTRODUCTION

The motivation behind the choice of the research topic is the recognized increasingly important role IT is playing in how business operations are conducted, and its significance as one of the determinants of strategic competitiveness. The researchers wanted to examine the complex process of implementing new systems owing to several interrelated issues involved that decide on the success or breakdown of implementation. The banking sector relies heavily on technology-based delivery channels and payment systems, such as automatic teller machines (ATMs), phone banking, call centers, mobile banking, and Internet Banking, to provide its service. But achieving strategic competitiveness through IT is challenging since in a modern economy, and with technology evolution,

DOI: 10.4018/978-1-60960-048-8.ch009

IT is necessary for doing business and creating value to different stakeholders. Moreover, achieving strategic competitiveness through IT alone is not easy as it entails an in-depth understanding of how to successfully deploy it. The researchers were guided by two trends of research, executive studies and reports, and academic research, theories and models. These interplayed in some extent in three areas of concern; namely, strategic management, management of change, and information systems, successful implementation was clarified by looking at the case of International Commercial Bank (ICB).

BACKGROUND

Successful implementation of IT and achieving strategic competitiveness has always been a challenging goal for banks. To understand the key success determinants of IT implementation in banking, the researchers examined previous literature on strategic competitiveness, literature on the impact of IT as one antecedent to competitiveness, as well as previous studies on mediating factors influencing the relationship between IT resources and strategic competitiveness; namely, challenges facing effective management of IT resources.

Previous researchers on antecedents of strategic competitiveness and value creation (Amit & Shoemaker 1993; Hatem, 2003; Hitt, Ireland, & Hoskisson, 2001; Hofstede, 1980, 1983; Penrose, 1959; Porter, 1985, 1990, and 1999; Rogers, 1983; Schumpeter 1934; Senge 1994; and Wah 2002), emphasized an organization's internal resources, capabilities and core competence, organizational innovativeness, learning organizations, the impact of cultural values and cultural competency on management systems and organizational practices, impact of nations, governments and industries, factors conditions and industry competition, as well as productivity levels. Research of quite a few scholars on the role of IT, being one antecedent of strategic competitiveness, on productivity and

profitability of organizations, and consequently competitiveness, has been substantial (Balmer, 1998; Balmer & Soenen 1999; Brand & Duke 1982; Kim & Weiss, 1989; Melewar & Navalekar, 2002; Morisi, 1996; Oster & Antioch, 1995; Swierczek, Pritam, & Bechter, 2005; Radigan, 1996).

Some literature on mediating factors influencing the relationship between IT resources and strategic competitiveness included issues on effective management of IT resources. In studying competitiveness, Ross, Beath, and Goodhue (1996), commented that in a business environment with a fast-evolving nature and technology-based operations, the challenge is not deciding upon the right technology; it is more on how to effectively address changes in organizational roles, structures, and processes on adopting a new IT, to eventually enhance organizations' competitiveness. A mounting literature highlighted the role of IT executives in effectively leading IT projects, different challenges facing organizations in finding mindsets that can effectively manage change, develop effective IT capability, synchronize strategy and IT, think about information technologies' strategic impact as a change agent, and in general, make things happen (Anol & Rudy, 1997; Barnes, Mieczkowska, & Hinton, 2003; Hill & Collins, 1999; Ian, 1997; Johnson, Fidler, & Rogerson, 1998; La Porte, Demchak, & Friis 2001; Matsui, 2002; Picket, 2004; Prahalad & Krishnan, 2002; Proctor & Doukakis, 2003; Ross, Beath, & Goodhue, 1996; Schultz, 2005; Weiss & Anderson, 2004). Therefore, to stay alive and face competition in the world of technology and to exploit speedy improvements in IT, organizations are bound not only to change but also to effectively manage such change. And accordingly, in an increasingly unstable and fast changing business environment, more pressure is placed on organizational leaders. They must be more alert to changes in the external environment to identify different forces such as breakthroughs in IT that might alter competition and provoke planned change (Branch, 2005).

The focus of previous studies on management of organizational change has been primarily on change and organizational development (Flower, 1996; French & Bell, 1978; French, Bell, & Zawachi, 2000; McNamara, 1999; Nickols, 2004; and Teare & Monk, 2002), the change process (Burton, Laridsen, & Obel, 1999; French, Bell, & Zawachi, 2000; Gersick, 1991; Lam, 2000; Lewin, 1950, 1958; Nickols, 2004; Teare & Monk, 2002). On top of the studies on management of organizational change is Kurt Lewin's force field analysis that lists competition, new equipment, high performance goals, employees with new skills, and the desire for increased influence as forces that stimulate organizational change. Lewin's (1951, 1958) three-phase unfreeze-change-refreeze model of change created a foundational framework for much of the literature that deals with planned organizational change. Studies on organizational transformation (French, Bell, & Zawachi, 2000; Hamel & Prahalad, 1994; McNamara, 1999; Teare & Monk, 2002; and Tosey & Robinson, 2002) focused on revolutionary and massive improvements of organizations.

Challenges to change and forces resisting change received considerable interest by researchers (Anderson & Anderson, 2003; Burnes, 1992; Collins & Porras, 1989; Dooley, 1998; French, Bell, & Zawachi, 2000; Hamel & Prahalad, 1994; Hamel & Valikangas, 2003; Kitchen, 1997, 2002; Kotter, 1995, 1996, 1998; Nickols, 2004; Teare & Monk, 2002; and Tichy & Sherman, 1993). However, there has been little research on best practices in managing IT and as a result, standards for IT governance do not exist. Nolan & McFarlan (2005) observed continuously changing IT strategies of hundreds of organizations for over forty years to find out that best practices for IT governance vary greatly among organizations according to a number of factors among which are a firm's industry, competitive pressures, financial status, sweeping changes in information technologies, IT expertise in managing IT risk and dealing with competitive forces. Accordingly, board members need to develop IT policies that strategically fit their firms' positions.

Strategic benefits of implementing IT solutions include improved service quality, speedy delivery, cost reduction, and risk aversion, all of which have both direct and indirect influence on output and factors of production and accordingly on productivity levels. IT implementation challenges, on the other hand, include issues pertaining to hardware and software contradictions, security, information distortion, and weak coordination between IT and operational managers, all of which have a negative impact on productivity, as well as on propensity to use further IT applications. Moreover, weak communication between IT and operational managers hinders good IT product development and reduces the benefits of the use of IT in banks. Therefore, while the good implementation of IT enhances productivity, challenges facing IT implementations reduces it (Swierczek, Pritam, & Bechter, 2005).

INTERNATIONAL COMMERCIAL BANK (ICB)

ICB is one of the largest private-sector banks operating in Egypt and provides an all-inclusive variety of banking and interrelated financial services throughout a network of 35 branches in addition to 108 ATMs in Cairo, 6th of October City, Giza, Alexandria, Sharm El Sheikh and Hurghada.

ICB provides a full variety of individual financial and assets management services. These embrace Egyptian pounds, US Dollars, as well as other currency products and services such as savings accounts, current accounts, personal credit services, time deposits, credit and debit cards, that are disseminated throughout channels such as its ATM network, (the second largest in Egypt), Internet Banking service, 24-hour phone banking centre, in addition to a network of over 35 branches and offices located in seven cities in Egypt.

ICB provides its most valuable customers with "Special Banking Packages," wherein they get highly personalized banking services, a committed liaison executive, as well as numerous advantages and privileges.

At ICB's core lies efficient technology that links domestic commercial banking and financial services to deliver a wide range of products and services (Website of ICB; ICB's Annual Reports 2002-2005).

ICB's Financial Performance

ICB performed well over the years 2002-2005 (see Table 1). Its improved financial measures were a tribute to major growth across its core business. Corporate banking expanded significantly both with regards to business volume and profitability. The achievement of investment banking is marked by its having been granted a number of key investment banking and finance consultative directives (ICB's Annual Reports 2002-2005). Personal financial services contributed significantly in further setting up the sound reputation the bank has put on in the market. Following a strategy implemented in 2001, the bank is increasing its personal financial services infrastructure. To fulfill its major goal of establishing diversified delivery channels for its customers, the bank launched the Automatic Teller Machine (ATM), and Internet Banking and a full-fledged call centre (Website of ICB; ICB's Annual Reports 2002-2005).

ICB's financial performance (See Figure 1) shows that ROA increased over the years in absolute terms showing management's efficiency in making earnings with funds available to them. ICB's ROE rates show an increasing earning power of common stock equity or stockholders'

Table 1. ICB's financial performance (Source: ICB's published annual reports)

Financial Measure		Year			
		2005	2004	2003	2002
Profitability Measures	Return on assets (ROA)	5.44%	5.14%	5.12%	4.98%
	Return on common stockholders' equity (ROE)	57.74%	43.16%	25.98%	17.06%
Banks' Ratios	Ratio of earning assets to total assets	62.68%	58.46%	75.84%	83.15%
	Return on earning assets	4.97%	4.33%	2.21%	1.43%
	Interest margin to average earning assets	4.86%	5.00%	3.93%	2.98%
	Deposits times capital ratio	15.68 Times per year	14.87 Times per year	13.65 Times per year	11.95 Times per year
	Loans to deposits ratio	37%	36.25%	42.80%	49.52%
	Equity capital to total assets	5.40%	5.86%	6.46%	6.98%
Investors & Stockholders	Earnings per share of common stock	57.69	49.02	24.17	21.18
	Leverage ratio. Also called the equity ratio, or funds to total assets	4.80%	6.15%	5.46%	7.92%
Analysis by Creditors	Interest coverage ratio	145.60%	141.72%	130.72%	114.68%
	Debt ratio	92.40%	93.80%	94.23%	91%

Figure 1. ICB's profitability measures (Source: ICB's annual reports)

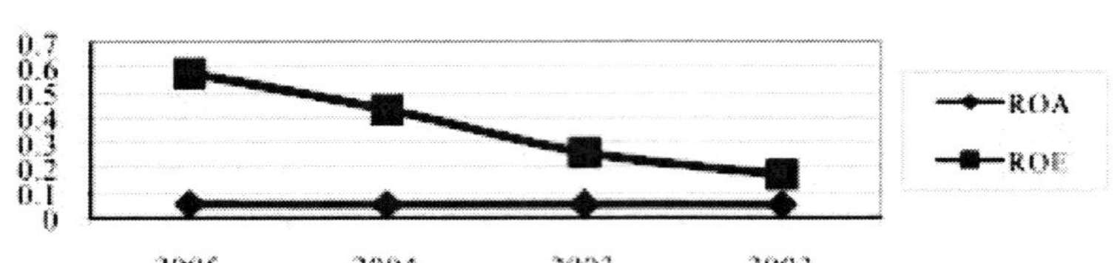

investments, and a much higher rate than the rate of interest paid on liabilities which is 4.449% as of end of year 2005.

High ratios of earning assets to total assets are indicative of high performance. Increasing return on earning assets indicates, considered along with the return on assets and the return on stockholders' equity, increasing profitability of the bank (See Figure 2).

A decreasing interest margin to average earning assets ratio indicates a lower profitability in 2005 despite a dramatically increasing one throughout the years 2002-2004. An increasing deposits times capital ratio reveals that the deposit base of ICB has increased over the years, indicating a profitable prospect. Increasing loans to deposits ratio is favorable because it means that the debt paying ability of the bank has also increased. However, the bank's decreasing eq-

uity capital to total assets ratio shows low safety measures against risk of debt and leverage usage. A low ratio is unfavorable from stockholders' viewpoint.

Increasing earnings per share of common stock (See Figure 3) indicates an increase in profitability of the bank, as well as an expected future growth. A low equity ratio shows a high use of leverage or borrowing of capital by creditors. An increasing interest coverage ratio in ICB (See Figure 4) indicates creditors' claims are becoming a greater percentage of the bank's total assets, which is not favored by long-term creditors who aim at keeping the claims at the minimum possible level so as to be protected if a bankruptcy happens. This ensures a margin of protection to creditors against diminishing assets. A lower debt ratio is more favorable to creditors since it means that there

Figure 2. ICB's ratios (Source: ICB's annual reports)

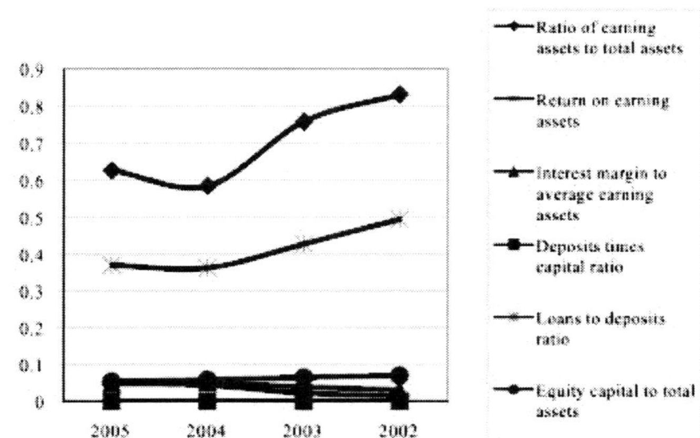

Figure 3. Investors & stockholders ratios (Source: ICB's annual reports)

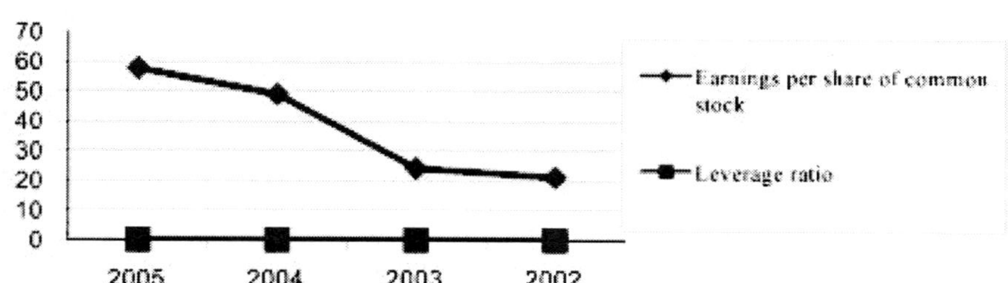

is little dependency on their funds while there is more use of stockholders' funds to the business.

Products and Services

The bank presents an assortment of banking and personal financial services, starting from current and savings accounts to time deposits, credit cards, loans as well as miscellaneous payment services. This case describes the implementation of two of the bank's products, namely ATMs and Internet Banking.

Numerous convenient ways to bank with the bank embrace the branch network, the widespread and handily located automated teller machines (ATM) in Egypt (about 123 ATM networks), the globally reachable Personal Internet Banking Service, the 24-hour Phone Banking Service and a direct sales group. Special assets management package are also available to meet the precise needs of the bank's most esteemed clients (Website of ICB; ICB's Annual Reports 2002-2005).

(A) ATM System

Functions and Transactions Supported

Through its easy-to-use, convenient, adaptable, and dependable Global ATM Network, ICB allows its customers to access their accounts and execute nearly all of their personal banking dealings around the clock without having to visit a branch. In 1993, ICB launched its network of ATMs (123 in Egypt, and 800,000 worldwide) which allowed customers to deposit cash and/or cheques and post mail to the bank in most of its ATM locations. ATM users can also make payment to their credit card(s), transfer funds between their accounts, withdraw cash up to L.E. 4000 (approximately US$697) per day, or up to L.E.6000-8000 (US$1045-1394) for medium and high net worth clients. Daily cash withdrawal limit is L.E.4000/ US$ 697 for MasterCard holders, and Standard MasterCard, whereas it is L.E 5000/US$871 for Classic Visa holders, and L.E. 6000/USD$ 1045 for Gold Visa holders, and L.E.8000/US$1394 for

Figure 4. Analysis by Creditors (Source: ICB's annual reports)

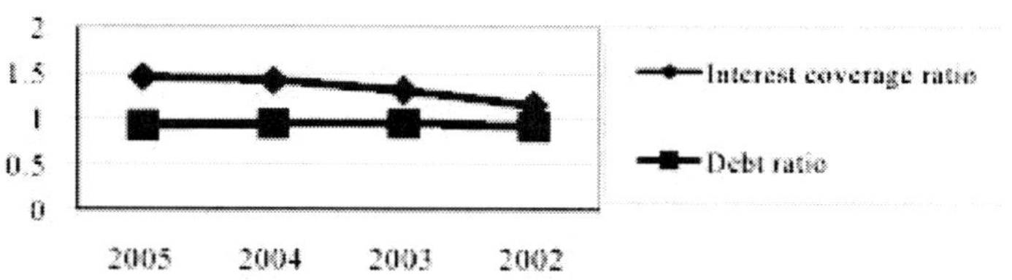

Premier Master Card holders. ATM users can also get a credit card cash advance, make enquiries on accounts' balances, and change their passwords or language of communication with the bank.

The bank is a recognized leader in building up and applying highly developed technology for the competent and convenient release of banking services. Apart from its other personal banking products and services, ATM and credit cards are among the most versatile and value added. The bank's ATM card allows customers to take out cash from more than one million ATMs globally. The bank provides its services around the clock throughout its automated teller equipment. A machine's downtime is less than 3 per cent, which is better than any of the bank's competitors.

ATM Fees and Charges

ICB charges an annual fee of L.E. 25 (US$4.35) on issuing local cards, and L.E. 60 (US$ 10.45) on issuing international ones. Replacement fees for local and international cards amount to L.E. 10 (US$ 1.74) and L.E. 30 (US$ 5.23) respectively, while re-issuance of a Personal Identification Number (PIN) costs L.E. 5 (US$ 0.87)

ICB exempts ATM card holders using the bank's network of ATMs in Egypt from paying usage charges. ICB levies charges though on other users at 123 Sharing Network (L.E.4/US$0.70 for cash withdrawal and L.E.2/US$0.39 for balance inquiry), the bank's group network (L.E.45/US$ 7.84), and at PLUS Network (L.E. 60/US$10.45) (Website of ICB).

(B) Internet Banking

Functions and Transactions Supported

An Internet Banking service was initiated in Egypt in February 2003. It is an award winning service extended by the bank's group with complete functionality and great concentration on user protection. The bank's e-banking structure provides clients with the right to use their accounts 24 hours a day, 365 days a year. A proprietary universal data system supports both worldwide electronic banking services and intra-Group communications. Internet Banking Service is a service that ICB provides to its account holders through the internet to enable them to electronically make and transmit information and/or instructions with regards to their accounts that can be accessed through the service by using a Personal Banking Number (PBN) and a Password.

Through its website, ICB provides it customers with Internet Banking Services terms and conditions in relation to the accounts accessed before using these services. To further ensure ease-of-use of the service, the ICB allows its clients to seek staff member assistance at any of the bank's branches, or call the bank's call center 24 hours a day, 7 days a week, or send an electronic mail to the bank, or finally visit any of the bank's branches. The bank provides a list of its products and services that can be accessed through Internet Banking Services. Through Internet Banking Service, clients of ICB can easily view their accounts and check their balances, transmit money to or from any account, make standing instructions or request payments relative to any account, as well as revise their personal details held by the bank for any of their accounts. Clients who access their accounts through that service can also make payment transfers from their accounts to another person's account with ICB within Egypt or with a different bank. The transfer can be to overseas branches of ICB or to another account they maintain with another financial institution. The quantity of such payments on any day cannot surpass the day by day service restrictions or the equivalent of such sum put by the bank, if being done in a currency different from the local one. ICB sets limits to clients' transfers to their accounts to 300,000 Egyptian Pounds; whereas, limits to transfers to a third party range between 30,000 and 60,000 depending on a customer's status within the bank. With Internet Banking Services, customers can also pay their credit card(s) bills and make online credit card(s) requests. They can further open Term Deposits or change existing maturity instructions.

The bank provides its clients with a "User Guide" that gives them an overview of the bank's Personal Internet Banking Services. This introduces clients to the service, informs them about minimum system requirements needed to use the Personal Internet Banking service, provides them with guidelines on how to register for the service and use it, and gives them security tips to proceed safely when using the service.

Security of PBN and Password

ICB helps clients guard against unauthorized access to their accounts, ensure security of their computers, and Internet Banking Service's Password or PBN, particularly when they know their PBN and/or Password is lost or stolen or they become aware or suppose another individual knows their PBN or Password or has made illegal use of the service. Simultaneously, ICB undertakes very sophisticated measures to protect its clients' transactions and accounts from any illegal access through its powerful 128-bit encryption system and automatic "time-out." Encryption is method that converts sensitive information into a series of unidentifiable characters prior to sending them over the Internet. Encryption assists in providing a protected channel for information transmission between the bank's computer system and clients' Internet browser to maintain data's privacy. The Internet Banking service has timed log-outs, which means that Internet Banking sessions will stop after some time of idleness. This assists in protecting customers against illegal access.

Internet Banking Services Fees and Charges

ICB provides Internet Banking Services free of charge. In addition, it provides a 50% discount on fees charged on transactions if these transactions are done through the net. However, clients incur fees and charges that are applicable to relevant transactions undertaken using other delivery chan-nels of the bank (Website of ICB; ICB's Annual Reports 2002-2005).

Phases of Implementation

A. **Planning**: Though ATMs were already implemented and tried in the Group bank, implementation of the product in Egypt went through the normal steps of implementing any new technology. ATM implementation in ICB started with gathering of all branch managers and key persons to be involved in the development of the product for brainstorming and making suggestions. Key executives made decisions on how to insert human factors, the new product manager, and key players to recruit for the new tasks required throughout the implementation phase. Implementation of ATMs started with bank branches first and then expanded with offsite machines. The new product manager also coordinated among the IT department and branches, as well as other departments concerned, such as the new Card Center to issue cards. The Project Manager also organized stages of implementation, some of which went in parallel. The product manager's accountabilities also included managing logistics. ICB followed the same procedures in planning for implementation for Internet Banking, and making decisions with regards to work on a website for customers, choosing and installing the right software (database, communication software) and modules to link the website (internet server) with the mainframe (customer database).

B. **Training:** To effectively implement the new system, the bank conducted training sessions for staff, especially those working in customer services. Sessions embraced training on implementation of hardware, software components, software use, maintenance and traffic management, and ATM machine replenishment. Training manuals and presentations clarified how to put the new system into operation. In conducting training sessions on Internet Banking, ICB further handled Security Issues. These included deciding upon and installing the right "Fire wall" tool to protect customers from others who could read their PIN numbers. ICB also installed a Virus Protecting Network (VPN) to protect against viruses, and installed a Host Secured Module to secure data transmission.

C. **Conversion:** ICB incorporated the ATM system into the bank's financial transactions with customers, and internal financial reporting and documentation. The bank's IT department conducted software runs to start the implementation phase and used a special module to operate the new systems simultaneously with the old system. Throughout conversion, staff interacted with peers and received technical support from the IT department, as well as the product development team.

D. **Acceptance Testing:** Acceptance testing of the ATM system started with global cards to all staff of the ICB's pilot launch to find out if there were any problems with the new product. After the product was tested on staff, it was then offered to the bank's clients. Internet Banking service was launched and tested first on staff members before it was released to customers. ICB announced availability of the service through flyers delivered with account statements, and on receipts of ATM transactions.

E. **Post-Implementation Audit:** ICB predetermines cut-off times and value dates for all its clients' transactions instructions. Instructions received from clients within the cut-off times will be executed the same day and given a value date according to the bank's preset value dates indicated. For example, there is no cut-off time for transfer applications to accounts held with the bank; accordingly the transaction is processed im-

mediately. For telegraphic transfers applications to local banks in Egypt, the cut-off time is 16:00hrs and the value date is same day; whereas, for a transfer to overseas branches of the group, the value date depends on the location worldwide. Requests for Demand Drafts and Cashier's Orders are processed and dispatched the same day if received before 14:00hrs; whereas, updating personal details is processed within one working day if instructions are received before 14:00hrs.

ICB's Target Market

The new ATM system targeted all clients who keep personal accounts in the bank. Both staff and clients were impressed by the efficiency of the technology. ATMs are used with great success with all people to the extent that an increasing number of organizations, including ministries, are now resorting to ATMs in their payroll payments. ATMs are most appealing to educated people who have skills with computers. A prospective target market could be executives, staff of multinational and joint venture companies, who found ATMs useful for them because their work requires them to work late, thereby preventing them from having time to stop by their banks to perform financial transactions on their accounts. Additionally, no difficulty is expected with regards to users of the new e-generation.

Internet Banking is also used with success by the same clients who use the new ATM system. Internet Banking is a selective product which is used with success among those who are educated and are computer oriented, have computer skills, and who own or have easy access to computers. This includes executives, staff of multinational and joint venture companies, who found the service useful for them because of the nature of their work which requires them to stay office hours at late periods, thereby preventing them from having time to stop by their banks. Internet Banking is a selective product, which could be appealing to executives, staff of multinational, and joint venture companies, university students, and high and middle class people of the society.

Handling and Overcoming Resistance to Change

In general, no challenges were encountered during implementation of new delivery systems introduced in ICB. Both ATMs and Internet Banking were sold properly and presented from top management to middle management (branch managers), who passed on the vision to staff, and then on to clients. The bank's employees had no time to rest though since the bank introduces many projects each one after the other and sometimes simultaneously, so staff is loaded with additional tasks over and above their normal work load. The bank faced some technical challenges such as hardware, software, and dropdowns too. To understand and overcome resistance to change, ICB divided work teams according to their members' experiences, whereby some did the hardware, some did the software, and so forth. Management also brainstormed with staff.

Leadership Measures and Best Practices

In implementing new systems, ICB appointed a new manager and formed a new team for each product. There was no specific way of forming each product's team. The bank made use of the best resources, including staff, according to their acquired talents. The bank's vision is to be the leading bank in Egypt, offering the best banking products and services comparable to other banks in the country. ICB's mission is to continuously develop its products and services in its dedication to best quality services. In support of this mission, ICB provides high quality electronic delivery systems, among which are ATMs and Internet Banking service.

Thus, to establish a sense of urgency to implement new systems, product managers conducted meetings with heads of branches to communicate the bank's goal of having an edge among other banks. By being a pioneer in introducing ATM cards that had two distinct features (global and triple access to three accounts), the bank wanted to be the best ATM card provider. The bank established a sense of urgency through meetings, brochures, training courses, and by creating awareness among staff to convince customers of new products and showing how these products are important to customers, as well as to the bank.

Product managers empowered branch heads to act on the vision. They in turn empowered selected staff and delegated to them authorities based on their experiences and their acceptance of new products. Though there were no direct rewards to employees who contributed to implementation, in staff annual performance evaluations for those who had a role in implementation were appraised, and management sent letters of thanks to team members who had participated in implementation of the new systems.

In consolidating improvements and producing more change, ICB tested new products on staff first to let them experience working on them before introducing them to customers. This helped the bank to detect the location of changing systems, structures, and policies. New approaches were easy to implement and institutionalize in the bank's culture since most of the bank's employees are from the young, computer-oriented generation and are well acquainted with computer applications and the internet. ICB gave a grace period of four months to start implementation.

Throughout the implementation process, management had the most important role since without their service-quality orientation and openness to change to update and improve technology, they could have never been able to pass on the bank's vision to employees and subordinates. An IT manager, for example, thinks of everything which makes the bank provide an easy and around-the-clock service to the client, using resources efficiently, safely, and reliably, creating value to its different stakeholders, and eventually contributing to a strategic competitive position. Management believes that 70% of the bank's strategic competitiveness is attributed to effective management of change.

Personal and Organizational Attributes

Finding a Common Base

To introduce new technology applications to employees, product managers conducted a 1-2 day information session with modules that revealed facts and concepts for both products. ATMs' information sessions were carried out on a limited scale; whereas, those on Internet Banking were conducted on a larger scale to explain all the functions of the product, by making presentations on its needs and infrastructure, and other facets of the changes. Some departments such as operations and customers service also contributed to the sessions. The purpose of these information sessions was to orient the bank's staff to the product by explaining the product, demonstrating its potential, showing how it will give the ICB an edge in the market and how it will be good for clients. Brainstorming was limited to suggestions about locations of machines in public areas, supermarkets, open areas, etc., as well as on how to secure transactions carried out on these new systems. These sessions allowed some input and a common base, but to a very limited extent because issues pertinent to new systems' implementation had already been based on business decisions and tested in other banks of the group before they were implemented in Egypt.

Organization of Implementation Activities Among Staff

The bank briefed all existing staff in customer service and the IT department about the product to be implemented and passed on the vision. Each staff member had his/her role defined according to his/her area of expertise. For example, those who had software experience were involved in software implementation. Usually in branches, staff members who are in charge of implementation have others act as backup in cases of their absence.

Adjustments to Work Assigned to People

With the introduction of new technologies, new tasks were added to existing employees' job descriptions with a target of raising the number of users as a percent of total clients. An exact figure was not specified, but the higher the rate of new users, the better. Staff got training and all acquired ATM cards and opened IB accounts. Human resources are dedicated to carrying out tasks of the new products/systems until tests are completed.

Adjustments to the Nature of Work and Job Content

During implementation, the bank's employees were exhausted because of 2-3 months of continuous work and normal work challenges. New business technologies reduced the number of clients who did business at the bank's counter, reduced vouchers/paper work for transactions, reduced work load/pressure and inquiries on employees, thus improving services provided to customers. New systems also made the bank's customer services more sales oriented and employees were convinced to only do the things that were productive to them, while encouraging clients to do transactions that can be carried out through other delivery channels like ATMs and IB.

New systems elicited new responsibilities in ICB. On the top level, the Product Development Manager and branches' heads were responsible for the implementation of ATMs. At lower levels, Customer Services Department (product marketing, and account opening), ATM Card Center (issuing and maintaining ATM cards), and the IT Department (technical problems, and system monitoring) was the team in charge. The Customer Services Department was heavily involved in implementation since it was responsible for the product development part in the change process, marketing it to the client, and onsite replenishment. With every account opened, the bank promoted the ATM card to clients. The ATM Card Center was responsible for the issuing and handling ATM cards, offsite machines' replenishment, handling complaints, detecting payment problems through journals, etc. The IT Department was responsible for software runs, as well as in all implementation phases, ATMs technical problems, and monitoring the machines to ensure they worked properly.

With Internet Banking, the Internet Banking Product Manager was responsible for implementation on the top level; whereas, at lower levels of the organization, the Bank's Call Center, IB Customer Services (marketing team), and internet service provider were accountable for carrying out the implementation phases. The Customer Services Department was heavily involved in implementation since it was responsible for the product development part in the change process and marketing it to the client. With every account opened, the bank promoted an Internet Banking account to clients. IB was also used as a marketing tool to attract clients to open accounts with the bank. The customer services department is responsible for opening and executing Internet Banking accounts, handling complaints, and detecting payment problems. The IT Department is responsible for maintaining and upgrading the electronic banking website and handling technical problems, to ensure the service is provided to customers around the clock. It also connects the new system with the bank's system and customers' database.

New Skill Requirements Involved
ICB recruits graduates and highly qualified and self-motivated Egyptians for different fields of banking such as Personal Financial Services, Corporate Banking, Operations, Financial Planning, Financial Control, Trade Services, Treasury, Information Technology, and Human Resources. The ICB appreciates and rewards individual per-

formance; encourages teamwork work environments and also promotes creativity and innovation. The bank believes that its human resources are its most important assets. Hence, it provides a career structure for its staff and invests heavily in training and developing to motivate them and keep those exceptionally talented.

The bank employs people with a good command of the English language and who are at ease dealing with numbers. Staff should also have decision-making abilities, as well as the capability to develop into future leaders. They should also have a proactive attitude to their work, a positive approach towards change, group spirit, and should be customer-oriented (Website of ICB; ICB's Annual Reports 2002-2005).

Therefore, and according to the responses of interviewed staff on implementing the new systems, no new skills were required. Staff involved in implementation had the normal banking skill requirements (staff should be a university graduate, well educated, and with good language skills). Some sales skills were needed though to sell the products. The Internet Banking product was very easy to sell since it is free of charge, but ATM cards needed some marketing since customers are charged fees on issuing an ATM card (Local ATM cards cost L.E.25/year whereas Global ATM cards cost L.E.60/year). In addition to the normal banking skills that the bank's staff had, they needed to acquire some background on the new systems' application to the bank's transactions, as well as get technical knowledge of the hardware and software involved.

Training and On-The-Job Learning Activities/ Programs
Efforts put forth to enlarge the bank's diverse backstage and front-stage activities with regards to ATM and Internet Banking Products and Services included further developing sales and service skills of frontline employees. Considerable improvements were made in operations processing, which provided the bank with an edge in managing work

volumes produced by the growth of the personal financial services business. The telecommunications and mainframe system infrastructures were improved to smooth the initiation of Internet Banking (Website of ICB; ICB's Annual Reports 2002-2005).

On implementing ATMs and Internet Banking, ICB conducted a one-day training session for each new product. Trainers then supported staff during both the pilot phase and after implementation of new systems with regards to both technical aspects, as well as on how to deal with customers and acquaint them with the new systems. On-the-job training was a crucial part of implementation since not everything was covered in training sessions. On-the-job training was carried out gradually as employees gained experience in their job.

ICB has an internal fully equipped Training Centre in its main branch, supported by training experts. All newly hired employees undergo an orientation course that offers them fundamental information on the bank, its record, principles, products and services. Entrants chosen to work in customer services are provided a particular Customer Services certification training. Likewise, the bank provides prospective Financial Planning Managers/Officers with special training courses to meet the requirements of the Certificate of Financial Advisors (CEFA), granted by the Chartered Institute of Banking (CIB) in the United Kingdom (Website of ICB).

Stakeholders Involved in the Implementation Process

Though customers were the end users of the products, they were not involved in the implementation phases. Mailings were sent to customers with their bank statements to promote the products. The bank considered complaints of customers to improve services provided.

Change Strategies

Specific Strategies in Managing Change

ICB invested heavily in IT and staff training. It conducted training sessions on local, regional and international levels. Locally, for example, it administered training sessions in the bank premises, AmCham in Egypt or Egyptian Banking Institute through its training programs or the American University's Institute of Banking. On the regional level, some training courses were conducted in Dubai. On the international level, training sessions were carried out in the Group's premises.

In general, the bank's strategy with regards to information technology products is to start small. After testing the product and finding it profitable and acceptable to users, it changes its distribution strategy and starts implementing the product in other branches.

Strategies Used to Transform ICB to the New Systems

The bank used a mixture of the following strategies: Empirical-Rational Strategies, Normative/Re-educative Strategies of Change, Power-Coercive Techniques to Effective Change, and Environmental-Adaptive Methods to Change, thereby indicating sufficient expertise in managing change. Management assumed people are rational since they were already of high caliber, which helped them adapt easily to new changes, especially for these two kinds of products. According to Nickols (2004), this indicates the bank is experiencing weak resistance to change. Managers also conducted meetings with employees to allow brainstorming and sharing of ideas. Managers also acted as role models for staff and empowered subordinates to provide outstanding customer services. In addition, management used power-coercive techniques to affect change, as suggested by short-term change plans of implementing IT-based delivery systems (Nickols, 2004: 10-11).

ICB's Organizational Structure

Changes in Organizational Structure

ICB appointed a new product manager from its existing staff for each product, and established new departments such as an ATM card center, a Personal Internet Banking Section (Call Center), and a new training center to conduct training sessions on ATMs and Internet Banking usage. A number of employees were accountable for monitoring peak hour's usage time and most commonly used transactions. An e-champion was responsible for communicating problems faced by customers using Internet Banking to related departments such as IT, the personal Internet Banking section, or the statistics department. Some staff were newly appointed. Others were transferred from other departments or were assigned additional tasks.

Task Specialization

In ICB there was no specialization in one job, as one employee can easily cover for the absence of another. The exceptions were the IT department and the cash area where individuals worked separately and specialized in one clearly defined task.

Integrating Mechanisms

Throughout implementation phases, staff acted as one team towards achieving targeted sales. Task forces or project teams were constituted, but were commanded by an established authority, in most cases the team leader. Hierarchy of authority was used in assigning a task to employees. Teams were used to carry out a common function for a defined time period and for a pre-specified purpose. Though team members might belong to different departments, they were all engaged in the same project.

Lines of Responsibility

New lines of responsibility were involved since development of each new product required a new product manager; a new head of ATMs/Internet Banking branches, head teams, and teams to carry out tasks pertaining to the new products. Hence the numbers of lines of responsibility increased as new positions were created that involved new responsibilities.

Authority to Control Tasks

Implementation of new delivery systems created a remote distribution department. Once products were in place, the bank had sections within the department, one for ATMs and the other for Internet Banking, and product development heads were responsible for communicating with branches' heads. The ATM product development head delegated authority to ATM teams who were responsible for tasks such as replenishing and balancing machines, and retrieving cards for offsite machines. Onsite machines were the responsibility of each branch. The Internet Banking product development head delegated authority to call centers to monitor statistics and report problems encountered by customers. The authority to control tasks in ICB, therefore, is decentralized to a certain extent, depending on the problem. A problem could be solved at low levels, while another might need higher-level intervention.

Coordination of Tasks and Work Process

ICB is going by both the Group's Instruction Manual and Desk Instruction Manual. Mutual adjustment and negotiation is only used in the way the tasks are carried out.

ICB's Organizational Culture

Extent of Employee Involvement in the Change Process

No feedback from employees was required with regards to new delivery systems since these were previously tested and implemented in other parts of the world before introducing them to Egypt. Employees were permitted to give their feedback on how tasks could be carried out, but not on what tasks are assigned or to whom.

Employee Empowerment

Employees' actions were controlled by bank's rules, policies, procedures, and regulations. Hence, minimum creativity was left for employees. Moving from command to empowerment throughout the day depended on the extent of experience the staff member has and was only related to how the job was done. An employee who is new differs from another who is about to be a branch manager, for example.

Values, Norms, and Behaviors of the Organization's Members

Implementation of ATMs made employees more sales oriented. It is doubtful that implementation of Internet Banking changed the bank's culture since all employees were already internet users when they joined the bank. In ICB, a "Best Service Bank" culture exists where all individuals believe themselves and their bank to be superior to other banks with regards to service delivery. Implementation of new delivery systems elicited a more customer driven culture as employees are constantly trained and business and customer philosophies are always reinforced.

Dominating Cultural Pattern

During implementation of ATMs, the dominating cultural pattern was aggressive and defensive since the bank was a pioneer in introducing the product. During implementation of Internet Banking, the dominating cultural pattern was constructive since the bank was not a pioneer or laggard as much as it was conservative in adopting the new system.

Dominating Cultural Style

In general, interviewees believed that ICB emphasized achievement, self actualization, encouragement, and affiliation.

ICB's cultural values are those of an accountable, ethically based, and traditional organization dedicated to long-standing relationships with its clients. The ICB aims to be greatly productive, team-oriented, and innovative and customer determined. The bank's group vision is to be the world's leading financial services corporation and is established on whole integrity, faith and outstanding customer service. The ICB aspires to guarantee that its brand name philosophy is constantly valid to all its activities. In this respect, ICB treats its clients with special care, is aware of their value, empowers them, is proactive to their needs; and offers them easy-to-use and useful products and services as well as reachable expertise (Website of ICB; ICB's Annual Reports 2002-2005).

Nature of Employees' Beliefs and Orientations

In the bank, people interacted with others and performed tasks that assisted them in meeting their top-priority satisfaction needs.

Business Processes

New Processes or Activities Involved

New services entailed additional activities and steps in the business process. Examples could be ATMs' replenishment and Internet Banking monitoring for availability of service around the clock. Staff orientation to the new products, training, implementation steps, and actual sale of the products were some of the activities involved. Newly added customer products and services also involved new experiences, support through on-the-job training or online training. New processes and activities involved increased the number of clients and services provided.

New Work Systems' Inputs and Outputs

Resources, technical skills, experience, hard work, challenges, and activities like daily replenishment and monitoring of machines 24 hours a day, even on weekends, were inputs to the newly installed ATM delivery system. On the other hand, webpage designs and images, as well as continuously screening the service, were inputs to Internet Banking work system. Outputs of new systems included better services that were provided to users whether new or existing ones, reduced stress on staff as the number of operations performed at the bank counter was reduced, and less paper work was involved.

Value Created to Stakeholders

With the implementation of ATMs, a certain segment, mainly employees and businessmen, found it more convenient, easier and more useful to use the machine and stopped coming to the bank. Internet Banking users are hardly ever seen at the bank ever since its implementation. Therefore, new systems increased customer satisfaction as it increased the bank's working hours from 30 hours per week to 24/7. It also reduced the number of times customers need to go to the bank, and it changed the type of customer to a one who is more demanding for technology and new products. Employees are satisfied with new systems because fewer clients are now using the bank counter, which reduces workload and cues. Shareholders are happy with new systems as they increase profits through commissions paid by customers and through cost reductions made in conducting transactions away from the counter.

Communication with Clients

New delivery systems increased communication with clients as they increased transactions' traffic. They reduced face-to-face interaction with clients though, thus reducing crowds and hassles, which allowed better communication. Clients' complaints switched to problems with regards to offline periods. In this respect, ICB created a complaint department and a call center to receive clients' suggestions and complaints.

Track of Transactions

Reports that ICB issued to monitor ATMs and Internet Banking transactions provided information on large and frequent transactions conducted, and Internet Banking account holders who ask to open a third party payment, thereby making clients feel more secure when transferring to a third party. Tracing of transactions was done manually before

implementation of systems. After implementation, the process was done automatically, which was easier as the human factor was only involved in some cases like ATMs replenishment or in cases of problems. This has reduced stress on staff as fewer loads were placed on them due to smaller queues or customers reporting at the bank counter.

Motivation of Employees

In general, new delivery systems motivated employees as they reduced stress by reducing the number of clients who come to the bank. One interviewee out of three thought that increases in salaries, bonuses, and promotions are more effective in motivating employees, however.

ICB's benefits and remuneration compensation packages are intended to promote the success of the ICB and to reward individual achievements. Through yearly income surveys, an improved payment program is applied to ensure that the bank is providing competitive salaries and stipends in contrast to similar banks in Egypt. In addition the bank offers a competitive benefits package embracing medical coverage and life insurance. The program further permits employees to be rewarded relevant to the point of their achievement (Website of ICB).

Post Implementation Challenges

In general, no problems were encountered. However, sometimes the bank experienced offline periods, complaints from clients, hardware problems, or fear from hackers.

ICB's Commitment Profile

Extent of Organizational Commitment to Change

In ICB, employees are highly committed. Evidence of great commitment is successful implementation and performance of both ATMs and Internet Banking products. Employees also use IB as a marketing tool to attract new customers to the bank, especially companies' accounts. Evidence to great commitment to change is also the increasing number of new branches opened, as well as the expansion of existing ones.

Prevailing Commitment Profile
Interviewees believed that the prevailing commitment profile, mindset nature, and decision to remain in the organization reflect a mixture of an affective and continuance profile. In their opinion, employees feel they work in a very prestigious establishment and rewarding environment, especially after recent salary adjustments (12-18 months earlier). The bank's environment is also secure with no hire and fire labor policy, and vast expansions allowed recruiting a large number of young people, who were almost at the same age. Reducing the age of managers thus created a common understanding between them and other young staff and also reduced the gap that could have existed between two different generations.

Strategic Competitiveness

Extent to which Organizational Commitment Profile Influences Bank's Strategic Competitiveness
The bank is very well organized in terms of authority. There are very clear-cut limits and authorities so everyone does his/her job effectively and efficiently, which eventually contributes to achieving strategic competitiveness. Organizational commitment improves customer services provided, which also contributes to strategic competitiveness. In fact, employees who are emotionally attached to the bank (70% of total number of staff members) contribute to 95% of the bank's strategic competitiveness. To retain the bank's most experienced staff who started leaving to other better paying banks, ICB revised its salary scale.

The Role of Technology in Achieving Competitiveness
Interviewees believed to achieve strategic competitiveness, the bank should have a sophisticated IT system and infrastructure, including hardware, and networks, which are user-friendly. In their opinion, the bank's IT system allows it to perform transactions faster, facilitates flow of work, and provides reliable accessibility to the bank's products and services all the time. Better service quality, speedy service, availability and accessibility have increased the number of "un-man bank", which with the technology revolution became a requirement for achieving competitiveness. "Technology is now a MUST; no product can be done without technology," declared one of the interviewees. In another's view, technology contributes from 40 to 50% to the bank's strategic competitiveness.

Other Sources of Strategic Competitiveness
The bank's interviewees perceived global existence and brand name as two other resources that contributed to the bank's strategic competitiveness.

The Role of Effective Management of Change through Information Technology in Achieving Competitiveness
In managing change, management planned, organized and controlled staff activities, and found solutions to problems of implementation. In this process, management thinks of everything that makes service to clients available all the time, easily accessible, useful, reliable, and so forth. Management's orientation and openness to change and technology solutions and passing it on to employees contributed to achieving strategic competitiveness. In their responses, interviewees believed that effective management of change contributed about 70% to the bank's strategic competitiveness.

Criteria that Measured the Impact of Management of Change through Information Technology on Strategic Competitiveness

A. **Response to customer:** Most products of the ICB are now available to customers all the time. Through its ATMs and Internet Banking services, ICB's services are now delivered to customers 24/7 instead of only 27.5 hours per week. "Service Level Agreement" guarantees a speedy service by measuring the rate of service response to customers. For example, it usually takes 3 days for a customer to receive an ATM card, and his/her password is available the second day in the branch. There is an agreed upon level of complaints; usually 2-3 complaints are considered bad.

For Internet Banking services, service quality control is guaranteed to customers through a default cut-off time set for Internet Banking transactions as follows:

1. Payment requests in the local currency that are made to another bank before the cut-off time (15:00) are processed the same day; otherwise they are processed on the next banking day.
2. Payment instructions between any accounts with ICB are executed immediately.
3. Payment instruction to make a transfer in the local currency to any account not held with ICB, either following the cut-off time of payment, or on a day which is not a working day, is not be carried out until the subsequent banking day. For transfer requests in foreign currencies, payments are executed on the next banking day in Egypt, the beneficiary's country, as well as the hub corresponding to the currency remitted.

Recording Telephone Calls: ICB records telephone calls made to its Call Centre for business deal confirmation and quality management purposes (Website of ICB).

ICB encourages its customers to consistently and cautiously provide their feedback on the service by verifying their ICB account statements when they receive them and reporting to the bank immediately through the bank's Call Center, electronic mails, or by visiting any of its branches in case they find an error in any business deal or an illegal operation. Customers are also encouraged to contact the bank if they have any concerns or complaints about the service. Normally, the bank resolves its clients' queries immediately. Customers are provided with an acknowledgement (either verbal or written) in 24 hours (across banking days) of receiving their suggestions and objections. Examination and handling of complaints start immediately after being received from clients. Until the concerned branch determines a proper solution, clients are normally updated with progress made. Complaints that are submitted in writing receive relevant responses that are approved by the branch manager, unit head of the branch, and the department to whom the complaint is directed (Website of ICB).

B. **Customers Working Online:** Statistical reports showing the number of customers, transactions' peak periods, durability on the net, and kind of transactions indicate increasing numbers of customers using both products. The traffic on ATMs usually reaches its peak throughout salary payment periods.

C. **Rate of customer to employee:** There is no direct relationship between rates of customer to employee because of two reasons: first, Egypt is still a cash society and bank tellers still serve corporate accounts which are not yet served by ATM machines. Further, the rate of customer to employee depends on branch location and services provided. But in general, the rate of customer to employee is going down as a result of new bank service channels (ATMs and IB) being opened.

D. **Waiting time to be served:** Waiting time for customers to be served has decreased after implementation of new delivery systems. This time differs though from bank to ICB and from one branch to another. Some branches are always crowded. But in general, waiting time is from 60 seconds to about 1 minute on ATMs; whereas, it is about 15 minutes on a bank counter with moderate crowd.

SUGGESTIONS FOR FUTURE RESEARCH

The study was limited to scanning and analyzing the independent variables in the model, which are internal to organization, and that accordingly might not be enough to give us an integrative explanation of the factors influencing key determinants of successful implementation of IT and banks' competitiveness in Egypt. Future research, therefore, might include external environment variables such as competitors and the economic, social, and political environments that might also have an impact on the dependent variable.

Research findings show that creating a conducive culture to change, involvement, and commitment were crucial for strategic competitiveness. Further investigation is recommended to find out the extent of involvement in the banking industry in general in light of dictated standardized rules and regulations. Comparison of involvement levels between the banking industry and other organizations that are less conservative have yet to be researched.

CONCLUSION

This case described ICB's activities and management practices throughout implementation of two IT-based delivery systems, namely ATMs and Internet Banking.

In implementing new technologies, ICB's managers were already aware of new developments through the parent bank. Still the bank went through all planning, training, conversion, acceptance testing, and post implementation audit phases of implementation.

Neither the bank's clients nor employees showed any resistance to the new products. In fact, they were, especially ATMs, appealing to most of the clients. Challenges to change included technical problems and work overloads. In this respect, the bank's management adopted several methods to overcome such challenges among which were dividing work among teams according to their members' experiences, brainstorming with staff, and conducting training courses.

ICB assumed several leadership actions to get people involved, adapt to change, support the change effort, and institutionalize new approaches. Management from several departments with different educational backgrounds and fields of experience cooperated throughout the implementation phases. Implementing new systems triggered re-organization of staff activities and each staff member had his/her role defined according to his/her area of expertise. Some work adjustments and new tasks were added to existing employees' job descriptions. New systems made people working in the customer services department more sales oriented and employees were convinced to do only things that are productive to them and leave the rest to clients through ATMs and Internet Banking. The bank used staff with the normal skills requirements of the bank, some sales skills, and a background on the new systems' application on the bank's transactions, as well as technical knowledge of hardware and software.

The bank's strategy was to invest heavily in information technology and training staff. It started small. Then, after testing the product, finding that it was profitable, and acceptable to users, it changed its distribution strategy and starts implementing the product in other branches. ICB used four change strategies simultaneously, namely Empirical-Rational Strategies, Normative-Re-educative Strategies of Change, Power-coercive techniques to effective change, and Environmental-Adaptive methods to change, indicating sufficient expertise in managing change.

Several changes were required in the bank's structure as a result of implementing these systems. The bank created new layers of management, new departments, and gave new job assignments to existing tasks of employees. There was no specialization in one job with the exception of the IT department and the cash area where individuals worked separately and specialized in one clearly defined task. Task forces or project teams were constituted, but were commanded by an established authority. Authority is decentralized as some tasks were delegated to experienced staff according to standardized rules and work procedures set by the parent ICBs well as the local one.

Employees' involvement in the change process was limited. Sometimes employees were allowed to give their feedback on how tasks could be carried out, but not on what tasks are assigned or to whom. Minimum creativity was left to employees. Employees' actions are controlled by bank's rules, policies, procedures, and regulations.

A culture, which is positive to change existed so that a "Best Service Bank" society existed in which all individuals believed they and their bank to be superior to other banks with regards to service delivery. The dominating cultural pattern was constructive, leading to higher levels of commitment, as employees believed that the bank's culture provided them with work experiences that fulfil their needs. Such work experiences stem from effective management plans and practices adopted. Sometimes, at times of implementing ATMs, the dominating cultural pattern was aggressive or defensive. People interacted with others and performed tasks that assisted them in meeting their top-priority satisfaction needs.

New processes or activities were involved with implementation of new delivery systems.

Inputs to the work system included resources, technical skills, experience, hard work, challenges, and activities. Outputs included better customer service, reduced stress on employees, and less paper work.

The new delivery systems implemented created value for stakeholders. They increased service to customers, as well as shareholders' profits, reduced workload stress on employees, facilitated work, and reduced costs and lead-time. And though new systems reduced face-to-face interaction with clients, they increased communication with clients in general as they increased transactions' traffic. New systems also improved tracking of transactions and motivated employees by reducing workload stress on them. They were also highly committed to change and emotionally attached to the organization. After implementation, the bank faced no problems though sometimes some technical challenges emerged.

Several factors contributed to the bank's strategic competitiveness. Organizational commitment improved customer services provided with implementation of new technologies in ICB. New technologies were also indispensable to achieving strategic competitiveness. Some other sources were global existence and brand name of the bank. Management's orientation and openness to change and technology solutions, including passing it on to employees, contributed to achieving strategic competitiveness.

The bank used several criteria to measure the impact of management of change through information technology on strategic competitiveness, such as response rate to customers, the number of customers working online, rate of customers to employees, and the waiting time to be served.

Strategic competitiveness and successful implementation of IT in ICB are outcomes of the interplay of some key determinant factors. Quickly responding to changes in technology, providing their customers with a variety of financial transactions through new systems and advancing efficiently through the change initiative process through a sequence of steps are some of those factors. A preset vision and strong leadership are essential. Management, however, must engage staff by conducting information sessions to reveal facts, concepts and models to permit all to find a common base in the technology applications.

Increased layers of management and even new departments were needed but employees had to be empowered to effect change themselves. All these had an impact on strategic competitiveness by influencing the bank's organization and values.

ICB achieved higher financial performance compared to its peer banks when assessing strategic competitiveness. This was true whether the assessment was measured by return on assets (ROA), return on equity (ROE), and earning assets to total assets or by other measures. Therefore, a successful implementation of IT means much more than just an IT department's initiative or technical endeavor and cannot be confirmed to terms such as "installation" or "execution" efforts of one party in an organization.

REFERENCES

Amit, R., & Shoemaker, P. J. H. (1993). Strategic assets and organizational rent. *Strategic Management Journal, 14*(1), 33–46. doi:10.1002/smj.4250140105

Anol, B., & Rudy, H. (1997). IT and organizational change: Lessons from client/server technology implementation. *Journal of General Management, 2*(winter), 31., Retrieved August 28, 2004, from http://www.aucegypt.edu/library/libdata/subject.cfm?classid=2

Balmer, J. M. T. (1998). Corporate identity and the advent of corporate marketing. *Journal of Marketing Management, 14*, 963–996. doi:10.1362/026725798784867536

Balmer, J. M. T., & Soenen, G. (1999). The acid test TM of corporate identity management. *Journal of Marketing Management, 15*, 69–92. doi:10.1362/026725799784870441

Barnes, D., Mieczkowska, S., & Hinton, M. (2003). Integrating operations and information strategy in e-business. *European Management Journal, 21*(5), 626. doi:10.1016/S0263-2373(03)00111-7

Brand, H., & Duke, J. (1982, December). Productivity in commercial banking: computers spur the advance. *Monthly Labor Review*, 19–27.

Flower, J. (1996a). *The five fundamentals of dealing with change*. Retrieved July 21, 2004, from www.well.com/user/bbear/change2.html

Flower, J. (1996b). *The root ideas in dealing with change*. Retrieved July 22, 2004, from www.well.com/user/bbear/change2.html

French, W. L., & Bell, C. H. (1978). *Organization development* (2nd ed.). Englewood Cliffs, N.J.: Prentice Hall.

French, W. L., Bell, C. H., & Zawacki, R. A. (Eds.). (2000). *Organization development and transformation*. New York: Irwin McGraw-Hill.

Hatem, T. (2003). *Understanding cultural differences between Americans and Arabs: The Egyptian case*. In the Proceedings of the tenth American University in Cairo Research Conference, Globalization Revisited: Challenges and Opportunities, April 6-7: 249-271. The American University in Cairo, Egypt.

Hill, F. M., & Collins, L. K. (1999). The quality management and business process re-engineering: A study of incremental and radical approaches to change management at BTNI. *Total Quality Management, 10*(1) (January): 37. Retrieved August 28, 2004, from http://www.aucegypt.edu/library/libdata/subject.cfm?classid=2

Hitt, M. A., Ireland, R. D., & Hoskisson, R. E. (2001). *Strategic Management: Competitiveness and Globalization*. United States: South-Western College Publishing.

Hofstede, G. (1984). *Culture's consequences: international differences in work-related values. Abridged editon.* Beverly Hills, CA: Sage.

Hofstede, G. (1991). *Cultures and organizations: software of mind.* London: McGraw Hill.

Johnson, P., Fidler, C. S., & Rogerson, S. (1998). Management communication: A technological revolution? *Management Decision, 36*(3), 160. doi:10.1108/00251749810208940

Kim, M. & Weiss, J. (1989). Total productivity growth in banking: the Israeli banking sector 1979-1982. *Journal of Productivity Analysis,* (1, 2), 239-153.

La Porte, T. M., Demchak, C., & Friis, C. (2001). Webbing governance: Global trends across national-level public agencies. *Association for Computing Machinery, Communications of the ACM, 44*(1) (January 2001), 63. Retrieved August 28, 2004, from http://www.aucegypt.edu/library/libdata/subject.cfm?classid=2

Matsui, Y. (2002). Contribution of manufacturing departments to technology development: An empirical analysis for machinery, electrical and electronics, and automobile plants in Japan. *International Journal of Production Economics, 80*(2), 185. doi:10.1016/S0925-5273(02)00317-1

Melewar, T. C., & Navalekar, A. (2002). Leveraging corporate identity in the digital age. *Marketing Intelligence & Planning, 20*(2). doi:10.1108/02634500210418518

Morisi, T. L. (1996). Commercial banking transformed by computer technology. *Monthly Labor Review,* (August): 30–36.

Nickols, F. (2004). A Primer. In *Themanager.org.* Retrieved July 21, 2004, from www.themanager.org/knowledgebase/management/change.htm

Oster, A., & Antioch, L. (1995). Measuring productivity in the Australian banking sector. In P. Andersen, J. Dwyer, & D. Gruen (Eds), *Productivity and Growth:* 201-12. Conference Proceedings, 10-11 July, Reserve Bank of Australia, Sydney.

Penrose, E. T. (1959). *The theory of growth of the firm.* London: Basil Blackwell.

Pickett, L. (2004). Focus on technology misses the mark. *Industrial and Commercial Training, 36*(6/7), 247. Retrieved August 28, 2005, from http://www.aucegypt.edu/library/libdata/subject.cfm? Classid=2

Porter, M. E. (1985). *Competitive Advantage* (pp. 11–15). New York: The Free Press.

Porter, M. E. (1990). The competitive advantage of nations. *Harvard Business Review, 90211,* 73–91.

Porter, M. E. (1990). *The competitive advantage of nations.* New York: Free Press.

Porter, M. E. (2001). Strategy and the internet. *Harvard Business Review, 75*(3), 63–78.

Prahalad, C. K., & Krishnan, M. S. (2002). The dynamic synchronization of strategy and information technology. *MIT Sloan Management Review 083, 43*(4) (summer), 23-34.

Proctor, T., & Doukakis, I. (2003). Change management: The role of internal communication and employee development. *Corporate Communication, 8*(4) (2003), 268. Retrieved on 28 August 2003, from http://www.aucegypt.edu/library/libdata/subject.cfm?classid=2

Radijan, J. (1996). Bring in danet, bring in da bucks. *US Banker, New York, 106*(9), 18.

Rogers, E. M. (1983). *Diffusion of Innovations.* New York: The Free Press.

Schultz, B. (2005). Out with the old, in with the new data center. *Network World*. Framingham, 22, no. 22 (June 6),38. Retrieved August 28, 2005, from http://www.aucegypt.edu/library/libdata/subject.cfm?classid=2

Schumpeter, J. A. (1934). *The theory of economic development: an inquiry into profits, capital, credit, interest, and business cycle*. Cambridge, MA: Harvard University Press.

Senge, P. (1994). *The fifth discipline field book: Strategies and tools for building a learning organization*. New York: Doubleday.

Swierczek, F. W., Pritam, K. S., & Bechter, C. (2005). Information technology, productivity and pofitability in Asia-Pacific banks. *Journal of Global Information Technology Management, 8*(1).

Wah, S. S. (2002). *Behavioral attributes of the transformational Chinese leader.* Unpublished doctoral dissertation. The Netherlands: Maastricht School of Management.

Weiss, J. W., & Anderson, D., Jr. (2004). CIOs and IT professionals as change agents, risk and stakeholder managers: A field study. *Engineering Management Journal, 16*(2) (June), 13. Retrieved August 24, 2004, from http://www.aucegypt.edu/library/libdata/subject.cfm?classid=2

ADDITIONAL READING SECTION

Abou-Musa, A. A. E. (2003). Egyptian banking industry: Its history and future. *Journal of American Academy of Business, 3*.

Ahadi, H. R. (2004). An examination of the role of organizational enablers in business process reengineering and the impact of information technology. *Information Resources Management Journal, 17*(4).

Alter, S. (1999). *Information systems: A management perspective*. Reading, MA: Addison-Wesley.

Argyris, C. (1990). *Overcoming Organizational Defensive Routines*. Boston: Allyn and Bacon.

Argyris, C., & Schon, D. A. (1978). *Organizational learning: A theory of action perspective*. Reading, MA: Addison Wesley.

Bosworth, D. (1996). Determinants of the use of advanced technologies. *International Journal of the Economics of Business, 3*(3). doi:10.1080/758539567

Brynjolfsson, E., & Yang, S. (1997). *The intangible costs and benefits of investments: Evidence from financial markets*. Paper presented in proceedings of ICIS, Atlanta, Georgia, United States.

Chin, R., & Benne, K. D. (1984). General strategies for effecting changes in human systems. In Bennis, W. G., Benne, K. D., & Chin, R. (Eds.), *The Planning of Change* (4th ed.). New York: Holt, Rinehart and Winston.

Competition: Driving innovation and consumer choice; in AMD Worldwide. Retrieved October 1, 2005, from http://www.amd.com/us-en/Weblets/0,7832_12670_13047,00.html

Covey, S. R. (2004). *The 7 habits of highly effective people*. United Kingdom: Simon & Schuster UK Ltd, a CBS Company.

Cox, E. B. (1986). *The Bank Director's Handbook. Dover*. MA: Auburn House Publishing Company.

Creswell, J. W. (2003). *Qualitative, quantitative and mixed methods approaches* (2nd ed.). Thousand Oaks, CA: Sage Publications.

Drucker, P. F. (1988). The coming of the new organization. *Harvard Business Review*, (January-February), Product number 88105.

Eisenhardt, K. M. (1989, Oct). Building Theories from case study research. [AB/INFORM Global.]. *Academy of Management Review, 14*, 4. doi:10.2307/258557

El Shenawi, N. (2004). *Credit cards and their impact on the development of the banking sector: the case of Egypt.* (DBA Thesis), Maastricht School of Management, the Netherlands.

Hall, J., et al. (2002). *Transformational leadership: The transformation of managers and associates.* A publication of the Department of Food and Resource Economics, Florida Cooperative Extension Service, Institute of Food and Agricultural Sciences, University of Florida, Gainesville, FL. Retrieved January 6, 2008, from http://edis.ifas.ufl.edu

Hamel, G. *Impetus for radical change: Interview with Gary Hamel.* Interview by Finnie, W.C. Retrieved July 12, 2004, from www.managementfirst.com/change_management/interviews/hamel.php

Hassanein, M. (2004). *The second wave of fiscal reforms in Egypt.* Presentation, Institute for International Economics, Washington DC. Retrieved February 2, 2005, from http://www.iie.com/publications/papers/hassanein0604.pdf

House, R. J. (1971). A path-goal theory of leader effectiveness. *Administrative Science Quarterly, 16*, 321–339. doi:10.2307/2391905

Kay, J. (1993). *Foundations of corporate success: How business strategies add value.* New York: Oxford University Press Inc.

Kitchen, P. J., & Daly, F. (2002). Internal communication during change management. *Corporate Communications, 7*(1), 46–53. doi:10.1108/13563280210416035

Kotter, J. P. (1990). *A force for Change: How Leadership differs from management?* New York: The Free Press.

Kotter, J. P. (1995). Leading change: Why transformation efforts fail?. *Harvard Business Review,* (March-April), Product number 4231.

Kotter, J. P. (1996). *Leading change.* Cambridge, MA: Harvard Business School Press.

Kotter, J. P. (1998). Winning at change. *Leader to Leader, 10*(Fall). Retrieved July 25, 2004, from www.pfdf.org/leaderbooks/121/fall98/kotter.html

Kotter, J. P. (2001). *What leaders really do?* Harvard Business Review OnPoint.

Kotter, J. P., & Schlesinger, L. (1979). Choosing strategies for change. *Harvard Business Review,* (March-April): 106–114.

Lahiry, S. (1994). Building commitment through organizational culture. *Training & Development,* (April): 50–52.

Levasseur, R. E. (2001). People skills: Change management tools- Lewin's change model. *Interfaces, 31*(4), 71–73.

McNamara, C. (1999). Basic context for organizational change. Retrieved July 21, 2004, from www.mapnp.org/library/mgmnt/orgchnge.htm

McNish, M. (2002). Guidelines for managing change: A study of their effects on the implementation of new information technology projects in organizations. *Journal of Change Management, 2*(March), 3.

Metwally, E. (2003). *Challenges facing foreign direct investment in Egypt.* In the Proceedings of the tenth American University in Cairo Research Conference: Globalization Revisited: Challenges and Opportunities, April 6-7, 292-325. The American University in Cairo, Egypt.

Metwally, E. (2005). *Understanding users' acceptance of Internet Banking in Egypt.* In the Proceedings of the twelfth American University in Cairo Research Conference: Reform in Egypt: Opportunities and Challenges, March, 261-290. The American University in Cairo, Cairo, Egypt.

Meyer, J. P., & Allen, N. J. (1990). Affective and continuance commitment to the organization: Evaluation of measures and analysis of concurrent and time-lagged relations. *The Journal of Applied Psychology, 75*(6), 710–720. doi:10.1037/0021-9010.75.6.710

Meyer, J. P., & Allen, N. J. (1991). A three-component conceptualization of organizational commitment. *Human Resource Management Review, 1*(1), 61–89. doi:10.1016/1053-4822(91)90011-Z

Ramsower, R. M. (1991). Competitive advantage with information technology. *Baylor Business Review*, (Fall), 9.

Rogers, E. M. (1983). *Diffusion of Innovations*. New York: The Free Press.

Yin, R. K. (2003). *Case study research: Design and methods*. London: Sage Publications.

Chapter 10
Integrating Information Technology into the Corporate Culture and Processes

Norita Ahmad
American University of Sharjah, UAE

Alanoud A. Alhaj
American University of Sharjah, UAE

ABSTRACT

Most of the Information Systems (IS) literature discusses the importance of managerial problems in the field such as how to evaluate Information Technology (IT) needed for effective intra-organizational communication and how to measure the value of changes influenced by IT. This paper uses a literature review detailing the importance of understanding the components of IS and how they affect one another and on the importance of the selection of IT by managers. This paper builds a case study on a large company in the United Arab Emirates (UAE) with approximately 35,000 employees world-wide represented in approximately 100 countries. We analyze the reasons behind the failure of integrating IT with the culture, people, and processes in a Market Research Department of this company. Finally, the case study discusses the consequences of failing to understand IS and provides recommendations for a better integration of IT within the department.

INTRODUCTION

Even though Information Systems (IS) has been around for centuries, only around four decades ago IS become very much dependent on Information Technology (IT). In order for the implementation of IT to be successful and effective in an organization, the IT must fit-in well with the system of the organization. Evaluating the changes that

are influenced by the implemented-IT will help understand the level of success or failure of the implementation process. Hence, it will be clear if the decision to purchase and implement a specific IT is a successful decision. To ensure the success of IT implementation in a given organization, it is important for the management and the people involved in the decision making process to understand what IS is and how it is related to IT.

This paper builds a case study in a Market Research department of a large company in the

DOI: 10.4018/978-1-60960-048-8.ch010

United Arab Emirates (UAE) with approximately 35,000 employees and with a vast operation worldwide represented in approximately 100 countries. This paper frames the consequences of failing to understand the best practices in selecting IT and of failing to understand what IS is and how IT implementation and upgrade might affect the various components of IS.

BACKGROUND

Literature related to Information Systems (IS) and Information Technology (IT) has been published for the past 50 years mainly in the area of Management Science simply because IS and IT affect all functions related to management (Banker & Kauffman, 2004). Most of the IS literature discuss the important managerial problems in the field such as how to evaluate IT needed for effective intra-organizational communication and how to measure the value of changes influenced by IT (Banker & Kauffman, 2004). In addition, many researchers focus on studying the link between information systems and managerial performance (Hamilton & Chervany, 1981a; Hamilton & Chervany, 1981b; Lucas, 1975).

Enterprise Information Systems are vital to the operation and management of every organization but the field is known to have constant technological change, therefore, organizations often find themselves struggling to keep up with changes when maintaining existing systems or evaluating new systems (Benbasat, Goldstein, & Mead, 1987; Díez & McIntosh 2009). Often, when investing in IS, organizations are interested in the benefit that they will gain from the investment. However, IS investments are hard to justify (Ragowsky, Ahituv & Neumann, 2000). Many researchers suggested that in order to justify an IS investment, organizations should look at specific functions instead of the entire entity (Gerstein & Reisman, 1982; Ragowsky, Ahituv & Neumann, 2000). In addition, it is also very important for

organizations to change the fundamental nature and structure of their work so that better use of IS can be realized (Keen, 1981). In fact, improving the efficiency and effectiveness of organizations, the traditional domain of the IS function (Bawden, 2008; Rockart & Morton, 1984). Keen (1981) predicts that IT and telecommunication systems will be the backbone of corporations and therefore they need to be properly managed. McFarlan and McKenney (1981, 1983) also point out the importance of proper management for the successful deployment of IT.

In the late 1990's and early 2000's we have witnessed the booming of IT. Many believe that information systems can help businesses use synergies, core competencies, and network-based strategies to achieve competitive advantage, justifying heavy investment in IT (Al-Gahtani, 2003; Al-Ashban & Burney, 2001; Blili & Raymond, 1993; Caldeira & Ward, 2002; Cooper & Zmud, 1990; Ein-Dor, Myers & Raman, 1997; Ives & Jarvenpaa, 1991; Yavas, Luqman & Quraeshi, 1992). However, in order to ensure the success of IT implementation, it is important for any organization to understand the difference between IS and IT and how they relate to each other. IS and IT adoption cannot be separated. There are four components to any formal IS used in an organization (Information Technology (IT), Process, People, and Structure according to Piccoli (2008). The four components of IS must work together in order for an organization to meet its information needs. IT is a very critical component in any modern IS. It is important to familiarize the people component of IS with the new IT implemented as studies indicate that there is a strong positive relationship between the perception of ease and the perceived effectiveness of any IT implemented to the actual amount of usage (Mahmood, 2001; Piccoli, 2008). It is also important to provide sufficient training to the people component of IS as this is evident to have a significant effect on the level of IT usage (Mahmood, 2001).

The components of IS are inter-related in that any change in one component could have an impact on the other components (Piccoli, 2008). IT is not the only factor that managers should consider. Even though IT's power has grown, its strategic importance has diminished, therefore, the way IT is managed should change dramatically (Carr, 2003). Studies show that in most cases the problem is in the inefficiency of the way non-IT executives are managing IT changes in the organization (Piccoli, 2008). It is important for managers to select an IT that integrates well with their organizations. Also, it is crucial for managers to view decisions from a user viewpoint (Ross & Weill, 2002). When making a decision on what IT to implement, managers should ensure that the IT to be implemented is user friendly and compatible with the structure of the organization. The failure of senior managers to select a manageable set of IT can lead to a disaster (Ross & Weill, 2002). Managers should also ensure that people within the organization or end users will be able to adapt to the new IT to be implemented. In other words, it is important for the organization to understand the impact of cultural differences on the IT adoption. Many studies have shown that in order for organizations to successfully implement IT, they have to understand the cultural differences that exist within their organizations (Applegate, McFarlan, & McKenney, 1999; Harris & Davison, 1999; Ives & Jarvenpaa, 1991; Shore & Venkatachalam, 1995; Tan, Watson, & Wei, 1995; Tractinsky & Jarvenpaa, 1995). Moreover, they should ensure that the IT to be implemented will fulfill the needs of the processes conducted within the organization and will meet the objectives of the organization.

THE ORGANIZATION

The subject organization for this case study is called EMGK for anonymity. EMGK is a travel and tourism organization with a well-known reputation of excellence services world-wide.

EMGK provides services in a number of industries such as tourism, holidays planning, safari tours and adventures, hotels and resorts, and training programs. The vision and mission of EMGK is to be the world-wide leading company in the services it provides. Service excellence is an important factor to this organization.

EMGK has a number of competitors within the region and world-wide. Therefore, in order for this company to retain its customers, attract new customers, and continue to deliver service excellence, it is crucial that it meets customer satisfaction in all its services. The strategy for service developments must remain aligned with the need of customers. Hence, there is a need to understand the customer satisfaction levels with the services provided by EMGK.

EMGK is a very large organization with approximately 50 specialized business divisions. About 35,000 employees constitute the family of EMGK. The headquarters of this company is based in the UAE with over 100 offices spread all over the world. The culture at EMGK is very diverse; employees come from almost a 100 nationalities. Approximately 45 divisions report to the Chief Executive Officer (CEO). Each division is responsible for a specific business function within the organization and has its own management. These include the Higher Management and Planning Division, the Marketing Division, the Organization Communications Division, the International Affairs Division, the Environmental Affairs Division, and others. The structure of EMGK is vertical with a steep level of hierarchy and the decision making is decentralized across the organization.

The Higher Management and Planning Division

The Higher Management and Planning Division is divided into two departments; the Planning Department and the Market Research Department. The Planning Department is composed of 29 employees. The main objective of this department, as its name suggests, is to work on the strategic plan of the company including identifying which new markets the company wants to expand its network into and identifying which current markets it wants to expand within.

On the other hand, the Market Research Department is a support department. It is composed of 7 employees. The main objective of this department is to manage market research projects for all divisions within the organization. Surveys play a very important role in the market research projects conducted. Market research projects help the organization in understanding the level of satisfaction of customers with the various services provided within the organization and in understanding the market needs and thus aligning the services provided with what the market desires. These projects also provide the higher management with recommendations for future strategic planning.

The Market Research Department

This case is focused on the Market Research Department of the Higher Management and Planning Division of EMGK. The Market Research Department is composed of seven employees with different levels of grades, indicating a vertical direction of hierarchy. The lower the job grade, the higher the position. The highest grade is grade 3 while the lowest is grade 7. The head of this department is the Market Research Manager with Grade 3 who reports directly to the Higher Management and Planning Division Manager (Grade 2). Four employees report directly to the Market Research Manager; one is the Research Manager who is a grade 4, while the remaining three are the Project Manager 1, Project Manager 2, and Project Manager 3, all at grade 4.

The roles of the Market Research Manager are to set the objectives of the department, supervise the implementation of these objectives, supervise all the projects conducted by the department, and

set the processes by which projects are undertaken in the department. One employee reports directly to the Research Manager, who is the Research Coordinator. The main role of the Research Manager is to overlook any IT related issues within the Market Research Department such as what software are needed in the Research Projects and what other software are needed to enhance the work procedures in the department. When there is an overload of work among the Project Managers, the Research Manager also manages some of the research projects. The Project Managers are all grades 5. There are three project managers. One employee reports directly to Project Manager 1, who is the Research Coordinator. No employees report to Project Manager 2 and Project Manager 3.

The main role of a Project Manager is to conduct research projects. The research project cycle involves writing proposals for suggested projects, developing time frames for conducting the project, writing the questionnaires on internet based or intranet based software tools, analyzing the outcomes using statistical software tools, and developing and sharing the reports and recommendations electronically with the respective higher managers. The three Project Managers perform the same role mentioned above except that each Project Manager oversees projects from different departments; Project Manager 1 oversees projects from the travel and tourism divisions, Project Manager 2 oversees projects from holidays planning and safari tours and adventures divisions, and Project Manager 3 oversees projects from hotels and resorts and training programs divisions as well as other support divisions within the EMGK (i.e. IT Division, Company Transportation and Logistics Division). Reporting directly to the Research Manager, the major role of the Research Coordinator is to assist the Research Manager with his tasks. Some of the tasks are working on IT related issues or taking a role in various research projects' tasks such as writing questionnaires to the online based software tool or analyzing data. On the other hand, the Research Assistant who is a grade 7 assists Project Managers with data entry. Decisions related to work procedures are mainly performed by the grades 3 and 4 employees. Grades 5, 6, and 7 employees are not involved in any decisions related to work procedures and processes. Decisions related to IT and software usage are done by grade 4 (Research Manager) with the approval of the grade 3 (Market Research Manager).

CURRENT CHALLENGES

This case is focused on the Market Research Department of the Higher Management and Planning Division of EMGK. One of the main goals of the Market Research Department is to conduct research projects for the departments across the organization. The outcomes of such research projects are important for the Higher Management of different divisions in the organization to understand the level of customers' satisfaction with the various services offered by the organization. These projects also help the management of the various divisions in understanding what services are needed. Based on the outcomes of these reports, upper level managers can align their strategy based on both the market and potential customers needs.

Information Technology at the Market Research Department

IT plays a major role in all research projects conducted by the Market Research Department. IT is essential at three stages during the progress of the research projects. IT is needed for writing the questionnaires to an online based survey hosting tool, for conducting the analysis of data phase, and for sharing the reports produced once analysis is conducted.

Information for this case was gathered via several face-to-face interviews. All end users at the Market Research department were interviewed

for this case. The end users interviewed are the Project Manager 1, Project Manager 2, Project Manager 3, and the Research Coordinator.

Software Used

One of the main goals of the Market Research Department is to conduct research projects for the departments across the organization. The outcomes of such research projects are important for the upper management of the different divisions in the organization to understand the level of customers' satisfaction with the various services offered by the organization. These projects also help the management of the various divisions in understanding the kind of services the general market would be most interested in. Based on the outcomes of these reports, upper managers can align their strategy based on the customers' needs and potential market.

The methods used by the Market Research department to conduct the research projects are quantitative methods mainly through questionnaires (or surveys). As mentioned earlier, the Research Manager is in charge of selecting software to be used for each process and on deciding whether renewals of software contracts and software upgrades should take place. The Research Manager has to receive the final approval from the Market Research Manager before implementing the decision.

The IT Division of the organization is in the front line when it comes to interacting with the supplier to purchase software. The installation of any software is done by the IT department of the organization. The IT personnel in charge of supporting the Market Research department are expected to learn how to use the software and then train the end users on the software.

The end users are not involved in any decisions even though they are the ones who will be using the software. They are usually not aware of whether or not contracts are to be renewed or software are to be upgraded and often times they only receive a very short notice prior to when implementation is scheduled to take place.

Three types of IT applications or software are needed throughout the project cycle; software to produce the questionnaires on an online (internet or intranet) hosting tool, software to analyze the data, and software to share the reports. The software used to write questionnaires on-line to a survey hosting tool is QuestionMark Perception, the software that is used to analyze the data of the questionnaires conducted is the statistical software SPSS, and the software that is used to share the reports with the stakeholders is QuestionMark Sharepoint.

Prior to purchasing QuestionMark Perception, the Market Research department conducted paper-based questionnaires. However, because paper-based questionnaires are time consuming and require data entry, the Market Research department decided to conduct the questionnaires online as this is a more efficient way of collecting the necessary data, leading to a shorter project cycle. As a result, reports are prepared in a shorter time and therefore recommendations can be communicated more efficiently and implemented within the strategy in a timely manner. QuestionMark Perception was brought to EMGK by the Training Programs Division to assist with the assessments conducted. QuestionMark Perception is a piece of software mainly used by instructors to write exams. Even though it can be used for writing surveys, it is mainly an assessment tool. After learning about the availability of this software in the organization through the IT department, the Market Research department decided to purchase this software and use it as a survey writing tool. The software has been used by the department for the past seven years.

SPSS is a statistical software that has been used by the department for data analysis for the past few years. After the period of survey fieldwork, this software is used to analyze the data using descriptive statistics commands and cross-tabulations. Even though SPSS is very similar to MS Excel,

it has some different features. Establishing a good understanding of the features of this software is necessary to make full use of it.

QuestionMark Sharepoint software is used to link to a portal server. This software has only been implemented in the department recently. It is used to share the PowerPoint reports produced after the data analysis is conducted, with all managers involved in a research projects.

Consequences of Deviating from Best Practices

Most of the problems encountered in the Market Research department arise when working with QuestionMark Perception due to its complexity. End users did not receive training on any of the software, instead they are expected to learn on their own and teach each other how to use the software. They are asked to contact the IT support personnel if they require any assistant or if they have any queries regarding any problems they encounter when using any of the software. It usually takes time for the IT support personnel to respond back to the end users. The software is not very easy to use since none of the end users come from an IT background. Moreover, the end users are not provided with any guidelines or any manuals on how to use the software.

As described by the end users within the Market Research department, "QuestionMark Perception is not user-friendly at all, it's a nightmare". For example, if a question is to be answered at five point scale such as "Excellent", "Very good", "Good", "Fair" and "Poor", a score has to be set for each choice. So, if there are five questions, the score has to be set 25 times, one for each question. This process is lengthy and time consuming. Furthermore, 'simple tasks' cannot be performed in a simple way when using this tool. A Project Manager stated that "the simplest thing when using any software today is to change font size, color, and type. However, QuestionMark Perception does not allow for that from the users side." The Researcher Coordinator added, "I had to spend hours playing around with the software and finally I was able to figure out how to change the color of the font using html. Even though I am not familiar with html, I performed a search in Google and learned how to do the coding. IT support is not giving us any directions on how to do these simple formatting changes from our side." Also due to the complexity of the software and the lack of proper communication between the end users and the IT support personnel, finalizing a questionnaire in this tool often takes a minimum of two weeks. End users cannot work on this process on their own because it involves the Development Server which is only accessible to the IT support personnel.

Furthermore, not all tasks can be performed from the end users side and this delays the procedure of writing the questionnaire to the tool. As described by one of the users, "Sometimes, even when contacting the IT support personnel for assistance, it turns out that they are not sure of how to use some features in the software and in some cases they don't know if specific features are available in the software".

The Research Manager at the Market Research department has also implemented some other improper decisions. An upgrade to the SPSS program was scheduled without notifying the end user. The notice of the upgrade came in a very short period of time. All of the end users were working on analysis using the old version of the software and had to switch to the new version. As is the case with any software at this department, no training was given to users on how to use the new version or to explain the new features of the new version of SPSS.

QuestionMark Sharepoint was purchased by the department to be used for sharing reports. However, because it also has the feature of updating reports, the Research Manager has decided to use it for the 'Progress Report' of the department. The 'Progress Report' of the department was in the format of MS Excel. Each week, the Project Man-

agers update the MS Excel file with the progress of their projects. However, just because updating reports is a feature available in QuestionMark Sharepoint, the Research Manager has decided to transfer the progress report to this tool. The Research Manager did not consult with the end users whether they would like to go about implementing this change or even how soon they will be able to work with this change. The decision to use QuestionMark Sharepoint for progress report created an additional problem instead of making it more convenient for the end users.

Deviating from the best practices of integrating IT with the people involved (end users), with the structure of the department, and with the processes conducted can lead to a number of consequences. The main consequence is that projects are delayed by weeks and this is not consistent with the objectives of the Market Research department. The main reason for purchasing QuestionMark Perception, for example, was to eliminate time wasted in data entry for paper-based questionnaires. However, QuestionMark Perception did not solve this problem, in fact it added to the problem due to the lack of support and training, the complexity of the software, and the lack of support from the respective IT personnel. Also, another consequence is the stress that is faced by the end users, especially the Project Managers, due to the inefficiency of QuestionMark Perception and due to the fact that they are not involved in any decisions regarding IT despite the fact that the decisions affect their work directly.

RECOMMENDATIONS

EMGK's previous and current IT initiatives illustrate the areas in which they stray from IS implementation and upgrade best practices documented in the literature. EMGK does not handle its organizational structure and people well, especially during IT upgrades and transitions. Upper and middle management at EMGK do not really understand IS and how IT implementation and upgrade might affect the various components of IS. In addition they do not foster support for the end users during new IT implementations or upgrades.

We can see that understanding what IS is and how IT relates to IS are important factors in the success of IT implementations within any organization. In the case study presented in this paper, it is very obvious that 'the cart is put in front of the horse'- software (IT) are selected without any consideration to whether or not these software will integrate easily within the department or even to whether or not they will help the department meet its objectives. Moreover, upgrades are implemented without looking at whether or not the upgrade is needed and without notifying the end users ahead of time with the scheduled upgrades. IS are needed in organizations to fulfill the organization's processing needs. In order for IS to be successful it should be built based on a defined set of objectives. Failing to understand the importance of IS and the importance of ensuring that implemented IT integrates very well with an organization's culture, people, and structure can lead to a mess as shown in the case of the Market Research department at EMGK.

There are a few things that can be taken by EMGK in order to help them solve the problem. First, they should evaluate more than one technology prior to deciding on what to implement because each technology has its strengths and weaknesses. In addition, they should assign a person who is not only familiar with the technology but also understand the current system in order to accurately evaluate the technology that is most suitable to the business needs. Whenever there is an upgrade to a technology, the IT support personnel should learn its new capabilities and features.

Secondly, they should ensure that there is enough training after implementing the technology. EMGK does not allocate enough time or money for end users training. The company expects the IT staff to provide full support to the end users. In addition, they also expect the end users to train each other and compile the training materials themselves. One of the end users interviewed said, "They should give training to users who have immediate need. In addition, they should have more training for end users that is scenario based because we (the end users) need to know what to do if something goes wrong." We can see that the bottom line is affordability. EMGK does not want to spend the money or the time for the training program.

On top of that, a very important factor that the management needs to consider is to ensure that IT (in this case the software) integrates well with the people, structure, and processes within the department and the company. Software that integrates people, structure and processes that collaboratively integrate the talents and insights of all users will increase value, reduce cost, and maximize company's efficiency (Piccoli, 2008). Integrating software with people means the software offers the right functionality and communicates the right information to the people. The software should facilitate people in their tasks by providing user easy access to a shared system and a single point of access through multiple channels at anytime, anywhere. By doing the right integration, the company's operational performance will improve and productivity will also increase. In the EMGK case, project management, users' participation, project reports, IT support, and many other tasks will all become easier if the software integrates well with people.

Integrating the organizational structure means that the software support the decentralized structure of EMGK, its function, and its culture. The culture of the company was such that senior management did not communicate frequently with the end-users. In addition, the two groups differ in terms of perspective, valuing new technology, and understanding of the systems' purposes. Therefore it is very important for senior management of EMGK to understand the impact of the company culture on IT related activities. Integration can be achieved if the upper management at EMGK communicates the priority and objectives with the IT support team and the end users. If employees are reluctant to spend time, are unaware of the priority, and are all still expected to support their normal workloads, it is very likely that the system will fail (Venkatesh, Morris, Davis, & Davis, 2003). Moreover, change in the organization resulting from a new system implementation is difficult because the company needs to get full support from its employees (Bakos & Treacy, 1986). Again, a big part of preparing for organizational change is training users on the new technology employees use to complete their jobs.

In addition to people and structure, the software should also include functionality for business process management, which greatly facilitates the integration of existing and new applications. The company should drive toward greater standardization because it will improve the systems' user-friendliness. Integrating all the different system modules and enabling personalization of the system will not only make the end users happy, but it can also provide their customers with better processes and products. We can clearly see from the EMGK case that IS success is affected by technological problems as well as people, process, and structure issues.

FUTURE RESEARCH DIRECTIONS

This research may be expanded in a number of different directions. First, it can include additional cases by including more departments in different locations covering a wider range of backgrounds rather than one specific department. Having cases from different departments with different structure provide more insights into the influence that tasks, organizational structure, and people have on the adoption and use of IS.

Future study can also focus on each individual component of IS and perform detail analysis of its impact on IS adoption in the organization. For example, a research that focuses on the role of societal culture which may influence the organizational culture as a whole. Culture is one of the most interesting factors to focus on because it is much more dynamic than has been assumed in the IS literature.

CONCLUSION

The study was practically motivated by the authors concern that the most common reason IT projects fail is not the failure of technology, but failure to understand that new systems involve changes to tasks, organizational structure, and people. This paper illustrates the importance of understanding the components of IS and how they affect one another. IS are more than just technology; they have management, organization, and technology dimensions.

In order to have a successful implementation of IT, it is not necessary for the management to have the knowledge of a technical person, but they do need enough knowledge and definitions to understand the role of IT and how it must support the organization's business strategy. They must also understand how IT can be used to help transform the business. In this study, we analyze the reasons behind the failure of integrating IT with the culture, people, and processes in a Market Research Department of a large company in the UAE. Finally, the case study discusses the consequences of failing to understand IS and provides recommendations for a better integration of IT within the department. The most important lesson that we learned from this case was IS must be built with a clear understanding of the organization in which they will be used. Technology is subordinate to the organization and its purposes and therefore needs to be carefully considered.

REFERENCES

Al-Ashban & Burney. (2001). Customer adoption of tele-banking technology, the case of Saudi Arabia. *International Journal of Bank Marketing, 19*(5), 191–201. doi:10.1108/02652320110399683

Al-Gahtani. (2003). Computer Technology adoption Saudi Arabia, Correlates of perceived innovation attributes. *Journal of Information Technology for Development, 10*(1), 57–69. doi:10.1002/itdj.1590100106

Bakos, J. Y., & Treacy, M. E. (1986). Information Technology and Corporate Strategy, A Research Perspective. *Management Information Systems Quarterly, 10*(2), 107–119. doi:10.2307/249029

Banker, R. D., & Kauffman, Robert J. (2004). The Evolution of Research on Information Systems, A Fiftieth-Year Survey of the Literature in Management Science. *Management Science, 50*(3), 281–298. doi:10.1287/mnsc.1040.0206

Bawden, D. (2008). Smoother pebbles and the shoulders of giants, the developing foundations of information science. *Journal of Information Science, 34*(4), 415–426. doi:10.1177/0165551508089717

Benbasat, I., Goldstein, D. K., & Mead, M. (1987). The Case Research Strategy in Studies of Information Systems. *Management Information Systems Quarterly, 11*(3), 369–386. doi:10.2307/248684

Blili, S., & Raymond, L. (1993). Information technology, threats and opportunities for small and medium-sized enterprises. *International Journal of Information Management, 13*(6), 439–448. doi:10.1016/0268-4012(93)90060-H

Caldeira, M., & Ward, J. (2002). Understanding the successful adoption and use of IS/IT in SMEs, an explanation from Portuguese manufacturing industries. *Information Systems Journal, 12*(2), 121–152. doi:10.1046/j.1365-2575.2002.00119.x

Carr, N. G. (2003). IT Doesn't Matter. *Harvard Business Review, 81*(5), 41–48.

Cooper & Zmud. (1990). Information Technology Implementation Research, A Technological Diffusion Approach. *Journal of Management Science, 36*(2), 123–139. doi:10.1287/mnsc.36.2.123

Díez, E., & McIntosh, B. S. (2009). A review of the factors which influence the use and usefulness of information systems. *Environmental Modelling & Software, 24*(5), 588–602. doi:10.1016/j.envsoft.2008.10.009

Ein-Dor, P., Myers, M., & Raman, K. S. (1997). Information technology in three small developed countries. *Journal of Management Information Systems, 13*(4), 61–89.

Gerstein, M., & Reisman, H. (1982). Creating Competitive Advantage with Computer Technology. *The Journal of Business Strategy, 3*(1), 53–60. doi:10.1108/eb038956

Hamilton, S., & Chervany, N. L. (1981a). Evaluating Information System Effectiveness - Part I, Comparing Evaluation Ap-proaches. *Management Information Systems Quarterly, 5*(3), 55–69. doi:10.2307/249291

Hamilton, S., & Chervany, N. L. (1981b). Evaluating Information System Effectiveness - Part II, Comparing Evaluation Viewpoints. *Management Information Systems Quarterly, 5*(4), 79–86. doi:10.2307/249329

Ives, B., & Jarvenpaa, S. L. (1991). Applications of global information technology. Key issues for management. *Management Information Systems Quarterly, 15*(1), 32–49. doi:10.2307/249433

Keen, P. G. W. (1981). Information Systems and Organizational Change. *Communications of the ACM, 24*(1), 24–33. doi:10.1145/358527.358543

Lucas, H. C. Jr. (1975). Performance and the Use of an Information System. *Management Science, 21*(8), 908–919. doi:10.1287/mnsc.21.8.908

Luqman, Y., & Quraesh. (1992). Facilitating the adoption of information technology in a developing country. *Journal of International Management, 23*(2), 75–82.

Mahmood, M. A., Hall, L., & Daniel, L. S. (2001). Factors Affecting Information Technology Usage, A Meta-Analysis of the Empirical Literature. *Journal of Organizational Computing and Electronic Commerce, 11*(2), 107–130. doi:10.1207/S15327744JOCE1102_02

McFarlan, F. W., & McKenney, J. (1983). *Corporate Information Systems Management.* Homewood, IL: Richard D. Irwin Inc.

McFarlan, F. W., & McKenney, J. L. (1981). *Information Systems Planning, A Contingent Focus. Harvard Business School Case* (pp. 181–128). Boston, Massachusetts: HBS Case Services.

Piccoli, G. (2008). *Information Systems for Managers. Text & Cases.* New York: John Wiley & Sons, Inc.

Ragowsky, A., Ahituv, N., & Neumann, S. (2000). *The* Benefits of Using *Information Systems. Communications of the ACM, 43*(11), 303–311. doi:10.1145/352515.352532

Rockart, J. F., & Scott Morton, M. S. (1984). Implications of Changes in Information Technology for Corporate Strategy. *Interfaces, 14*(1), 84–95. doi:10.1287/inte.14.1.84

Ross, J. W., & Weill, P. (2002). Six IT Decisions Your IT People Shouldn't Make. *Harvard Business Review*, 84–91.

Venkatesh, V., Morris, M. G., Davis, G. B., & Davis, F. D. (2003). User Acceptance of Information Technology, Toward a Unified View. *Management Information Systems Quarterly, 27*(3), 425–478.

ADDITIONAL READING SECTION

Abdul Rashid, Z., Sambasivan, M., & Abdul Rahman, A. (2004). The influence of organizational culture on attitudes toward organizational change. *Leadership and Organization Development Journal, 25*(2), 161–179. doi:10.1108/01437730410521831

Bostrom, R, P. & Heinen, J. S. (1977). MIS problems and failures, A socio-technical perspective,

Davis, F. D. (1989). Perceived Usefulness, Perceived Ease of Use, and User Acceptance of Information Technology. *Management Information Systems Quarterly, 13*(3), 319–339. doi:10.2307/249008

Davis, F. D., Bagozzi, R. P., & Warshaw, P. R. (1989). User Acceptance of Computer Technology, A Comparison of Two Theoretical Models. *Management Science, 35*(8), 982–1002. doi:10.1287/mnsc.35.8.982

Davis, F. D., Bagozzi, R. P., & Warshaw, P. R. (1992). Extrinsic and Intrinsic Motivation to Use Computers in the Workplace. *Journal of Applied Social Psychology, 22*(14), 1111–1132. doi:10.1111/j.1559-1816.1992.tb00945.x

Goodhue, D. L. (1995). Understanding User Evaluations of Information Systems. *Management Science, 41*(12), 1827–1844. doi:10.1287/mnsc.41.12.1827

Goodhue, D. L., & Thompson, R. L. (1995). Task-Technology Fit and Individual Performance. *Management Information Systems Quarterly, 19*(2), 213–236. doi:10.2307/249689

Harris, R. & Davison, R. (1999). Anxiety and involvement, Cultural dimensions of attitudes toward computers in developing countries. *Journal of Global Information*

Harrison, D. A., Mykytyn, P. P., & Riemenschneider, C. K. (1997). Executive Decisions About Adoption of Information Technology in Small Business, Theory and Empirical Tests. *Information Systems Research, 8*(2), 171–195. doi:10.1287/isre.8.2.171

Hartwick, J., & Barki, H. (1994). Explaining the Role of User Participation in Information System Use. *Management Science, 40*(4), 40–465. doi:10.1287/mnsc.40.4.440

Hofstede, G. (1980). *Cultural Consequences, International Differences in Work Related Values.* Beverly Hills: Sage.

Hofstede, G. (1983). National culture in four dimensions. *International Studies of Management and Organization, 13*(2), 46–74.

Hofstede, G. (1991). *Cultures and Organizations, Software of the Mind.* New York: McGraw-Hill.

Hofstede, G., Neuijen, B., & Ohavy, D. D. (1990). Measuring organizational cultures, a qualitative and quantitative study across twenty cases. *Administrative Science Quarterly, 35,* 286–316. doi:10.2307/2393392

Hofstede, G. J. (2000). You must have been at a different meeting, Enacting culture clash in the international office of the future. *Journal of Global Information Technology Management, 3*(2), 42–58.

Karahanna, E., Straub, D. W., & Chervany, N. L. (1999). Information Technology Adoption Across Time, A Cross-Sectional Comparison of Pre- Adoption and Post-Adoption Beliefs. *Management Information Systems Quarterly, 23*(2), 183–213. doi:10.2307/249751

Keil, M., Tan, B. C. Y., Wei, K. K., Saarinen, T., Tuunainen, V., & Wassenaar, A. (2000). A cross-cultural study on escalation of commitment behavior in software projects. *Management Information Systems Quarterly, 24*(2), 299–325. doi:10.2307/3250940

Laudon Kenneth & Laudon Jane. (2009). *Management Information Systems.* Upper Saddle River, NJ: Prentice Hall.

Leonard-Barton, D., & Deschamps, I. (1988). Managerial Influence in the Implementation of New Technology. *Management Science, 34*(10), 1252–1265. doi:10.1287/mnsc.34.10.1252

Management, 7(1), 26-38.

Morris, M. G., & Venkatesh, V. (2000). Age Differences in Technology Adoption Decisions, Implications for a Changing Workforce. *Personnel Psychology, 53*(2), 375–403. doi:10.1111/j.1744-6570.2000.tb00206.x

Part, I. (nd). —The causes. *Management Information Systems Quarterly, 1*(3), 17–32.

Pavlov, P., & Fygenson, M. (2006). Understanding and predicting electronic commerce adoption, An extension of the theory of planned behavior. *Management Information Systems Quarterly, 30*(1), 115–143.

Peppas, S. C. (2001). Subcultural similarities and differences, An examination of US core values. *Cross Cultural Management–An International Journal, 8*(1), 59–70. doi:10.1108/13527600110797209

Rogers, E. (1995). *Diffusion of Innovations.* New York: Free Press.

Rogers, E. M., & Shoemaker, F. F. (1971). *Communication of Innovations, A Cross-Cultural Approach.* New York: Free Press.

Straub, D. (1994). The effect of culture on IT diffusion, E-mail and FAX in Japan and the U.S. *Information Systems Research, 5*(1), 23–47. doi:10.1287/isre.5.1.23

Straub, D., Limayem, M., & Karahanna, E. (1995). Measuring System Usage, Implications for IS Theory Testing. *Management Science, 41*(8), 1328–1342. doi:10.1287/mnsc.41.8.1328

Watson, R. T., Ho, T. H., & Raman, K. S. (1994). Culture, A fourth dimension of group support systems. *Communications of the ACM, 37*(10), 44–55. doi:10.1145/194313.194320

Chapter 11
Factors of Culture Affecting ICT Adoption in an Arab Society

Ali Nasser Al-Kinani
Abdulaziz Medical City, KSA

ABSTRACT

The world is witnessing vast and rapid developments in Information and Communication Technology (ICT), which can affect all aspects of our daily lives. This chapter aims to investigate cultural factors that influence the Saudis adoption of ICT. A questionnaire was used to gain insight related to business language, communication language, culture and ICT R&D, support of the Saudi Arabia government of production development, consistency of Internet with local culture, openness of the culture of the country to foreign influence, the impact of employees' culture on their work, and the protection of Saudi culture.

INTRODUCTION

In line with Arab governments' efforts to overcome the digital divide barriers, the Saudi government has contracted foreign companies to develop new applications and web sites for various government departments and had spent about $2270 million in 2002 on training staff. The government is also encouraging the private sector to participate in software development, in addition to being proactive in sponsoring and hosting ICT related conferences for the last few years (Al-Maliki 2005).

Saudi Arabia had Internet connection since 1994 when King Faisal Specialist Hospital and Research Centre (KFSHRC) established a satellite link to Bethesda, Maryland, via the International Medical and Educational Data link (IMED) (Shafi, 2002; Burkhart, 998), then, a US based company managed the Saudi connection and its Internet infrastructure, as well as hosting the official Saudi government website in the United States. In 1995, Saudi Arabia transferred to the GulfNet Academic network to connect academic institutions, research centres, hospitals and public libraries (Shafi, 2002).

In 1997, after lengthy debates, discussions and consultations within the Saudi authorities,

DOI: 10.4018/978-1-60960-048-8.ch011

the Government approved a resolution giving the coordination, introduction and management of initial Internet services to King Abdul Aziz City for Science and Technology (KACST) (Shafi, 2002; Al-Furaih, 2002). This authorized KACST to provide Internet services to customers under certain conditions and controls in order to preserve the values and the Islamic traditions of the Saudi society,

By 1999, the Internet service was made available to the general public in 1999 through domestic services, with some regulations put in effect to restrict Internet use. One of the most important regulations of those was the introduction of censorship on the Internet contents on a national scale. The aim was to block access to websites which were deemed non-acceptable sites; those which conflict with the religious, cultural, legal and traditional values of the Saudi society (Shafi, 2002). At the early stages of introducing the Internet publicly, there were certain fears and anxiety within the Saudi society that this new medium will be misused to negatively influence their religious and cultural values and therefore, it was expected that the penetration rate of the Internet will be slow.

Several studies (Choucri et al, 2003; Ives & Jarvenpaa, 1991; Shore & Venkatachalam 1995; Dean et al, 1997; Palvia, 1998) have investigated the relationship between the 'national culture' and ICT. These studies have emphasized the importance of culture to the success of ICT adoption. Other researchers such as (Hasan & Ditsa, 1999; Tricker, 1988) conducted multinational surveys that resulted in identifying a relation between culture and effective use of ICT applications. Levitt (1983) predicted a global 'coming together', a convergence and eventual elimination of distinct cultures based upon improvements in communications and increases in global trade. Examination of the literature reveals that culture is an important factor to be considered in any e-assessment framework. In this chapter, the issues and questions that have been considered

to reflect on the impact of culture on ICT in the Saudi context were:

- What language used in daily business
- What language used in daily communications
- Does the culture support research and development
- Does the government supports the development of ICT culture in the country
- Is there consistency of Internet technology with organisation beliefs and business needs
- Is the nation's culture open to foreign influence
- What is the impact of an employee's culture on his/her work
- Can new technology assist in protecting a nation's culture

The author has also investigated the critical factors that impact local contents and cultures in developing countries, which can potentially impede or foster the adoption of ICTs. This study aims at analysing cultural issues and the use of ICT at Saudi organizations. At the end of chapter some recommendations are put forward related to the adoption and adaptation of ICT applications for organisations in the developing countries.

RESEARCH METHOD

This study uses a survey as an instrument for collecting data in order understand and evaluate the impact of cultural factors on e-society in terms of ICT adoption and acceptance, which is considered as an important factor to determine the level of e-readiness of Saudi Arabia to adopt and cope with new technologies.

The study uses real data which was collected directly from public and private Saudi organizations rather than using already published data from international organizations e.g. the World

Bank (Rizk 2004). A total of 200 questionnaires were distributed to Saudi organizations across the country. A total of 87 organizations (48 public and 39 private sectors) responded. The selected sample includes small, medium and large organizations that are mainly in the healthcare business in order to assess their level of e-readiness. The questionnaire includes a set of cultural questions which are used as an e-readiness assessment instrument (al-Solbi A. & Mayhew, P. 2006). These questions were answered by ICT managers who are working as executive mangers or decision makers in their respective organizations.

DATA ANALYSIS

As mentioned earlier culture is an important factor in determining the e-readiness of a country. Language, national heritage and beliefs constitute an important part of a nation's life. Eight questions were designed to investigate the impact of this issue on ICT adoption and use in Saudi organisations. Descriptive statistical methods were used to analyze the collected data using calculated means, and percentages in order to determine the differences between the public and private sectors. The findings are discussed next.

Language Used in Daily Business

Result shows that Arabic and English are the main languages used in Saudi business organisations, with Arabic taking the lead in conducting business activities. The study shows that 60% of public sector organisations are using Arabic as the main language in their business compared to 40% of organisations in the private sector. This was expected since the official language of Saudi public sector organisations is Arabic and their business activities are mainly internal activities within their society. The result also shows that 51.8% of the public organisations and 48.2% of the private are using English in their business activities. In the private sector, slightly more companies are using English as the official business language. This is because of their external business activities. Many do business with foreign companies where English is used as the communication medium. Therefore, the language influence is limited since the majority of Saudi participants, whose mother tongue is Arabic, also familiar with English, which happens to be the main language of ICT applications such as software and Internet websites. It is an advantage for any nation to be able to communication in more than its own native language, this can assist nations to understand each other and facilitate the exchange of ideas and experiences much easier.

Language Used in Daily Communications

Result shows that 57.9% of the public sector organisations and 42.1% of the private sector organisations use Arabic in their social communications. The result also shows that 40% of the public organisations and 60% of the private organisations use English in their social communications while 57.1% of the public organisations and 42.9% of the private organisations use other languages such as French or Urdu. It can be said that the main language used in social communications is Arabic; however, English is also used in some public and private organisations as a communication medium and also occasionally other languages.

A Culture that Supports Research and Development

This variable was designed to investigate the role of societal culture and working culture in the ICT industry in supporting Research and Development (R&D) in Saudi Arabia. The results indicate that both societal and working culture slightly support R&D in the ICT field. However, this issue needs to be investigated further in future studies. Cultural factors, which include values, beliefs, heritage values and religion, are important and have impact on employees, which could affect their attitude towards the adoption of ICT. It is clear from the result that there is slight difference between the responses of both public and private sector organisations.

Government in Support of Development of ICT Culture in the Country

Results show that the general views of the respondents to this question are (23% very supportive, 34% fairly supportive, 43.7% a little supportive and 2.3% not at all). This result shows that 43.7% of the managers surveyed believed that their government has limited support for the development of an ICT culture in the country. This indicates that Saudi Arabia as a developing country does not satisfy the local ICT requirements and more support should be provided to their local organisations.

Consistency of Internet Technology with Organisation Beliefs and Business Needs

This variable was chosen because it is an important factor to assess the e-readiness of Saudi Arabia. Majority (74.7%) of the surveyed organisations believe that Internet technology is consistent with their values, beliefs and culture, which means the Internet technology can support the local culture. The result indicates that society is willing to accept new technology, which is a positive sign on the e-readiness of the country. During interviews, two of the interviewees said, "We believe that Internet technology might help much any conservative society such as Saudi Arabia. While women are not allowed to drive cars, this technology might allow them to shop online or work from home." The results, illustrate that most of the surveyed organisations encouraged the use of the Internet and believed that it would assist their business and support their values.

Nation's Culture Open to Foreign Influence

This variable was selected to seek information from the ICT managers as to whether society accepts new ideas from foreign countries (e.g. cultural interaction, anything related to the new technology e.g. videophones, Internet etc.). Results show that there is some division of opinion over this issue with 21.8% and 33.3% 'disagreeing' and 'strongly disagreeing' respectively but a large percentage 19.5% and 14.9% 'Strongly agree' and 'agree' respectively. It can be said that the Saudi culture is less likely to be open to foreign influence and interact with other cultures without the fear of any harm to their own culture.

Impact of an Employee's Culture on His/Her Work

This variable was designed to investigate to what extent the employee's culture has a positive impact on his/her work, in order to examine if this variable affects the e-readiness assessment. The results indicate that 50.6% of respondents believe that the culture of employee has medium impact on his/her work. The respondents indicated that the employee's culture could have a positive influence if they were working in their own country e.g. the culture of Saudi employees, could have a good impact on their work in Saudi organisations but not when they work abroad.

New Technology as a Help in Protecting a Nation's Culture

This variable was used to investigate to what extent new technologies such as the Internet can assist in preserving a nation's culture. This variable shows the extent of people's perception towards electronic preparedness, in order to improve the overall e-readiness of a country. The results indicate that the majority of the surveyed Saudi organisations believe that the new technologies and the Internet could help in preserving their culture. During interviews, the managers mentioned that they believe the Internet can have positive impact on the livelihood of Saudi women since they are not permitted to drive, or travel unaccompanied. With the Internet, they will be able to access a variety of services from home. The Internet can also be used to promote religious observance and reinforce the message of the Holy Quran through the use of e-mails, chat rooms and to remain in touch with families, which is an essential part of the culture of the country. The researcher argues that this result could encourage the diffusion of ICT within the Saudi society. The results also show that the managers both agree in principle that the new technology and the Internet do protect Saudi culture. The researcher argues that the investigation of this issue is useful to be included in assessing the e-readiness of a country.

Culture Selected Variables vs Manager and Organization Variables

To investigate the impact of both the managers and organisation variables, a One-Way ANOVA statistical test was used to examine the variance between group means of manager and organisation variables in response to e-readiness variables. In this test, the selected variables were considered as 'dependent variables', and both the manager and the organisation variables were considered as 'factors' or independent variables. The tests were carried out using the SPSS statistical package.

One-Way ANOVA can distinguish between selected variables which have significant relationships with independent variables. As a result the researcher can identify those variables that have significant relationships with independent variables and have them examined further to understand the nature of the relationships and their Impact on ICT. This means that the researcher can neglect variables, which have virtually no significant relationships with independent variables. This

process assists in creating a new improved indicators to measure the culture influence on e-society.

The statistical testings revealed 5 e-readiness variables have no significant relationships with any of the manager or organisation variables and they were removed from the original e-readiness tool. By contrast, the variables which have shown significant impact are listed below:

The test results indicate that 7 out of the 12 culture variables have significant relationships with some of the managerial and organisational variables. However, it is clear from the above list that the numbers of managerial and organisational variables, which have impact on the e-readiness variables, vary from one e-readiness variable to another. The highest number of significant relationships can be observed in Q4 and Q1, which scored 5 and 4 significant relationships with managerial and organisational variables respectively.

DISCUSSIONS

The study has revealed that the use of Arabic is the official business and communication language: 87.5% of the public organisations surveyed use Arabic in their daily business activities compared to 71.8% of the private organisations surveyed. However, 60.4% of the public organisations surveyed and 69.2% of the private organisations surveyed were also found to use English in their business activities. These results are as expected since Arabic is the official language in Saudi Arabia and English is often used in communications with the foreign companies.

Concerning the language of communication, the study revealed that 91.67% of the public organisations surveyed and 82.05% of the private organisations surveyed use Arabic, and 16.67% of the public organisations surveyed and 30.76 of the private organisations surveyed use English in their daily communications. The results also revealed that 8.33% of the public organisations

Table 1.

	Significant Questions	**No**	**Variable**
Q1	The Arabic language is mainly used in daily business	4	Nationality, Industry type, ICT department, Resources
Q4	The Arabic language is used in daily communications	5	Nationality, Qualifications, Field of study, ICT experience, Resources
Q5	The English language is used in daily communications	1	Qualifications
Q6	Other languages that you use in your daily communications	1	Qualifications
Q7	The role of culture in supporting R&D	1	ICT department
Q9	Is the Internet technology consistent with your organisation's values, beliefs, culture and business needs	2	Age, Industry type
Q10	Whether the nation's culture is open to foreign influence	1	Industry type

surveyed and 7.69% of the private organisations surveyed use other languages such as French or Urdu. Arabic is obviously the main language being used in business and communications in addition to the English language for some of the organisations from both public and private sectors, and a few use other languages too.

The role of culture and working in the ICT industry to support R&D in Saudi Arabia was also investigated. The study revealed that the organisations surveyed believe that culture and the working environment support the R&D field. This could be related to the fact that the government has started to encourage scientific research and sufficient funds have been allocated.

The study also found that 74.7% of the organisations surveyed believe that there is a consistency between Internet technology and the organisations' values, beliefs and culture, and business needs. This indicates that Internet technology is not unwelcome in their organisations, but the study also found (views of both public and private organisations) that the Saudi culture is unlikely to be open to foreign influences.

This could be related to the fact that Saudi society is conservative and it fears the negative impact of the Internet on their culture. The interviewees also highlighted this point, they asserted having regulations to govern the use of the internet, is a positive factor to encourage its adoption.

Another vital finding from this study, suggests that employees' background, previous work experience and way of life have an impact on his/her work. The majority of the organisations surveyed, both public and private sector organisations, believe that the new technology and the Internet may promote the protection of the Saudi culture and values, for example, use of the Internet for education and shopping means that the separation of the sexes required by the Saudi culture (Islam religion) will be maintained, and women will be able to participate in different ways with the outside world via their computers.

The results also show the following findings:

- Each of the following variables was found to have a significant impact on 2 e-readiness variables of the culture factor: Nationality, ICT department and Resources variables
- Each of the following variables was found to have a significant impact on 3 e-readiness

variables of the culture factor: Qualifications and Industry type variables.

- Each of the following variables was found to have a significant impact on 1 e-readiness variable of the culture factor: Field of study, Age, and ICT experience.
- Each of the following variables was found to have a significant impact on any of the e-readiness variables of culture factor: Decision-Maker, Position, Sector, Size, Time being in business and Training programs.

In general it can be concluded that the most important e-readiness variables are Q4 and Q1 since they have 5 and 4 significant relationships with managerial and organisational variables respectively, while the most important managerial and organisational variables are Qualifications and Industry type since each has 3 significant relationships with e-readiness variables.

Finally, the study confirms that the openness of the Saudi culture has a direct impact on the e-readiness outcome.

CONCLUSION AND RECOMMENDATIONS

This chapter demonstrates that cultural factors have impact on the development of any society, and this calls for more efforts to encourage the Saudi society to be more open and to increase the interaction with other societies for mutual benefits. As a result, the following recommendations are made:

- The government should encourage the media to concentrate on the benefits of ICT systems to the Saudi society.
- Learning English should be encouraged since it is an international language, and the majority of ICT systems have been developed in English speaking countries and the majority of websites are designed using English.

- The government should establish ICT-oriented social clubs to enhance ICT knowledge in the Saudi society.

REFERENCES

Al-Furaih, I. S. (2002). *Internet regulations: The Saudi Arabia experience*. Retrieved (month,date), from: http://inet2002.org/CD-ROM/lu65rw2n/papers/u05-a.pdf

Al-Maliki, S. G. Y. Al-Khalidi, (2005). *Information Systems Evaluation Process in Private Organisations in Saudi Arabia*, Unpublished doctoral thesis, University of East Anglia, UK.

Al-Solbi, A.& Mayhew Pam, (2006, December). A strategic Framework for Electronic Readiness Assessment. In *Proceeding of the Internet and Information System in the Digital Age Conference,* Brescia, Italy.

Al-Solbi, A., & Mayhew, P. (2004, June.). Development of Information and Communication Technology Indicators in the Kingdom of Saudi Arabia. In Proceedings *of the 17th European conference on combinatorial optimization*, Beirut.

Al-Solbi, A., Mayhew Pam, P., & Al-Badi, A. (2005a, July). An Alternative Model For Measuring E-Readiness for Developing Countries: Applied to Saudi Organisations. In *Proceedings of the International Business Information Management Conference,* Lisbon, Portugal.

Choucri, N., Maugis, V., Madnick, S., & Siegel, M. (2003). *Global E-Readiness For What? Centre for E-Business at MIT*. Sloan School of Management.

CSPP. Org, (2000). *The CSPP readiness guide for living in the networked world*. CSPP.Org, Retrieved May, 2010, from: http://www.cspp.org

Delone, W., H. (1998). Determinants of success for computer usage in small business. *Management Information Systems Quarterly*, *12*(1), 51–67. doi:10.2307/248803

Dutta, S., Lanvin, B., & Paua, F. (2003, January). *The Global Information Technology Report 2002-2003: Readiness for the Networked World*, Oxford University Press Published.

Hasan, H., & Ditsa, G. (1999). The impact of culture on the adoption of IT: an interpretive study. *Journal of Global Information Management*, 7(1), 5–15.

ICT, (2002), Information and Communication Technology and Development in the Arab Countries Report.

Ives, B., & Jarvenpaa, S. L. (1991). Applications of Global Information Technology: Key Issues for Management. *Management Information Systems Quarterly*, 15(1), 33–49. doi:10.2307/249433

Palvia, P., Shailendra, C., Palvia, J., & Whitworth, J. (2002). *Global information technology management environment: representative world issues, in Palvia and Roche (eds.)* Global information technology and electronic commerce: Ivy League publishing.

Rizk, N. (2004, September). E-Readiness Assessment of Small and Medium Enterprises in Egypt: A Micro Study. *Topics in Middle Eastern and North African Economies, electronic journal*, (6), Middle East Economic Association and Loyola University Chicago.

Shafi, I. M. (2002). *Assessment of the impact of Internet technology use among Saudi business organisation*, Unpublished doctoral thesis, faculty of Mississippi State University, Mississippi State, Mississippi.

Shore, B., & Venkatachalam, V. (1994). Prototyping: a metaphor for cross-cultural transfer and implementation of IS applications. *Information & Management*, 3(27), 175–184. doi:10.1016/0378-7206(94)90045-0

Tricker, R. I. (1988). Information resource management - a cross-cultural perspective. *Information & Management*, 15(1), 37–46. doi:10.1016/0378-7206(88)90028-6

Chapter 12
Electronic Readiness of Saudi Health Organizations

Ali Nasser Al-Kinani
King Abdulaziz Medical City, National Guard Health Affairs, KSA

ABSTRACT

In this chapter, the author uses a questionnaire as an instrument s to evaluate the use of Information and Communications Technologies (ICT) in Saudi health organizations. Information and Communication Technology has become an important tool for improving the efficiency of health organizations. E-Health applications are increasingly being drawn into evaluating the Internet as a useful source of information on health by end-users. This chapter is an attempt to explore E-Health applications and related implementations issues in developing countries, and in particular Saudi Arabia.

INTRODUCTION

Electronic health or e-health is increasingly becoming an important factor in our daily lives. E-health is an emerging technology that benefits from the emergence of new powerful databases capable of storing, retrieving and managing different health-related information, the effectiveness of the Internet as a tool for communication and collaboration within and beyond the hospital premises, and the high-tech medical and disease diagnosing equipment. These advancements represent the main pillars for innovative new

DOI: 10.4018/978-1-60960-048-8.ch012

technology applications in the health industry. The availability of technology investment funds and the adoption of appropriate ICT health-strategy are considered vital to promote ICT in the health sector. There are several studies (CSPP, 2000a; Edward, 2003; Al-Omari, 2004) that examined different e-health variables in specific countries' contexts using statistical data from international organisations rather than real data. In contrast this study adapted several e-health variables or indicators to assess the direct use of ICT in health organisations.

The ICT and Development in the Arab countries report (2002) states that e-heath needs a proper infrastructure, with adequate institutional arrangements and long-range planning, ICT applications can help advance health services in terms of the quality and the extent of dissemination. Notable efforts have been made in Saudi Arabia, where one of the channels on a communication satellite has been employed in setting up a decentralised health information system linking a number of health centres. Several telemedicine initiatives have also been launched in most Arab countries but the majority still lack maturity. In Saudi Arabia, the first organisation to introduce the Internet service was King Faisal Specialists Hospital and Research Centre (KFSHR).

In order to assess the e-health status in a specific country, it is essential to use a comprehensive e-readiness assessment framework that can examines the availability of a range of services and strategies, such as:

- Internet connection at the hospital/health centre
- Databases for use by employees to support the dissemination of telemedicine in health centers
- Internet to communicate/collaborate with other local hospitals
- Internet to communicate/collaborate with other external hospitals
- Telehealth/telemedicine services in hospitals/health centres
- Digital databases to be used in maintaining records
- Websites for major health care facilities
- Interactive websites for major health facilities
- On-line health services for doctors and patients needs
- The Internet to provide on-line consultation service with doctors using e-mail
- ICT health standards and strategies

In this study the author explores key issues or services that are related to electronic health implementations, and specifically investigates critical factors that have potential impact on e-health applications and implementations. The study had the following objectives:

- Evaluate the level of use Information and communications technologies at health organizations in Saudi Arabia as a developing country by examining the above services
- Provide some recommendations for those who are involved in the design or use e-health applications.

BACKGROUND

There is no doubt that Internet creates new opportunities for both public and private sectors. The Internet has the potential to revolutionize healthcare by providing unprecedented access to information as well as health products and services on "e-health sites". Electronic Health applications utilizing Internet websites should be given high priority. E-Healthcare applications are increasingly being drawn into evaluating the Internet as a source of end-user information for health and medicine. World Health Organization defines electronic health as "digital data transmitted, stored and retrieved electronically in support of health care, both at the local site and at a distance" (WHO, 2004).

E-Health as a term is commonly used to describe the application of information and communications technologies for health purposes and when it is used in the context of hospitals' services.E-Health also refers to electronic patient administration system laboratory and radiology information systems, electronic messaging systems, telemedicine, telepathology and teledermatology. While at the domestic end, e-health refers to teleconsults and remote monitoring systems used for diabetes medicine, asthma monitoring and home dialysis systems. The term e-health when used with primary health and clinics refers to the use of computer systems by general practitioners and pharmacists for patient management, medical records and prescription handling.

One of the key objectives of any Hospital Information System (HIS) is to use a network of computers to collect, process and maintain information from the various departments covering a range of hospital activities. The availability of digitized information will assist in the development of a decision support system for the hospital authorities in order to plan and develop cost effective and efficient comprehensive healthcare policies.

Few of the existing e-readiness assessment tools (CSPP, 2000; Al-Solbi & Mayhew, 2005 a,b, 2006c; Al-Solbi, 2007a) considered healthcare as one of the variables in its assessment of e-readiness. Jennett (2002) stated that Harvard presented definition of telehealth readiness, as "the degree to which a 'community' is prepared to participate and succeed in the telehealth." Telecommunication health or electronic health are important e-readiness factors that have impact on the level of assessment of e-readiness of a country. Therefore, it is essential in order to design a comprehensive and useful e-readiness tool to investigate the e-health factors and the use of ICT in health organizations.

RESEARCH METHOD

The main research tool used in this study was a survey for the assessment of e-readiness in the health sector. In order to design and test alternative e-readiness models, the researcher collected actual data from public and private Saudi organizations rather than using published data from international organizations e.g. the World Bank (Rizk, 2004). The questionnaire was distributed to 200 Saudi organizations across the country. A total of 87 organizations (48 public and 39 private sector organizations) responded. The selected sample included small, medium and large organizations which have healthcare as their main business ac-

tivities in order to gauge the level of e-readiness nationwide. As a result, the researcher designed 11 e-health questions to cover e-health applications in health organizations (Al-Solbi, 2006b). The respondents were not only ICT managers who worked for the health organizations, but also ICT managers who were users of ICT applications.

DATA ANALYSIS

Descriptive statistical methods were used to determine the differences in e-readiness status between public and private sectors, using representative means and percentages. The study has used a set of questions to assess the readiness of use Information and communication technologies in health organizations in Saudi Arabia as a developing country. These questions were the result of an extensive review of relevant literature, three pilot studies and two workshops with expertise from the health industry and academia. The selected factors were validated for their importance as determinant of effective use of ICT in health organizations. The outcomes of the survey are discussed in the following sections.

Connection to the Internet

The survey results related to this factor show that 70.1% of the hospitals and health centres surveyed are connected to the Internet. This indicates that there is a significant level of connection to the Internet, which is considered a good indication of e-readiness.

Database to Support Telemedicine

This related question is not only concerned with measuring ICT applications in telemedicine but also to reflect the level of skills of people working there. The survey results related to this factor show that two thirds (66.7%) of the hospitals and health centres surveyed are using databases to support their activities in the dissemination of telemedicine. This is important since it indicates that those health organizations are aware of the benefits of ICT systems to their daily activities.

Internet Communications/ Collaboration with Local Hospitals

This question was considered to investigate to what extent these organisations communicate/ collaborate to improve their services by utilising ICT in e-health activities. This factor shows that only 28.7% of the surveyed organisations are using the Internet to communicate/collaborate with other local hospitals/health centres. The results indicate that communication/collaboration via the Internet has limited practice.

Internet Communications/ Collaboration with Foreign Hospitals

In this area the results show that a limited number of health organisations (14.9%) have communication/collaboration with similar foreign health organisations. This may indicate that only large health organisations have connections with foreign international health organisations.

Telehealth/Telemedicine Services

The survey results related to this factor indicate that only 11.5% of the surveyed organisations are using the telehealth/telemedicine. The results show that 8.3% of public and 15.4% of private health organisations are using the telehealth/telemedicine service. Although the result is very low, it does give an indication that some health organisations are aware of the usefulness of this new field of service.

Digitising Medical Records

This factor was included to study whether medical records were maintained in a digital format in a

readily accessible database. The results show that nearly half (44.5%) of the surveyed ICT managers use digital medical records within their organisations. The results indicate that the utilisation of the ICT in health service is still at an early stage. The result shows that 48.7% of public and 51.3% of the private health organisations are digitizing their information.

Community Website

The results show that only 13.94% of the major health organisations have community websites, 6.25% of the public and 7.69% of private health organisations. This is a very low percentage indicating that ICT utilisation and the Internet service in health organisations are still at early stages.

Community Interactive Website

This factor was considered to seek information on whether the major health organisations (e.g. KFSHR) have interactive websites, offering e-services (enquiries, bookings, etc,) to the staff, patients and other stakeholders. The results show that only 13.94% of the surveyed health organisations have interactive websites, 6.25% of the public and 7.69% of private health organisations. This indicates limited numbers of e-health organisations have websites. Much work is still needed to leverage on the use of the Internet to improve services and attract new patients and customers.

Line Services to Communicate Between Doctors and Patients

The results show that 35.6% of the surveyed organisations indicate that the Internet service is used by doctors to communicate with patients. It shows that 29.17% of the public and 43.59% of the private health organisations have Internet services for doctor-patients interaction. This result indicates that the private sector is more active in this area than the public sector, and probably due to the competition on services between the private organisations.

Internet Communication with Doctors

The results show that 49.4% of the surveyed organisations are using the Internet to communicate with their doctors. This is significant since it shows that people have started to utilise advanced technologies to improve efficiency. It also shows that 47.9% of public and 51.3% of private health organisations have Internet facilities to enhance communication between customers and doctors. This factor indicates the extent of using new technology e-health activities and services; however, this factor requires further investigation.

Standards in Health Care Information System

The results show that 43.7% of the surveyed organisations indicated that there was a governmental ICT strategy and guidelines in setting standards in e-health. It also shows that 37.5% of public and 51.2% of the private health organisations agree in principle that the government has an ICT strategy to regulate and encourage the utilisation of ICT systems in the health sector.

It was found in the literature that the e-Health readiness assessment depends on the managers and organisation variables. The manager profile comprises of eight variables: nationality, qualifications, field of study, age, ICT experience, decision-maker, position and ICT specialist, while the organisation profile includes seven variables: sector of the organisation (public or private), size of the organisation in terms of their employees, industry type, time being in business, existence of ICT departments, availability of resources and training programmes.

This section was designed to explore the impact of both manager and organisation variables on e-readiness using a One-Way ANOVA.. In

this test, the e-Health variables were considered as 'dependent variables', and both the manager and the organisation variables were considered as 'factors' or independent variables. The tests were carried out using SPSS statistical package.

One-Way ANOVA can distinguish between the e-readiness variables which have significant relationships with independent variables. As a result the researcher can select those e-readiness variables which have significant relationships with independent variables to be included in the new e-readiness assessment tool.

Impact of the Managerial and Organisational Variables on E-Health

After the statistical tests on the impact of the managerial and organisational variables on the e-health variables it was noted 7 variables have not shown any significant relationships with either managerial or organisational variables. The e-readiness variables which have significant relationships with either managerial or organisational variables or both are shown in Table 1.

The test results indicate that 7 out of the 11 e-readiness variables were found to have significant relationships with some managerial and organisational variables. However, it is evident from the above list that the numbers of manager and organisation variables, which have impact on the e-readiness variables, vary from one e-readiness variable to another. The highest number of significant relationships can be observed in Q7 and Q8, since they scored 4 and 3 significant relationships with managerial and organisational variables of the e-health factor respectively.

The results also show the following findings: the Field of Study, Age, ICT Specialists and Training Programs were found to have a significant impact on one e-readiness variable of the e-health factor. Moreover, the ICT Department and Resources were found to have a significant impact on three e-readiness variables of the e-health factor.

Table 1.

Q	Significant Questions	No	Variable
Q4	Internet Communication to collaborate with foreign hospitals	1	ICT department,
Q6	Digitising medical records	2	ICT specialists, Resources
Q7	Community has a web site	4	Age, ICT department, Resources, Training programmes
Q8	Community has an interactive website:	3	Field of study, ICT department, Resources

In general it can be concluded that the most important e-readiness variables are Q7 and Q8 since they each have 4 and 3 significant relationships with managerial and organisational variables respectively, while the most important managerial and organisational variables are ICT department and Resources since each was found to have 3 significant relationships with the e-readiness variables of the e-health factor.

DISCUSSION

The results show that 70.1% of the surveyed organizations have Internet services and that 66.7% of the use databases to support their daily activities. As far as using the Internet for communication/collaboration with local hospitals, 28.7% of the surveyed organizations use this activity.

The study also found that less than 15% of the surveyed organizations use the internet/website for supporting their activities such as communications/collaborations with external hospitals, having telehealth/telemedicine facilities, having health websites that are either static or interactive. The study also found that 35.6% of the sample was found to have on-line Internet services for doctors and patients.

The surveyed sample also shows that there were some implementations of applications in the field of e-health. However, the health sector needs a clear ICT health plan and adequate budget to promote its development.

The study considered the following eleven questions; hospital/health centre connection to the Internet, availability of databases in hospitals, using the Internet to communicate/collaborate with other local and external hospitals, availability of telehealth/telemedicine, availability of digital format to save hospital records, availability of a health website, availability of an interactive website, availability of Internet for doctors in the hospitals, on-line consultations with doctors, and availability of governmental health strategy for ICT systems. With regards to the Internet connection of the health organisations, 70.1% of the sample, 61 organisations (33 public and 28 private) were found to be connected to the Internet, and 66.7% of the organisations i.e. 58 organisations (30 public and 28 private) were using databases to support their dissemination of telemedicine. The results are encouraging since they indicate that the majority of the health organisations are connected to the Internet and use electronic databases for the dissemination of telemedicine.

This study has also highlighted some negative results related to the use of the Internet to communicate/collaborate with other local hospitals. Only 28.7% of the sample surveyed (14 public and 11 private) use the Internet to communicate/collaborate with other hospitals in the country. As far as communications/collaboration with external hospitals is concerned, the study found only 14.9% of the organisations surveyed (6 public and 7 private) communicate/collaborate with other external hospitals. The reason could be related that the Internet is considered as an emergent technology in Saudi organisations. The results of the survey of telehealth/telemedicine services were also found to be low, as only 11.5% of the sample (4 public and 6 private) had this type of service. This could be related again to the fact that this type of technology is considered new in Saudi Arabia.

As for maintaining medical records in a digital format, the study found that 44.5% of the surveyed sample (19 public and 20 private) have this facility. This finding reflects the fact that in Saudi Arabia, organisations are not familiar with digital technology for maintaining documents. The researcher believes that this result is expected from a developing country. When examining the availability of a health website, the study found that only 13.8% of the organisations surveyed have this type of facility but only 6.9% of the organisations surveyed (3 public and 3 public) have health-care interactive websites. This could also be related to the above-mentioned reason, i.e. the Internet and website technologies are emerging in Saudi organisations.

The study also found that 35.6% of the organisations surveyed (14 public and 17 private) have on-line services for doctors and patients, 49.4% of the sample surveyed (23 public and 20 private) use the Internet for on-line consultation services with doctors. When asked whether the government should provide guidelines or standards in health-care information systems, the study found that 43.7% of the surveyed organisations (18 public and 20 private) agreed.

In general, this study is indicative of the strengths and weaknesses of the ICT utilisations in the surveyed health organisations.

CONCLUSION

This study has revealed that e-health is not well developed in Saudi Arabia and that electronic health is an emerging technology only introduced recently to the health sector and it will take some time to be diffused. The overall e-readiness index for e-health was found to be at a low level Therefore, the government should encourage the utilisation of ICT in e-health, and an ICT strategy and adequate funds should be allocated for this important venture.

REFERENCES

Al-Omari, A. (2004). *International Electronic Transactions Measurement Scale System.* Digital publish doctoral thesis, School of Information Technology and Engineering, the George Mason University.

Al-Solbi, A. (2007, May), *Evaluating E-Health and Use of ICT: Learning From the Experiences of Saudi Health Organisations*, ECCO 2007 XXth Anniversary Conference European Chapter on Combinatorial Optimization, Limassol, Cyprus, May 24-26, 2007

Al-Solbi, A. & Mayhew, P., (2004, June.). *Development of Information and Communication Technology Indicators* in the Kingdom of Saudi Arabia, ECCO XII'04, Beirut.

Al-Solbi, A., & Mayhew, P. (2005b, July). Measuring E-Readiness Assessment in Saudi Organisations: Preliminary Results From A Survey Study. In *Proceeding of the First European Conference on Mobile Government*, 10-12 July 2005, University of Sussex, Brighton, UK

Al-Solbi, A., & Pam Mayhew, P. (2006, December*), A strategic Framework for Electronic Readiness Assessment*, Proceeding of the Internet and Information System in the Digital Age Conference, Dec. 14-17, 2006 in Brescia, Italy. ISBN: 0-9753393

Al-Solbi, A., Pam Mayhew, P., & Al-Badi. A., (2005a, July), *An Alternative Model For Measuring E-Readiness for Developing Countries:Applied to Saudi Organisations*,The proceeding of the International Business Information Management Conference (IBIMA 2005) on July 5, 6, and 7, 2005 in Lisbon, Portugal.

CSPP. Org, (2000). The CSPP readiness guide for living in the networked world. *CSPP.Org*. Retrieved October, 2000 from http://www.cspp.org

Edwards, J. E. (2003). *The relationship of e-commerce readiness to technology acceptance: The case of Barbados*. Digital publish doctoral thesis. NOVA South-eastern University.

Jennett, P. (2002, September). *A readiness Model For Telehealth*. In Proceedings of International Society for Telemedicine (ISFT) in the seventh International Conference on Telemedicine, Regensberg, Germany Sep. 22-25th, 2002.

Ramani, A. (2006, March). Hospital information system: PULSE [Implementing IT in health-care]. In *Proceeding of 18th National Computer Conference* (NCC18), March, 2006, Riyadh, Saudi Arabia.

Rizk, N. (2004, September). *E-Readiness Assessment of Small and Medium Enterprises in Egypt: A Micro Stud,* Topics in Middle Eastern and North African Economies, *Electronic journal*, Vol. 6, Middle East Economic Association and Loyola University Chicago, Retrieved Apil, 2010, from http://www.luc.edu/orgs/meea/volume6/Rizk.pdf

World Health Organization, (2004), *Strategy 2004-2007 eHealth for Health – care Delivery, Report 2004.*

Chapter 13
E–Learning Acceptance and Challenges in the Arab Region

Nasim Matar
Anglia Ruskin University, UK

Ziad Hunaiti
Anglia Ruskin University, UK

Shahid Halling
Talal Abu Gazala Group, Jordan

Šadi Matar
Ministry of Communications and Transport, Sarajevo

ABSTRACT

This chapter discusses the status and quality of e-learning in Arab Universities located in the Middle East. The first objective of the study was to provide an analytical overview of the use of e-learning and its quality in these universities located in the Middle East region. Another objective is to fill the gap in literature in this particular topic regarding the Middle East region. Also draw into different solutions and recommendations in order to make a successful match that will result in a better adoption and serving of the e-learning technology. The study was based on two different approaches that include a survey to navigate the official web sites of universities in the region plus a questionnaire to fetch for the current stand of e-learning quality in the region. The results of each approach have been analyzed, and the outcomes and recommendations have been presented which can be used for future adoption and other related studies.

BACKGROUND

E-Learning

Electronic Learning or what is referred to as "e-Learning" is defined as the delivery and acquisition of education or training in electronic format using electronic media. The latest technology to lend itself to e-learning has been the internet which is now the largest single resource repository for educational establishments (Smith and Hardaker, 2000). During the last few years, e-learning has witnessed growing interest from different

DOI: 10.4018/978-1-60960-048-8.ch013

directions, which helped in fast promotion and prosperity of these electronic educational activities. Different organizations that were involved in distance education have adopted this new technology and started to offer online learning that was treated as a logical extension to their distance educational activities. Educational institutions have also been involved in promoting for higher interest towards e-learning through incorporating this educational technology to improve access towards their educational programs and to provide a better merge with the current market.(Bates, 2005). The growth of e-learning has also been regarded to the advances in the field of information and communication technologies which has helped in lowering the coast of adopting e-learning services and technology in a better and much effective way.

Despite this massive interest in the e-learning technology, it is not without constraints and limitations. The primary obstacles towards this e-learning technology are the lack of proper infrastructure of information and communication technology (William et al., 2004).

Middle East Scenario

In The Middle East scenario, many Arab universities are taking gigantic steps in their use of e-learning to enhance higher education (Abouchedid and Eid, 2004). The countries surveyed in this investigation were found to be heading in the same direction as far as implementing this technology was concerned. It was also found, however, that the process was rather slow and still at initial stages in some cases due a number of factors such as, political peculiarities; rigidity of government agendas; levels of economic development and technological challenges. The political situation in the region has been facing many challenges. These challenges have had a significant bearing on the different sectors of human activity with education being a case in point. So influential has their impact been that it has affected other areas especially those that have a close relationship with education such the country's economy, technology and financial status (George, and Davoodi, 2003).

Brief Insight on the Status of Arab Countries in the Middle East

The Arab World is a region which has suffered both economically and financially compared to the rest of the World which has thus affected the level of literacy and adoption of new education systems and methods. The high level of illiteracy and poor technology infrastructure has meant that good educational opportunities are typically reserved for those who are societies wealthiest, which as a proportion of the overall population in Arab states is relatively small (Cassidy and Matthew, 2002).

The 'digital divide' as it is known, is the difference in access to technology between the countries located in the economically richer Northern Hemisphere compared to those in the poorer Southern Hemisphere. Developed nations mainly based in the richer northern hemisphere, have long enjoyed access to fast, reliable technology and communication media due to their greater economic prosperity and financing of IT infrastructures and educational initiatives. This in turn has provided them more opportunities for education and development for their citizens (Elango and Selvam, 2008). The need for e-learning solutions for the Arab world is essential for bridging the digital divide as there are over 130 million illiterate Arab children with no formal education. The adoption of e-learning systems is of paramount importance as currently the average yearly population growth in the Arab States is 2.5%, compared to a figure of 1.5% worldwide, which equates to having approximately 110 million 5-18 year olds by the year 2010. This will drastically change the demographics of the region and place a huge burden on the limited resources and the infrastructure it presently has (Internet World Statistics, 2009). This will also place extra strain on the traditional schooling and education methods which will further reduce the chances for nations to break the cycle of poverty and illiteracy they currently face. Providing just the same level of education for future generations will be extremely challenging unless new education methods are employed. An added advantage of e-learning over other solutions is that it is far less expensive. Building new schools or establishing evening classes within the universities' programs is obviously a great deal costlier. Thanks to all this knowledge, implementing e-learning technology in the Arab world has now become the natural solution of choice for institutions seeking to leverage educational standards and pedagogical outcomes (Arab Knowledge Report, 2009). Although e-learning systems can help to address these imbalances by providing high quality educational resources through technology, such an approach suffers from many different challenges which need to be addressed, if in the long term, e-learning is to be a successful combat to illiteracy (Clarke, 2008).

Regional ICT Infrastructure in the Middle East

Today the most popular media for providing e-learning is the internet. By using the different facilities that this technology provides such as web pages, e-mailing, discussion groups, blogs, chatting, Learning Management Systems (LMS), Multimedia and VoIP (Bates, 2005), the way of delivering education in the Arab world has changed. The internet was first used in Arab countries sometime between 1993 and 1995 (Dewachi, 2001). Since then, its usage has grown exponentially as can be seen in Table 1.

After the internet's introduction in Arab countries, many universities adopted the technology. They made changes to their IT infrastructures in order to adapt them for use with this new technology. The purpose was to make the best use of it in providing services to their faculty members and their student communities. Most, if not all of

Table 1. Middle East Internet Usage and Population Statistics (Internet World Statistics, 2009)

Middle East Internet Usage and Population Statistics						
MIDDLE EAST	**Population (2009Est.)**	**Usage, in Dec/2000**	**Internet Usage, Latest Data**	**% Population (Penetration)**	**User Growth (2000-2009)**	**(%) of Table**
Bahrain	728,709	40,000	**250,000**	34.3%	525.0%	0.5%
Iran	66,429,284	250,000	**23,000,000**	34.6%	9,100.0%	48.0%
Iraq	28,945,569	12,500	**275,000**	1.0%	2,100.0%	0.6%
Jordan	6,269,285	127,300	**1,500,500**	23.9%	1,078.7%	3.1%
Kuwait	2,692,526	150,000	**900,000**	33.4%	500.0%	1.9%
Lebanon	4,017,095	300,000	**1,570,000**	39.1%	423.3%	3.3%
Oman	3,418,085	90,000	**469,000**	13.7%	421.1%	1.0%
Palestine(West Bk.)	2,461,267	35,000	**355,500**	14.4%	915.7%	0.7%
Qatar	833,285	30,000	**436,000**	52.3%	1,353.3%	0.9%
Saudi Arabia	28,686,633	200,000	**7,200,000**	25.1%	3,500.0%	15.0%
Syria	21,762,978	30,000	**3,565,000**	16.4%	11,783.3%	7.4%
United Arab Emirates	4,798,491	735,000	**2,860,000**	59.6%	289.1%	6.0%
Yemen	22,858,238	15,000	**320,000**	1.4%	2,033.3%	0.7%
Gaza Strip	1,551,859	n/a	**n/a**	n/a	n/a	n/a
TOTAL Middle East	202,687,005	3,284,800	**47,964,146**	23.7%	1,360.2%	100.0%

them, set up official web sites through which they could disseminate essential information to current and prospective students (Albirini, 2006).

Based on the presented data in Table 1, it can be seen that Arab countries in the Middle East are suffering from low internet penetration. The reason for this is the weak regional Information and Communication Technology (ICT) infrastructure. As revealed at the Beirut summit (1988) most Arab countries lack adequate ICT infrastructure (UNISCO, 1988). The percentage of internet users in the Middle East is still very small when you compare it with some of the other regions in the world. This probably explains why most Universities and other educational establishments have been slow to adopt e-learning and thus, have contributed to depriving the region's population of reaping the full benefits that emanate from using this mode of delivering education (Bates, 2005).

Table 2, taken from (www.internetworldstats. com), shows growth levels per country in the region between 2000 and 2007. Comparing these figures to the rest of the world and summarizing, the position of the Middle East is as shown in Table 2.

It is clear from these figures that the rate at which the number of internet users is growing in the Middle East is far higher than that of the rest of the world. This is because of the fast-paced development of internet infrastructures in the region (Soumitra et al., 2006). Despite this fast development, remote areas in Arab countries have not been well served with the advances in the ICT, as still the concentration of such efforts is within urban areas. This has led to poor adoption of e-learning in the region when compared to advanced nations (Soumitra et al., 2006). This study's purpose is to evaluate the state of e-learning and its quality in the Middle East and provide possible solutions as to how e-learning adoption can be increased.

METHODOLOGY

The objectives of this study were two fold. Firstly, to evaluate the state of e-learning in the Arab educational establishments, and secondly, to evaluate the quality of such e-learning endeavors.

Two surveys were used to achieve these objectives.

Survey 1: Evaluate the State of E-Learning in Arab Educational Establishment

The approach was based on content analysis methodology (Krippendorff, 2004), which was conducted by navigating 171 official university web sites across the Middle East region. The lists of universities were obtained from the Ministry of Higher Education in each country surveyed. The aim was to:

- Identify the use and incorporation of e-learning services within universities web site

Table 2. Middle East Internet growth against the rest of the world (Internet World Statistics, 2009)

INTERNET USERS IN THE MIDDLE EAST AND IN THE WORLD						
MIDDLE EAST REGION	Population (2009 Est.)	Pop. % of World	Internet Users, Latest Data	% Population (Penetration)	User Growth (2000-2009)	Users % of World
Total in Middle East	202,687,005	2.9%	47,964,146	23.7%	1,360.2%	2.9%
Rest of the World	6,565,118,203	97.1%	1,620,906,262	24.7%	353.1%	97.1%
WORLD TOTAL	6,767,805,208	100.0%	1,668,870,408	24.7%	362.3%	100.0%

- Define the accessibility towards e-learning services in terms of internet / Intranet
- Define the type of e-learning service provided in terms of blended and online courses.
- Define the learning management system used within each university.

The information obtained was considered vital in order to make better judgments for solutions and future recommendations.

Survey 2: Evaluate the Quality of E-Learning in Arab Universities that have Adopted E-Learning

This approach was adopted from a number of national educational agencies across the world including Canada, Australia, UK, which relied (UNISCO INTERNATIONAL SCIENCE, 2003), and others. The questionnaire that was developed was tailored to fit the socio-economic nature of the Middle East to make it more relevant to this study. A summary of the questionnaire can be seen within the table of results (Table 5).

Prior to the distribution of the questionnaire, a pilot study was undertaken in order to evaluate the credibility of it. The pilot was conducted on IT staff working for the University of Jordan and provided information regarding their attitudes prior to conducting full scale. The results were encouraging, it gave us a value of 93 which is beyond 70, which considers that the survey questions are reliable for an in- depth examination of issues related to this research based on "Alpha Cronbach's" formula for measuring consistency of reliability (Zoltán, 2003).

The questionnaire was sent to 70 universities that have adopted e-learning and 26 of those universities answered the survey. The results were analyzed using statistical software package (SPSS) and have been presented in the next section.

STUDY OUTCOMES

This section will present the compendium results of the previous steps that were performed in the methodology section. The results will be summarized into facts about the region, which will

Figure 1. Apportions LMS adoption by universities per Middle-Eastern country

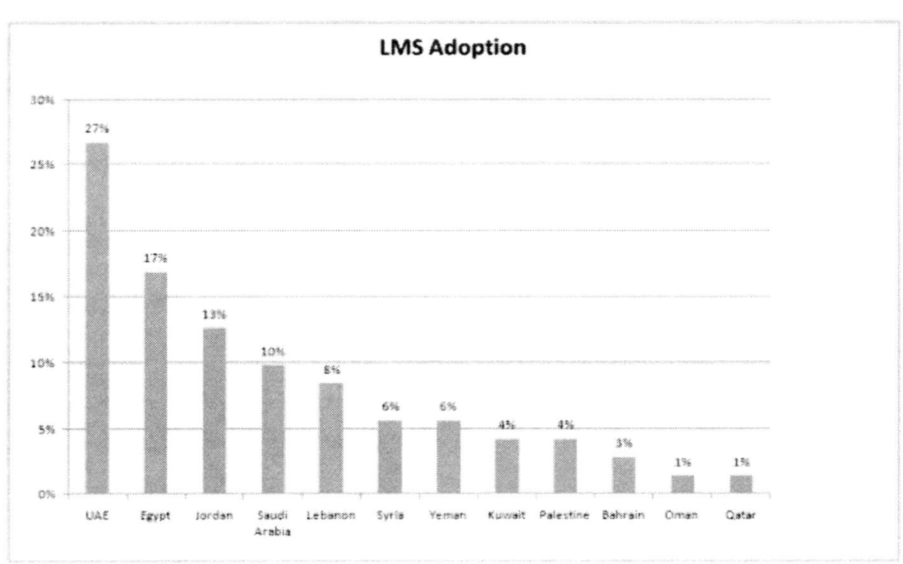

help to update the current literature regarding e-learning adoption and its quality in the Middle East region.

Study Outcomes: Part 1

The first survey study that was initiated includes 171 universities in the region. Each university's web site was visited and carefully analyzed to get the targeted answers regarding e-learning adoption and what type of services they are offering. The study revealed that the e-learning adoption rate among Arab universities in the region is 41%, which represents 70 universities of the 171 universities sampled. The revealed rate is considered relatively low in comparison with the European Union countries that reached 75%. (LTAC Project Group, 2009). Through analyzing the data further, it was found that out of the 41%, 58.5% accounted for public universities with the remainder being private universities. The United Arab Emirates (UAE) had the highest adoption of LMS among Arab countries (Figure 1), with 19 universities

having installed a LMS, representing 27% of all LMS usage in the region's universities (Matar,N. & Hunaiti et al., 2007).

From this chart, we can rank (from highest to lowest) each country's universities' LMS adoption.

To give an overview of adoption levels per country according to the number of universities using e-learning, Figure 2 is presented.

The survey has also revealed that out of the total of 71 universities providing e-learning in the region, 15 also run 20% of their programs online. Table 3 shows the number of universities running online programs in each country.

In terms of e-learning management systems used, it was revealed that there are 3 preferable types of software used for providing the management of learning and the learners. These software applications are Moodle, Blackboard and WebCT (Matar et al., 2007). Table 4 shows how many universities in each country have chosen a particular type of LMS.

Figure 2. Percentage of universities using e-learning in each country

Table 3. Number of universities running online programs in each country

Country	Egypt	Kuwait	Jordan	Lebanon	Palestine	Saudi Arabia	Syria	*Yemen*
Online University	*3*	*1*	*1*	*2*	*1*	*1*	*4*	*1*

Table 4. Universities and LMS

Country	Moodle	WebCT	Blackboard	Not Defined
Egypt	12	×	×	×
Kuwait	1	×	1	×
Jordan	3	×	4	2
Bahrain	1	1	×	×
Lebanon	2	2	1	3
Oman	×	×	×	×
Palestine	3	×	×	×
Qatar	×	×	1	×
Saudi Arabia	3	4	×	×
Syria	×	×	×	4
UAE	1	14	4	×
Yemen	4	×	×	×

According to this table, **44%** of the region's universities are using licensed programs while **41%** use open-source programs and **14%** remain undefined. By further breakdown the data we obtained, we were able to determine what percentage of universities using LMS were government run, and what percentage were in private hands. This is depicted in Figure 3.

Second Survey Outcomes

The second survey was a questionnaire which was sent to 70 Arab universities in the region. The actual response came from 26 universities, and the analysis of their responses was used to update the literature towards e-learning services and technology in the Middle East region. The first survey approach showed that 41% of Arab universities have adopted e-learning services and technology while the remaining 59% did not. This survey study was initiated to evaluate the quality of e-learning in those 41% Arab universities. After gathering and analyzing the data that was filled in these questionnaires, the following results in Table 5 came to light.

DISCUSSION OF RESULTS

Based on the previously discussed methodology, we had two different survey methods to define the status and the quality for e-learning in the region, for which the results have been presented. This section will discuss each result with respect for each survey type undertaken.

Figure 3. Percentage of governmental versus private adoption for e-learning in each country

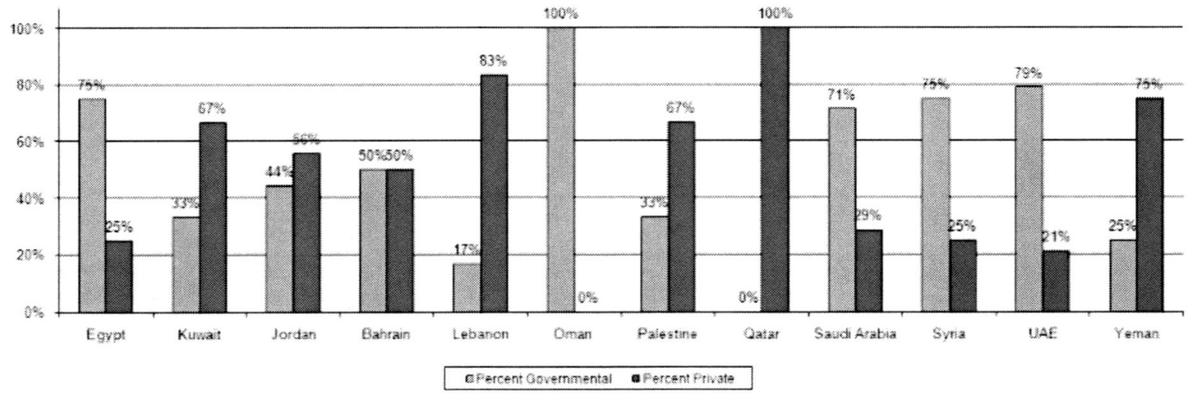

Table 5. Results of second survey

N.O	Question	Comments (The following Percentage are based on the universities that have adopted e-learning)	Percent
1	What method of e-learning do you provide within your university	• Are providing blended e-learning courses	71%
		• Are providing partial online courses	33%
		• Are providing Full e-learning courses	4%
2	When did your university initiated e-learning services and technology?	• Between 2000 and 2002	4%
		• Between 2002 and 2004	38%
		• Between 2004 and 2006	46%
		• Between 2006 and 2008	13%
3	Which benefits does your university seek through implementing e-learning in your organization	• Just in time learning	50%
		• Substitute instructor shortage	17%
		• Provide more educational resources for learners	83%
		• Effectiveness of e-learning and raising educational standard	96%
4	What were the main challenges for implementing e-learning at your university	• Cost of implementing e-learning system and services	50%
		• Time for implementing e-learning system and services	67%
		• Effectiveness of using e-learning system and services	38%
		• Employees resist towards this new technology	50%
		• Computer literacy among users	33%
		• Vendor issue	21%
		• Management resist towards this new technology	8%
5	Is your local e-learning system and materials accessible from outside the learning campus any time and any where	• Yes	96%
		• No	4%
6	Which of the following e-learning services do you provide within your university	• Learning Management System (LMS)	96%
		• Content Management System (CMS)	21%
		• Video Conferencing	29%
7	What are the resources for providing e-content for the electronic courses taught at your university?	• Instructors	100%
		• Computer Centre	25%
		• Companies	17%
		• Internet	58%
8	Which training courses does your university provide for instructors to assist them in interacting with e-learning services and technology?	• Creating web pages	58%
		• Creating Multimedia files (Audio, Video, Interaction,...etc)	17%
		• Interacting with Learning Management System (LMS)	96%
9	Is there a repository system for saving and categorizing different e-contents to be used by another instructor or courses?	• Yes	33%
		• No	64%

continued on following page

Table 5. continued

N.O	Question	Comments (The following Percentage are based on the universities that have adopted e-learning)	Percent
10	Is there a committee or center for supervising and advising e-contents creation and distribution	• Yes	54%
		• No	46%
11	How do you rate the success of implementing e-learning within your university	• Rates of participants and user satisfaction	63%
		• Assessment of learning that has taken place	21%
		• Impact on students performance	42%
		• Impact on organizational performance	21%
12	What is the percentage of courses that are supported with e-learning technology and services?	• 1 → 5% Courses	38%
		• 6 → 15% Courses	42%
		• 26 → 35% Courses	4%
		• 46→50% Courses	4%
		• Above 50% Courses	13%
13	What will determine your use of e-learning in the next two years?	• Coast effectiveness	42%
		• Ability to reach more learners	67%
		• Improved e-learning platforms	50%
		• Resolution of vendor issues	21%
		• Students' performance and satisfaction	71%
		• Attitude of Faculty member towards e-learning	93%

Survey Approach Discussion

From the previous results it shows that e-learning adoption is low in the Middle East region compared with many Western Universities which is due to the lack of adequate ICT infrastructure services. This is considered the main backbone for supporting current e-learning services and technologies. The ICT services in the Middle East are commonly concentrated in cities, where, in some countries, only 20 to 30% of the people live, whilst the vast majority (around 70% to 80%), are scattered in smaller rural communities where basic essentials such as telephone lines and electricity are either irregular or nonexistent (Arab Knowledge Report, 2009). In some countries, up to 75% or more of the country's phone lines are found in the capital city alone. Other reasons cited included the high cost of using internet services, which depend on the availability of phone lines, and the cost of using the phone lines themselves. In Arab countries, the cost of using a phone line is considered expensive and when you add the cost of using the internet itself which is provided by a third party companies, the whole thing becomes rather astronomical (Dewachi, 2001). Internet technology use in Arab countries is minimal, but nonetheless, the number of users is growing exponentially. This can be pinned down to advances in other areas of human endeavor such as improved income levels, increased awareness and a more educated population.

In terms of adopting e-learning services and technology, the study showed that 58 public universities were keen to adopt new technological approaches in the field of education. The United Arab Emirates has shown a leading position in the current adoption of e-learning technologies, which is due to the huge economical and technological advancement that has affected the country, which resulted in a better support for educational technologies. The UAE has provided an open and transparent environment for local and foreign investments, through a careful revision of their legislations and making the appropriate adjustments to attract many foreign companies and corporations. It has also improved the phone lines and cable infrastructure and technology to create a robust information and communication technology infrastructure to better serve the needs of its population (Matar et al., 2007).

In terms of online learning, the survey revealed that out of the total of 70 universities providing e-learning in the region, 15 also ran 20% of their programs online. The biggest challenge faced by the online method of learning in this region is educational-policy hostility. Most countries still refuse to recognized let alone accredit online degrees. In fact, there appears to be on-going effort by authorities to discourage students from enrolling in online degree programs offered by many European, Australian and north American universities. The first Arab Online University (AOU) was established in Kuwait in the year 2000 (www.arabou.org/). Being the pioneer, it was greeted with scorn in many countries. It was only after it had proven itself that it started gaining acceptance. It has been so successful in pushing its charter that some countries in the region such as Kuwait, Jordan, Bahrain, Saudi Arabia and Lebanon have even accredited its degrees (UNESCO Regional Bureau, 2009). The huge step taken by Arab Open University has paved the way for other universities to follow. Nowadays the number of universities accepting a blend of traditional and online learning is on the rise. The Al-Baath University, the University of Aleppo and the University of Damascus in Syria; the University

of Science and Technology in Yemen; and the Al Quds Open University in Palestine have all taken to online education. In the next few years, we are likely to see more countries and universities accrediting e-learning degrees (UNESCO Regional Bureau, 2009). In terms of using different LMS's, the survey results shows that Universities differ broadly in their choice of LMS and the choice is predominately tied to the University's financial standing as well as the support it receives from government.. The study shows that countries that are hard up financially and are bogged down by financial and economic burdens usually opt for open-source systems such as Moodle. More affluent countries such as the United Arab Emirates, Saudi Arabia, Qatar and Bahrain, on the other hand, prefer licensed programs like WebCT and Blackboard (Matar et al., 2007).

It is important to state that most licensed software is provided through companies that provide training and support for their products. Such support is an added cost and should deliver good value in order to make their offering lucrative.

Questionnaire Results Discussion

This section will discuss the result with respect to each question posed in the questionnaire.

What method of e-learning do you provide within your university?

Arab Universities are still in favor of traditional education and e-learning is acting as a support for adding resources and additional references. They have also accepted partial online courses for general subjects that have a large number of well designed quality resources, with assessment for students in the university campus. The number of online universities accounted for 4% of the total sample. Each had a different mode of assessment that mostly rely on reports and a final assessment done in the university campus.

When did your university initiated e-learning services and technology?

Results between 2002 and 2006 have shown a large adoption of e-learning in Arab universities due to the technological advancement in the field of ICT and the development web 2.0 applications. It is believed that the reason of low adoption during 2006 to 2008 was due to lack of proper utilization of e-learning contents and resources. It seems as though that the one size-fits-all approach has had negative impact on students' performance and adoption to this educational service.

Which benefits does your university seek through implementing e-learning in your organization?

The purpose of most Arab universities to introduce e-learning services was to raise education standard and to provide more educational resources for students. According to the results, the region is suffering from lack of instructors', with 17% of the universities sampled participating in such an approach. Such a serious shortage comes from the fact that many instructors are leaving universities to work in other countries which provide more facilities for research or better income. In many cases, women colleges are lacking female instructors, which is a prerequisite for the teaching of females in such countries such as Saudi Arabia.

What were the main challenges for implementing e-learning at your university?

The results show that the main challenge for implementing e-learning is the time needed. This is calculated as the accumulative time for installing and running the services, training faculty members on the use and utilization of such applications and services and supporting courses with a sufficient number of electronic resources. Employee resistance has also been a great challenge, and from the survey it was obvious that the cost was a challenge for non-gulf countries, while employee resistance was mainly from employees working in gulf countries. As another challenge, vendor issues were voted at 21%. This figure should drop in coming years as many agencies and organizations are working on providing a common framework for e-learning standards.

Is your local e-learning system and materials accessible from outside the learning campus anytime and anywhere?

E-learning is accessible at all universities, but only a small number of universities are providing this service through an intranet. This is due to various reasons, such as security, copyright issues, and the cost of required maintenance.

Which of the following e-learning services do you provide within your university?

Most universities are providing learning management systems for managing courses and educational resources. Video conferencing is also popular in some countries where there is a requirement for female instructors, but a lack of them. In such cases male instructors are utilized through video conferencing. LCMS had the lowest rate of 21% due to the special training requirements in order to utilize such applications.

What are the resources for providing e-content for the electronic courses taught at your university?

All universities in the region are relying on instructors to create e-learning content. With minimal support and training for creating e-content, it is clear why the rate of adoption in Arab universities that is less than 15%. The Internet is considered the second source for information in the e-learning course in Arab universities, and instructors should be trained on the use of such contents and copyright issues. The involvement of computer centers and companies is considered minimal. Most University computer centers are not supported with a sufficient number of qualified staff which is why many of them are unable to support all the University's requirements. This puts an additional load on instructors to fix the errors, or find a way around to the solution. E-learning companies have the lowest rate of adoption for such initiatives, due to the high expenses of creating electronic courses. The cost is calculated by paying the company for its work and paying or releasing faculty members from their teaching duties in order to supervise the creation of such e-content.

Which training courses does your university provide for instructors to assist them in interacting with e-learning services and technology?

Most universities provide training on the use of LMS and creating static web pages, and minimal resources are provided to develop effective multimedia courses that provide a better pedagogical approach. Multimedia courses are considered specialized courses and they need a firm computing background. Also in terms of expenses, they need a special applications, equipment and training.

Is there a repository system for saving and categorizing different e-contents to be used by another instructor or courses?

Most universities in the region do not hold a repository system for storing or categorizing the e-contents that have been used during the course. Such policy is considered a negativity as it will bring the process of creating e-contents in the reinvention wheel concept.

Is there a committee or center for supervising and advising e-contents creation and distribution?

A percent of 54% of universities sampled proclaimed that they have a committee or center for supervising and advising e-learning contents and distribution. Many questions should be investigated towards the efficiency and procedures of such centers or committees due to the previously discussed results, but such investigation is beyond the current study aims.

How do you rate the success of implementing e-learning within your university?

Most universities rate the success of e-learning based on student satisfaction and performance. Less attention is paid towards the assessment of learning and impact on educational performance due to lack of expertise in the field of e-learning pedagogy, with educational organizations not having e-learning within their mission statement as was clear from the outcomes of question 3.

What is the percentage of courses that are supported with e-learning technology and services?

Most universities have between 1% and 15% of their courses supported with some form of e-learning. This minimal support rate is considered the fruit of the previous obstacles that have been identified.

What will determine your use of e-learning in the next two years?

Faculty member attitudes and students performance are considered the corner stone for the continuing process of using this technology. The main purpose of such technology is enhancement of student learning, so if such initiatives don't show any progress towards their performance, they will be neglected as a service. Also, if instructors' attitudes are against this technology, noting the fact that universities are depending 100% on instructors for creating e-contents, this technology will be useless in such scenarios.

SOLUTIONS AND RECOMMENDATIONS

The study has presented many obstacles towards adopting e-learning technology, and different factors that are affecting the quality of electronic courses. This section of the chapter will try to present some of the solutions that can be embraced in order to provide better services towards adopting e-learning technology and solutions. The solutions and recommendations are going to be presented in sequence of the discussed obstacles that have been previously presented.

Obstacle 1: Lack of Proper ICT Infrastructure

Most Arab countries in the Middle East region rely on broadband connections as their major choice for providing information and communication technology (Soumitra et al., 2006). This technology has provided wide options for internet connections such as (DSL, ADSL, T1, T3) connections. Such availability for fast connections in comparison with the traditional phone lines has raised the usage of internet technology and access. But also it has major deficiencies which are:

a. Limited in its reach
b. Reliant on residential cable infrastructure
c. Expense of setup and use.

For many commercial zones, offices, universities, schools and houses (DSL or such) connections are not available. Many customers are outside of DSL reach, and mainly cable connections are installed in commercial zones (Soumitra et al., 2007). Lately a new standard has been provided as a promise for many obstacles that are currently present with the current ICT infrastructure. This technology has been called (WiMax) which stands for (Worldwide interoperability of Microwave Access). This technology is believed that it will standardize and promote wireless broadband as a wireless alternative to digital subscriber line (DSL) and cable that can help remove barriers to broadband access (Soumitra et al., 2007). Setting up a wireless broadband access is similar to setting a cellular station using a base station that service a radius of several kilometers.. A customer has a unit similar to a satellite TV setup to connect to the base station. The signal is then routed via standard Ethernet cable either directly to a single computer, or to an 802.11 hot spot or a wired Ethernet LAN. Such networks are considered very

Figure 4. WiMax base connection (Source: Techwarelabs, 2009)

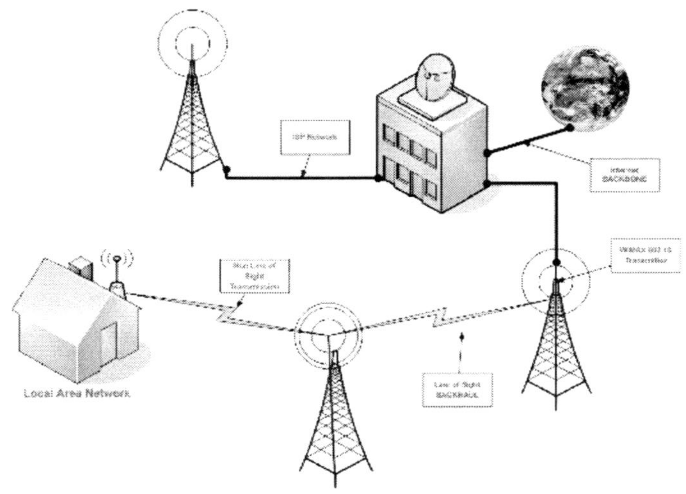

effective since one base station sector can provide enough data rate to simultaneously support more than 60 businesses with T1-type connectivity and hundreds of homes with DSL-type connectivity (Jeffrey et al.,2006). There are many benefits of WiMax through enabling standards-based products with fewer variants and larger volume production. This will drive the cost of equipment down and encourage competition making it possible to buy from many sources. For areas poorly served by a wired infrastructure, including many developing countries, WiMax will be important as it eases Internet implementation at a relatively low cost. Figure 4 shows a typical WiMax base station connection.

Obstacle 2: Lack of Proper Electronic Resources to Satisfy the Educational Context

In many Western and European universities, different approaches have been initiated towards providing proper e-learning, as many universities have suffered the lack of proper electronic contents. The solutions developed by them can be adopted to raise the educational quality of e-learning initiatives in the Middle East. The most favorable solution that many universities have adopted is working in collaboration. A readymade model has been provided with the name of CAMEL that stands for (Collaborative Approaches to the Management of E-learning) (Thumbria University, 2006). This mode facilitates the collaborate efforts of many parties which enhances e-learning utilization through presenting and sharing experiences, expertise, solutions and resources among the participating universities. This model has well defined standards for initiating an effective collaborative effort and suggests many solutions through a set of templates for effective use and implementation (Thumbria University, 2006). Figure 5 shows the CAMEL model.

In terms of technological support for collaborative efforts in order to share resources and information between universities and faculty members, many unified e-learning system approaches have been established. The following list shows the various categories of these systems.

Figure 5. The CAMEL model (Source: Thumbria University, 2006)

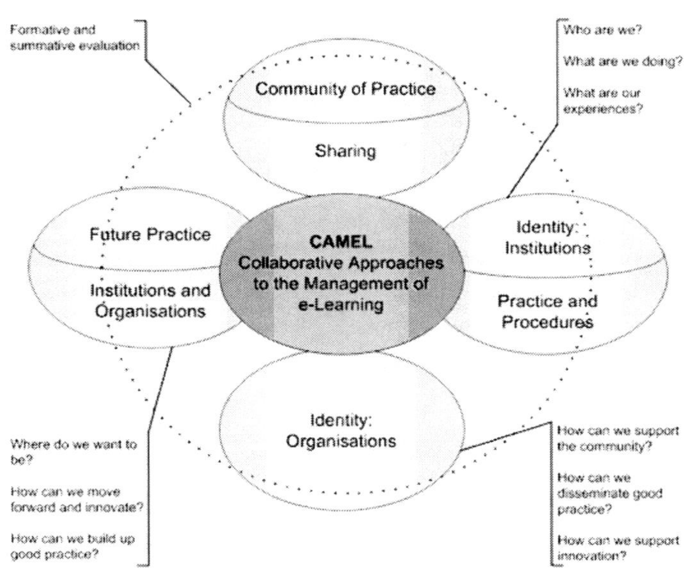

1. **Centralized LCMS:** Uses one main LMS and each university can collaborate by adding its readymade complete courses to the system, with other university members able to use the same course for their teaching. Such an approach can be performed using WebCT, Blackboard, Moodle, ATutor, etc.
2. **Networked Unified E-learning System:** Uses special software and hardware which are used for the collaborative creation of course materials. An example of such systems is LON-CAPA, which is an open source system and stands for Learning Object Network with CAPA (http://www.lon-capa.org/).
3. **Unified Repositories systems:** Uses special software and a specific structure for creating the courses with each university participating by adding learning resources or what is known as learning objects. In Australia and New Zealand, such project has been initiated under the name of "The Le@rning Federation" (TLF). It provided learning content for all kindergarten to grade twelve teachers and learners in addition to Australia's distance learning schools. TLF have provided more than 4000 interactive learning contents in different discipline such as math; science; literacy; languages. TLF ensured that the learning contents are based on educational soundness, copyright standards, and technical standards that reflect interpretability, flexibility and access for teachers through specific design and development criteria (Le@rning Federation) more information about this project is available from (www.thelearningfederation.edu.au). Different systems have been developed that provide the same concepts such as "The Co-Operative learning object exchange" (CLOE), Fedora, plus many more different systems that are developed and used for different disciplines and uses (Matar et al., 2009).

Most of these systems have the disadvantage of one-size-fits-all, which has proven to affect the adoption and interaction level between students and learning material. A new approach is being investigated towards providing a unified e-learning system with flexible approach to support students learning style. This approach is called E-ticket, and it has been tested and provided promising results towards raising the collaboration and adoption of e-learning in the Middle East region. For more information check the following site (http://www.u-elearning.net) (Matar et al., 2009).

CONCLUSION

Arab universities in the Middle East have a low adoption level of e-learning with a small percentage of electronic courses that are suffering from a low quality and low utilization of different services that are provided with current revolution of web 2.0 applications and services. Different solutions are available which can help promote better ICT with a lower cost of installation. WiMax should be considered in many Arab countries in the region as many residents in those countries are located in scattered areas, where wired broadband connectivity would be expensive to install. Regarding the quality and availability of educational resources, a collaborative solution should be considered. Unified e-learning systems provide tools for effective collaboration and sharing resources among different universities. Arab universities should also focus on establishing e-learning research centers that can collaborate with each other. This should be supplemented by using expertise from western university's to implement what has been tried and tested, in order to deliver effective e-learning environments whilst taking into consideration the cultural sensitivities of the region.

REFERENCES

Abouchedid. K., & Eid. G. (2004). E-Learning Challenges in the Arab World. New York: *Emerald Group Publishing Limited.*

Albirini, A. (2006). Cultural perceptions: The missing element in the implementation of ICT in developing countries. *International Journal of Education and Development using ICT, 2*(1).

Arab Open University. (2007). *Arab Open University.* Retrieved, April 2007, from: http://www.arabou.org/

Bates, A. (2005). *Technology, e-learning and distance education.* Boca Raton, FL: Routledge Taylor & Francis Group.

Cassidy, T., Matthew, M. (2002). Higher Education in the ARAB states: Responding to the Challenges of Globalization. *AMIDEAST conference on Higher Education In ARAB Countries*

Clarke, A. (2008). e-Learning Skills, 2nd edition. New York: *Palgrave Macmillan.*

Dewachi, A, (2001). Overview of Internet in Arab States. *International Journal of Education and Development using ICT, 2*(1).

Elango, R, and Selvam, M. (2008). Quality of e-Learning: An Analysis Based on e-Learners' Perception of e-Learning. *The Electronic Journal of e-Learning, 6*(1), 31-44.

George, A., & Davoodi, H. (2003). *Challenges of growth and globalization in the Middle East and North Africa.* International Monetary Fund, Publication Services.

Internet World Statistics. (2009), Retrieved November, 2009, from http://www.internetworldstats.com/

Jeffrey, A., Arunabha, G., & Rias, M. (2007). *Fundamentals of WiMAX: Understanding Broadband Wireless Networking.* Upper Saddle River, NJ: Prentice Hall Communications Engineering and Emerging Technologies Series.

Krippendorff, K. (2004). *Content analysis: an introduction to its methodology.* New York: Sage Publications Ltd.

Learning Object Network with CAPA project. (2008). *Learning Object Network with CAPA project.* Retrieved April, 2008, from http://www.lon-capa.org/

LTAC Project Group. (2009). Student Induction to E-Learning (SIEL), *IMS Global Learning Consortium*, Retrieved November, 2009, from http://www.imsglobal.org/siel.cfm

Matar, N., Hunaiti, Z., Huneiti, Z., & Al-Naafa, M. (2007). E-Learning Status in Arab Counties. In *Proceeding of the International conference on Information Society (i-Society 2007),* Indiana.

Matar, N., Hunaiti, Z., & Matar, S. (2009). Promoting Flexible Unified E-Learning Structure. *The 4th International Conference on Mobile and Computer Aided Learning*, IMCL2009

Northumbria University. (2006). *The CAMEL Project.* Higher Education Funding Council For England & JISC InfoNet.

Report, A. K. (2009). Towards Productive Intercommunication for Knowledge. *Mohammed Bin Rashid Al-Maktoum Foundation & United Nations Development Program,* Retrieved April, 2007 From: http://www.arabstrategyforum.org/asf2009en/attachments/144_programme-english.pdf

Smith, D., & Hardaker, G. (2000). e-Learning Innovation through the Implementation of an Internet Supported Learning Environment. *Journal of Educational Technology & Society, 3*(3), 2000.

Soumitra, D., Augusto, L., & Irene, M. (2006). The Global Information Technology Report 2005-2006: Leveraging ICT for Development, 5th Edition. New York: *Palgrave Macmillan*

Soumitra, D., Shalhoub, Z., & Samuels, G. (2007). Promoting Technology and Innovation: Recommendations to Improve Arab ICT Competitiveness. *World Economic Forum: The Arab World Competitiveness Report*, 2007

Techwarelabs. (2009). *Techwarelabs*. Retrieved November, 2009, from http://www.techwarelabs.com/articles/other/wimax_wifi

UNESCO Regional Bureau. (2009). A Decade of Higher Education in the Arab States: Achievements & Challenges. *UNESCO Regional Bureau for Education in the Arab states,* Beirut.

UNISCO International Science. (2003). Science and Technology Education in the Arab world in the 21st Century. Technology & Environmental education Newsletter, XXVIII (3-4).

William, D., Pauline, C., & Namkee, P. (2004). The Social Shaping of a Virtual Learning Environment, *Electronic Journal of e-Learning. 2*(1), 69-80.

Zoltán, D. (2003). *Questionnaires in second language research: construction, administration, and processing.* New York: Routledge.

Chapter 14
The Adoption of the Internet to Support Arab Academics

Abdelnasser Abdelaal
Sur College of Applied Sciences, Sultanate of Oman

ABSTRACT

This chapter addresses the adoption of Internet applications by an Arab Student Association in North America (ASANA)[1]. ASANA uses the Internet to integrate its members, promote the Arabic culture, bridge with the American society and transfer knowledge to its native country. It delivers these services through websites, email service, electronic payment systems, online conferencing, file sharing tools and other digital resources. These online services build social capital, accrue intellectual capital, and cement mutual understanding between Arabs and the American society. However, these e-services are not widely adopted due to the lack of awareness of their usefulness, the absence of suitable IT culture, poor service quality, instability of leadership, and inadequate incentive system. Improving the adoption of ASANA e-services requires stable leadership, supportive IT culture, assessing provided services, and providing incentives to members to participate. This chapter provides insights and conceptual details that help Arab academic communities to use the Internet to participate in the overall socioeconomic development of their societies.

INTRODUCTION

The Internet is perceived as the second best invention, after the written word, for encoding, saving, disseminating, and sharing human knowledge. In fact, its applications have revolutionized the way individuals, communities, organizations and governments conduct their business. Academics, in particular, benefit significantly from Internet applications and online resources. They can access open-content journals, articles, textbooks, tutorials, presentations, and simulators. In addition, most colleges and universities have adopted information technology in their education process and offer electronic learning opportunities through distance or online courses. In fact, you might have

DOI: 10.4018/978-1-60960-048-8.ch014

taken one of these online courses. A noteworthy utilization of the Internet in the realm of academia is Wikipedia which is a collective authorship site developed and revised voluntarily by individuals (Forte and Bruckman 2008). The novelty of Wikipedia stems from providing individuals the opportunity to be learners, authors, editors and educators at the same time. These enormous learning opportunities have transformed the Web into a virtual dictionary, encyclopedia, reference, library and classroom. These opportunities help learners to develop faster, teachers to teach better and scientists to explore swifter. International students are one of the academic communities that benefit extensively from online resources. They use the Internet to search for reputable schools, scholarships, admission requirements, tuition fees and more. They can also connect with their families and peers through email tools, chatting venues and VoIP applications.

The Internet has become a venue where individuals and organizations can build intangible assets such as social capital, intellectual capital, and human capacity. *Social capital* is a broad term that encompasses the norms and networks that facilitate collective actions (Woolcock 1998). Social capital has become a key concept in social science, politics, economics, organizational behavior, and information systems. The term *intellectual capital* refers to the sum of intangible assets that determine the value and competitiveness of an organization (Johnson 1999 and Magrassi 2002). This includes know-how, knowledge, information, intellectual property, competences, and relationships. The term *human capacity*, in this context, describes the ability of individuals to provide, learn, communicate and lead.

The Arab society has severe shortage in durable natural resources. Therefore, they need to adopt online resources and electronic facilities to invest in intangible assets such as intellectual capital, know-how, knowledge, trust, reputation, and social capital. Investing in these intangible assets is important for them to sustain socioeconomic

development. Fortunately, Arabs are growingly adopting the Internet for delivering and/or accessing a wide range of social, political, personal, and commercial services (Hofheinz 2005).

The question is what factors affect the adoption of these emerging information technologies by Arab academics?

Scholars do agree that the adoption of information technology is a function of technical, behavioral, cultural, and economic factors (Davis 1989, Zakour 2004; Agarwal 1999; and Brynjolfsson and Hitt 1998). Yet, there is a lack of agreed upon wisdom on this topic that applies to the Arab context. To shed more light on the mentioned factors, this chapter uses a case study approach to provide anecdotal evidence on the motivations and the challenges of the adoption of the Internet by an Arabic Student Association in North America (ASANA) to deliver a number of services to its members. In so doing, we describe the managerial structure of ASANA, identify its services, and explore their usefulness. We will also discuss how ASANA serves as a venue where members broker and develop intangible assets and resources. In addition, we discuss the challenges facing the adoption of these e-services and provide suggestions for improvement.

Before we carry on, it is important to remember that the realm of the Information Systems discipline lies at the confluence of people, organizations, and technology. Specifically, it is concerned about the roles, capabilities and characteristics of people. Regarding organizations, this emerging field brings in strategy, structure and culture of organizations. It also crosses over infrastructure, applications, communication architecture, and capabilities of technology. We will show how the case of ASANA presents a unique and smart confluence between the three pillars of information systems. We will also provide some details about the business domain of ASANA and articulate on the specific problems it solves. This chapter, however, does not venture into detailed description of the structure, technology, culture, services and challenges of ASANA. Instead, it shows how Arab academic communities can use the Internet to achieve collective actions.

BACKGROUND

There is a large body of research that addresses the behavioral, cultural, organizational, technical and economic factors affecting the adoption and acceptance of technology by organizations and individuals (Davis 1989, Zakour 2004; Agarwal 1999; and Brynjolfsson and Hitt 1998). Acquiring a new technology is a necessary but not sufficient condition for using it effectively (Agarwal 1999: p.85). The organization in hand developed an innovative collaborative infrastructure but it failed to fully utilize it. The Technology Acceptance Model (TAM) is a profound theory that relates the adoption of innovations to its Perceived Ease of Use and Perceived Usefulness (Davis 1989). Perceived Usefulness (PU) is defined by Davis as "the degree to which a person believes that using a particular system would enhance his or her job performance." Perceived Ease of Use (PEOU) refers to "the degree to which a person believes that using a particular system would be free from effort." TAM is considered to be the most acceptable model for explaining IT adoption at the individual level. We will explore the usefulness of ASANA services provided to its members. Segars and Grover (1993) introduce a new factor that affect the adoption of technology which is its effectiveness. Tornatzky and Klein (1982) find that compatibility, relative advantage, and complexity of the technology are significant factors for adoption of innovation. Hage and Dewar (1973) note that high levels of centralization and formalization in the organization have negative effects on adoption of innovations. The term innovation refers to a new way of doing something or new thing that is made useful (Barras 1984).

This may be incremental or revolutionary changes in thinking, production, processes, or structural of organizations.

Prior research also shows that the national and organizational cultures affect the development and adoption of IT innovations (Png, Tan, and Wee 2001). According to Kluckhohn (1962), "Culture consists of patterns, explicit and implicit, of and for behavior acquired and transmitted by symbols, constituting the distinctive achievement of human groups, including their embodiments in artifacts" (p.73). Zakour (2004) presents the most common culture dimensions. These dimensions include hierarchy, power distance, conservatism, uncertainty avoidance, and nature of people.

Hofstede (1980) categorizes four main dimensions of national culture: power distance, uncertainty avoidance, collectivism versus individualism, and femininity versus masculinity. Uncertainty avoidance is the extent to which individuals feel threatened by uncertainty or unknown situation. Power distance describes the relationship between the higher-class and lower-class members of a society and how differences in power and wealth are dealt with in a particular culture. This may include the hierarchy in an organization, the political centralization, and the expected respect members may gain.

Individualism versus collectivism describes the extent to which members of a culture rely on and have loyalty to either their self-interest or the collective interest. The term masculinity refers to a culture in which social gender roles are distinct, whereas femininity refers to a culture where there is an overlap between social gender roles. Several studies have tested the relationship between these four cultural dimensions and IT adoption (Png, Tan, and Wee 2001; Pfeil, Zaphiris, and Ang 2006).

Png, Tan, and Wee (2001) show that uncertainty avoidance negatively impacts the adoption of the frame relay technology by businesses. In addition, the authors show that power distance is not significantly correlated with adoption of this technology. A study by Pfeil, Zaphiris, and Ang

(2006) revealed cultural differences in the style of contributions to Wikipedia according to the cultural dimensions identified by Hofstede. Straub (1994) noted significant differences between knowledge workers in America and Japan with respect to their perception about the usefulness and use of e-mail and fax services. He relates these differences to the uncertainty avoidance. We will discuss how uncertainty avoidance is a challenge facing the adoption of ASANA e-services. For an in-depth critique of research on the relationship between information systems and culture, we refer readers to Mayer and Tan (2003).

Dadashzadeh (2002) points out that technology is ethnocentric, or culturally, biased as it favors the social context and the culture where it is developed. As a result, developing societies face obstacles when they try to bring in new technology into practice at home, according to Dadashzadeh. There is evidence showing a positive relationship between the use of information technology and the following work practices: the existence of a self-directed work teams, more authority of individuals, and increased investments in training and education (Brynjolfsson and Hitt 1998). Teng and Nelson (1996) investigate a number of organizational and human factors that may influence the adoption of the CASE technology by organizations. The investigated organizational factors are centralization, formalization, size, and professionalism. Among these factors authors found that the existence of a champion, a human factor, who enthusiastically promote the new innovation positively impact the adoption of innovation. One of our research focuses is exploring the relationship between leadership and centralization of the organization in hand and technology adoption.

Another line of research relates the adoption of IT innovations to economic factors such as productivity (Brynjolfsson and Hitt 1998). One of the benefits of the Internet is its ability to accrue social capital (Abdelaal, Ali, and Khazanchi 2008). We treat ASANA as a value network and explore the productivity of its e-services by indentifying the values generated, shared, and/or transferred among its members. A value network is defined as a group of people who work together via relationships to create public goods or economic values (Allee 2008).

THE CASE OF ASANA

Most Arab students attending educational institutions in North America do not get adequate exposure to the American culture and educational system before they move there. This leaves them vulnerable to academic, cultural, legal, and social problems. ASANA was established in 1971 as a non-profit organization to virtually integrate and support its members in this new environment. ASANA members, mostly graduate students, have moved from one of the Arab countries to North America to pursue graduate studies. Their membership exceeded 1,500 in 2008. They are distributed in more than 50 units all over Canadian and American universities. The majority of these graduate students are funded by the Ministry of Higher Education and supervised by the Cultural and Educational Bureau (CEB) of their country of origin. Its membership is open not only for their fellow academics but also for all their countrymen living in North America. Non students, however, can only have an honorary membership.

ASANA has elucidated the following four goals:

1. Syndication goal: integrating and supporting its members living in North America. This includes providing them with necessary social, financial, legal, and academic support.
2. Propagation goal: introducing the true Arabic and Islamic culture to the American society.
3. Scientific goal: addressing the academic deficiencies of their country of origin.
4. Political goal: expressing the opinions of its members regarding ASANA, national and international affairs.

ASANA Management

ASANA has federal and regional administrative hierarchies. The federal hierarchy includes the General Assembly (GA) and the Executive Committee (EC). The GA is composed of presidents of local chapters who convene once a year to discuss the performance and challenges facing ASANA and its members. It is the highest authority according to its constitution. The EC is the governing body and is elected on an annual basis during the GA meeting. It is composed of the president, vice president, treasurer, social officer, science officer, unit affairs officer and media officer. The president represents ASANA in front of other organizations, manages its meetings, implements the EC decisions, and informs member about its decisions. The vise president assists the president, substitutes for him as necessary, manages ASANA projects, compiles quarter reports, and improves ASANA resources. The media officer manages media activities such as newspapers, mobile exhibitions, media campaigns, and the national day. The unit officer manages ASANA chapters. The science officer manages and promotes the scientific activities of ASANA. The treasurer manages its finances.

Regarding finances, ASANA has limited internal and external sources of funding. The Ministry of Higher Education of their home country sponsors the GA and pays for all related expenses. This includes travel tickets, hotel reservations, food and local transportation. External funds also include a contribution of $3000 from the CEBs of their country in Washington D.C. and Canada. Another external fund comes from domestic universities which pay for the shipping expenses of the received books collected by Book Campaign project. Internal funds, however, come as a membership fee of $12 for single members and $18 for families. As with most non-profit organizations, ASANA lacks stable financial resources and it is always struggling to guarantee fund for the GA and finding a sponsor for its Book Campaign project. Therefore, the regular and irregular voluntary contributions of members and donors are also a significant resource for such organizations.

The Adopted Information Technology

Internet-based communication systems give participants, of this type of organization, the opportunity to work together in a virtual workplace and participate in shared tasks and projects regardless of the organizational structure and geographical locations of members. In addition, they could collectively exchange information, edit documents, communicate with each other, and share experience. These resources and facilities help organizations to successfully achieve their goals in a timely manner.

ASANA adopts a wide range of online resources and Internet-based communication tools to perform its activities. For instance, members use email to interact and communicate with each other. In addition, they use online conferencing, Wikis and VoIP. There is also an electronic payment system used to collect membership fees and reimburse members. Moreover, members use online functions for registration, accessing scientific resources, submitting reimbursement forms and more. Furthermore, ASANA publishes meeting minutes, guidelines, performance reports and other related documents on its webpage to make it publicly available to all members. It also runs an online poll so that members can participate and vote on political and community issues. To achieve its scientific objectives, ASANA utilizes online resources such as scientific references, open-source software, simulators, information and knowledge.

The usefulness of these online resources and Internet facilities stems from the following factors:

1. ASANA is a geographically dispersed and loosely structured socio-academic organization. Therefore, face-to-face communication, in this case, is very costly if

not impossible. As a result, the Internet is the only viable and practical innovation to consolidate these distributed groups into one single organization.

2. Unlike traditional academic communities, ASANA provides a wide range of academic, scientific, social, cultural and political services. Delivering these different services requires a broad spectrum of resources and telecommunication facilities.

3. ASANA lacks financial resources and; therefore, free online resources and communication tools are important assets for ASANA.

Certainly, it is difficult to imagine how ASANA could integrate its members and achieve its objectives without Internet applications and online resources. The following is more details on its services delivered through the Internet:

ASANA Services

ASANA provides a number of services and e-services to its members and other irregular beneficiaries. The term *e-services* refers to the transformation of services into digital media and transport them through electronic networks for the purpose of driving revenue streams or improving efficiency (Henten 2001).

The following is a discussion of ASANA's major services and e-services facilitated through the Internet:

- **Research Support:** The target audience of this project is the researchers at universities in their country of origin. It guides them to useful research materials such as scientific references, simulators and open-source software available on the Internet. In addition, the project also facilitates a number of research clubs and blogs that discuss different research areas in North American Universities. Blogs, a Web 2.0 technol-

ogy, are characterized of ease of use, rapid deployment, efficient information sharing, and minimal technical requirements (Parker and Chao 2007). Perceived ease of use and usefulness are main reasons for technology adoption as discussed before. However, these blogs and online research clubs have failed to attract a significant number of members for reasons we will discuss later. The Research Support project also directs scientists at their native universities to their peers in North America. Moreover, it publishes a quarterly newsletter about the scientific achievements of ASANA members and native scholars.

- **The Link Project:** The Link project was found in 1998 for the purpose of linking academics in North America (students and faculty members) with their fellow researchers back home. In other words, it facilitates collaboration and knowledge sharing among them. It also provides services to native students pursuing a degree from North America. These services include providing estimation of living expenses in North America. In addition, it helps newcomers find suitable accommodations before their arrival. The link project also arranges for pickup service from airports for newly arrived students. Moreover, it publishes a number of booklets on its website that provide guidelines and detailed directions on common issues. Furthermore, it invites a native scholar every year to attend the annual meeting and provide his/her insights into topics of interest. These topics may address the American educational system, culture shocks, scientific progress, and the major problems facing their home country.

- **Study in America:** The Study in North America project assists those who are seeking admission to North American universities. It provides them information about graduate programs scholarships. It also helps them develop competent applications, resumes and statement of intent. In addition, it maintains a database of academic advisors, departments, and schools recommended by members. This database is available to members upon request after approval from the governing body.

- **Social Fund:** ASANA provides members with limited financial assistance through the Social Fund project. For instance, members are entitled to 30% of the unpaid medical bills and $50 for a newly born baby. The EC also facilitates online fundraising for a particular member in emergency cases such as death or financial difficulties. For instance, a member lost his scholarship and got laid of his job after the severe financial crises that hit the globe in 2009. ASANA called for a fundraise campaign that collected around $10,000 for him.
- **The National Day:** ASANA members have the potential to serve as peacekeepers that cement friendship, install tolerance, acceptance, and mutual understanding between Arabs and Americans. Building such understanding is coming to the forefront of international relations particularly after September 11. For that purpose, the Media Officer manages the National Day event. This event is a social and cultural function where local units invite American students to discuss Arab culture, Islamic beliefs, traditions and history. For example, ASANA chapters organized such events at Rice University in 2008 on March 21st, Purdue University on April 6th, West Virginia University on August 15th, and Penn State University on April 10th.
- **The Book Campaign:** The purpose of the Book Campaign project is to supply the libraries at native universities with updated textbooks and scientific references. Through this project, ASANA members collect donated books, references and scientific periodicals from all disciplines and ship them to one of the native universities. They select a university each year to receive the collected materials and charge it for the shipping expenses.

In addition to these projects and activities, ASANA publishes a magazine that serves as the voice of its members in North America.

THE USEFULNESS OF ASANA SERVICES

ASANA serves are best described as a value network, or a network of aid, where members develop, provide, obtain, share and/or broker a wide range of social, financial, cultural, scientific, political, and intellectual values. A value network is a social structure where social, technical and/or intellectual resources are produced, reproduced, shared or brokered among members. These values are accrued by ASANA members, its society back home and/or the American society. ASANA also serves as a venue where academic resources such as used textbooks and scientific journals are recycled or shared among members. We can also describe ASANA as a *community of practice* (CoP). A CoP is usually formed for the purpose of sharing knowledge, creating innovative ventures or solving problems. Its members can openly brainstorm projects, share ideas, and discuss problems which can lead to new capabilities (Dalkir 2005).

The following is a summary of the values produced or brokered within ASANA as a value network or a CoP:

1. **Conserving resources:** The success of an organization generally depends on its ability to provide consumers with more real values without using extra labor, capital or other inputs (Brynjolfsson and Hitt 1998). ASANA

saves its members, and other stakeholders, a lot of time money and efforts. This is because it provides them access to information about housing options, advisors, research directions, scholarships and academic programs in North America. For example, one member may share his/her experience on solving a particular problem. This specific type of information may not be easily obtained from books, online resources, or academic advisor. Information sharing may help member avoid mistakes, increase learning, and improve their academic achievements. In addition, these services could be accessed by members from anywhere and anytime. ASANA also assists the CEB staff to do their job.

2. **Building bridges:** ASANA activities build bridges and mutual understanding between Arabs and the American society. In other words, this highly intellectual group serves as culture brokers who create opportunities and arrange activities where Americans can gain a close understanding of the Arab culture, traditions and beliefs. They also build bridges and collaborate with scholars who live in North America. Other organizations could also build partnerships with ASANA to achieve their own objectives.

3. **Generating social capital:** one of the key functionalities of ASANA is integrating and supporting its members. While do so, its members solicit feedback from volunteers and community members; hold discussion forums and awareness events; plan and coordinate logistic activities; and manage volunteers. These activities are venues for social capital building. The concept of social capital refers to the resources and values embedded in social networks that could be mobilized to achieve collective actions. The type of information that could be shared and the number of problems that could be solved in ASANA are boundless. Scholars consider the knowledge shared and learned within social structures as a form of social capital. ASANA members communicate at various levels without the constraints of geographical barriers or time zone. They communicate internally, within their local unit, and externally, across states, to share their expertise and learn from each other.

4. **Producing intellectual capital:** intellectual capital is defined as the collection of intangible assets that utilize human intellect and innovation to create wealth (Johnson 1999). This includes know how, innovative ideas, leadership, expertise, intellectual properties, and internal relationships. ASANA members transfer knowledge to their home country.

5. **Building the capacity of members:** ASANA also serves as a site for building the human capacity of its members. They get the opportunity to build communication, leadership, political, problem solving and personal skills. They also obtain practical experience related to American ideals such as democracy, freedom, gender equity, openness, diversity and tolerance. This practical exposure would build their personal, professional, and political capacity. It would also improve their personal, academic and career growth.

CURRENT CHALLENGES

Most ASANA services, except the Book Campaign, are not widely accepted by academics from their home country. For instance, the Research Support project received only 11 requests for assistance in year 2008. In addition, the link project received only 11 requests for assistance in the same year. The Study in North America project handled only 11 requests. The vice president wrote 4500 emails assisting members. Given the capacity of ASANA and the number of its members, this is a rather poor performance. The following are our observations on the challenges facing the adoption of ASANA services:

Inadequate Organizational Culture

ASANA does not adhere to a well-defined organizational culture or institutional rules that govern its activities and services. For instance, there are no rules to guide who has the right to send information and how this information could be directed, shared and interpreted. The the culture of an organization encompasses the values, beliefs, norms, symbols, language, rituals and myths shared among its members (Sannwald 2000). Understanding the organizational culture helps us to comprehend how this organization evolves and identify reasons of failure. Successful online organizations, in particular, should have a distinctive culture that guides behavior and give meaning to membership. It also helps leaders to determine where and how changes of culture should occur. Virtual communities, as with ASANA, usually do not have a single monolithic culture; instead, they have ecosystems of subcultures (Rheingold 2000:pp xviii).

The main cultural concern with ASANA is the lack of trust towards outsiders. The term outsiders, in this context, refers to fellow scientists who are not funded by the Ministry of Higher Education of their country or supervised by the CEB. For instance, the science officer refused to forward an email from a fellow researcher who was collecting data through a survey for his research. He refused the researcher's request claiming security and uncertainty reasons. Another student had a problem with law enforcement agency in the U.S.A. Instead of supporting him, ASANA leaders acted unprofessionally and asked him to leave ASANA to avoid any trouble with this agency. Avoiding uncertainty is one of the cultural dimensions that negatively affects IT adoption within a specific organization (Hofstede 1980). Professionalism is another factor that affects technology adoption (Teng and Nelson 1996). In addition, ASANA does not have a strict privacy policy and there was a big problem when leaders violated the privacy of one member. ASANA deals with social, financial, and legal issues of problems and protecting the privacy of members while addressing these problems is important to gain their trust. The existence of a self-directed work team with more authority for individuals increases the adoption of IT innovations (Brynjolfsson and Hitt 1998). In this regard, the EC have the tendency to coordinate with, and implement the directions of, the CEB rather than considering the feedback of members. For example, ASANA runs a poll so that members can vote on political, academic and community issues. The CEB advised them to avoid political issues. Once more, most of these graduate students are supervised by the CEB. We observed that lack of professionalism; interventions of the CEB and uncertainty avoidance have negative impact on the adoption of ASANA's e-services.

Assessing the Usefulness of its Services

ASANA generate and broker tangible and intangible outcomes. The tangible ones include money and used books. Intangibles include knowledge, software, leadership, ideas, and experience, reputation, trust, social capital and intellectual capital. Tangibles, particularly books, could be evaluated according to their opportunity cost. For instance, the manger of the Book Campaign project assesses the collected books in year 2009 by more than one million dollar. However, it is difficult to convert intangible assets into negotiable forms of values (Allee 2008). Therefore, ASANA should find a way to evaluate the values of the delivered and produced intangibles in terms of marketable values in order to engage and attract a broad spectrum of beneficiaries.

Poor Incentive System

The fact that this non-profit organization does not provide adequate reward to contributors for their contributions (e.g. time, money, efforts, and ideas) makes participation less favorable. During the

2009 elections, no one was interested in running for treasurer because this position is time and effort consuming. Instead, most of competitors were interested in being the president or vice president for the recognition they obtain. We believe that the reason for that could be the lack of incentive system necessary to motivate a wide range of individuals to contribute and/or benefit from ASANA services. Adopting an adequate incentive system is important for effective online scientific cooperation (Forte and Bruckman 2008). This is because incentives explicitly attribute the contributions of participants. The provided incentives should accommodate diverse facets of motivations in order to sustain their contribution. The CEB gives awards to EC members and project managers during the annual meeting. Unfortunately, it does not provide units' leaders or regular members any awards.

Instability of Leadership

ASANA's governing body changes frequently as elections run every year to select EC members, project managers and units' presidents. The success of online organizations usually requires a stable executive champion or leader who has the capacity to make the project. Prior research shows that the existence of a motivated and motivating leader has positive impact on the adoption of these online services (Teng and Nelson 1996). This is because leaders plot the directions of the organization and craft its online culture.

SOLUTIONS AND RECOMMENDATIONS

ASANA needs to implement a number of managerial, cultural, and technical changes to improve the adoption of its e-services. The following are some suggestions that may help in this regard:

1. **Improving the quality of service:** email service is the most common tool used by ASANA to deliver its services and interact with its members. However, relying only on emails and Internet communication tools is not sufficient in this mobile age. Mobile phones have the capabilities to provide ubiquitous assistance. In the digital economy, the value of a product or a service depends increasingly on its quality, timeliness, convenience, customization, and other intangibles (Brynjolfsson and Hitt 1998). While email service is an asynchronized communication tool, mobile phone is a synchronized communication one. Therefore, ASANA should provide support by mobile phone to assist those who have limited Internet access such as travelers and mobile individuals. In addition, it needs to develop more online functions that enable members to browse popular ideas, gather feedback, vote on common issues, add comments, submit new ideas, and promote the overall functionality of its website. They may adopt emerging social media such as Facebook and Linkedin to achieve this task. It can also post video streams online to effectively present common issues and promote the Arabic culture. They could also develop a video clip that provides orientations about the American culture and educational system and post it on their website. These are some facilities or informal ways through which all members can contribute, benefit, and learn.

2. **Crafting a novel strategy:** ASANA should adopt an organization strategy that foretells how it can grow, engage more members, detect future problems, and identify resources required to deliver its services. This strategy should include an online quality assurance or a feedback mechanism. Feedback is important to dynamically reconstruct its goals based on previous experience or suggestions from members. In addition, the adopted strategy should find online and offline instruments to promote ASANA's

services. One of these ways could be building a win-win partnership with entities that may benefit or contribute to its mission. These entities may include American academic institutions, developing agencies, and civil society. This partnership would enable members to benefit from the resources and experience of these organizations. For that purpose, ASANA alumni, who graduated from American universities, have opened a chapter in their home country to expand its base. One of the key missions of the new unit is to provide newcomers orientations about the American culture and educational system. Such preparatory program would help them to avoid culture shocks, reduce problems, and increase the possibility of their success. Furthermore, ASANA leaders should also realize that engaging a wide range of members in its services may need programs, events and activities that respond to their needs. For example, they may organize more seminars, debates, meetings, and shows about the scientific developments and academic life in North America. They can also arrange summer camps with American students for the purpose of sparking conversations, building mutual understanding and cementing friendship between Arabs and Americans.

3. **Installing an attractive incentive system:** another way through which ASANA can engage more participants is adopting an attractive incentive system. This system should satisfy their different needs such as social satisfaction, success, security, recognition and expectations. One of these needs is protecting their security and safety which have grown to be a big concern after September 11. The provided incentives may also include financial benefits, award certificates and/or recommendation letters that honor their contributions to the community.

4. **Adopting a suitable IT culture:** ASANA also lacks an IT culture that identifies shared values and rules. Therefore, crafting a well-defined IT culture is necessary for the progress of ASANA and adoption of its online services. There is also a need for standardized best practices and learning lessons related to the use of these services. This culture is important for driving its online functions and objectives.

5. **Reaching out:** one more suggestion that may promote the adoption of ASANA e-services is to increase the awareness among native citizens, and other stakeholders, of the usefulness of its services. The author of this chapter has published two newspaper articles about ASANA services in a major Arab newspaper to achieve this goal.

These are some suggested organizational, technological and cultural changes needed to increase the adoption of ASANA services.

RESEARCH IMPLICATIONS AND FUTURE DIRECTIONS

Ubiquitous computing and pervasive communications have grown to be crucial instruments to conduct all aspects of our daily life affaires. Yet, the Internet penetration rate in the Arab world was 28% in year 2009 (Internet World State 2009). Therefore, there is a need for ways to motivate Arabs to use the Internet and benefit from its opportunities.

We can elicit a number of implications for policy makers, student leaders, academic reformers, and scientist from this case. This case directs our attention to the social, academic, intellectual, cultural, and economic values that may be accrued from scientific online resources. It also draws our attention to the capacity of academic communities and student organizations as *knowledge brokers*. They may include student leaders, volunteers, academic advocates, educational reformers, and intellectual anchors. The importance of these knowledge brokers stems from the fact that they are rooted in academia and the society at large. In addition, their careers will reshape the future of their societies. Moreover, these resourceful individuals are more aware of the immediate challenges, available resources, capacity and common needs of academics. Therefore, fostering and supporting these creative communities is important for the overall progress of the Arab society.

This chapter also provides detailed information about the problems that Arab students may face while studying overseas; and how collective actions can assist them to tackle these problems. We also hope that the discussed martial would expand the problem solving abilities of academic practitioners, student leaders, and educational reformers. We can also conclude that exchange students, if prepared well, can play as bridges between the West and the Arab World. Another important implication is that online academic communities could serve as venues for social capital creation. This concept is so intellectual that it captures key moral and social factors of an organization (Abdelaal and Ali 2008). Arabs, in particular, should invest in social capital because of its ability to alleviate poverty and sustain human and economic development.

The topic of technology transfer and adoption by Arab organizations is a rich research area because of the complexity of the Arab culture and social setting. For the scientific community, this chapter provides a rich set of conceptual insights and theoretical details that could guide current and future related research. Our work could be extended to empirically replicate the TAM model with respect to ASANA e-services. In addition, practitioners or researchers may devise a method to measure the values, or usefulness, accrued for academicians from adopting Internet services. One way is to measure the intangibles produced or brokered within ASANA using the social network analysis approach. This analytical approach uses the network theory to study the structures, relationships, and the transactions between members of social systems at all scales (Peter, Scott, and Wasserman 2004). In particular, it could be used to measure intellectual capital, social capital, expertise, trust, and reputation built in the domain of ASANA. This problem could be diagnosed also as a productivity problem as noted before. In other words, researchers may investigate the productivity of the Internet in the domain of online academic communities. It is important to note that the topic of IT productivity has sparked a big debate among scholars (Brynjolfsson and Hitt 1998). It would be also interesting to collect data and empirically examine the cultural factors that affect the adoption of IT by Arab academic groups. Another interesting topic that deserves the attention of the scientific community is exploring the incentives that should be provided to members and business partners in order for them to participate. Our future work will use the social network analysis approach to categorize the stakeholders of ASANA based on their attributes, roles, contributions, and benefits.

CONCLUSION

This chapter shows that the Internet has become an effective virtual workplace for academic communities to prosper and participate in the overall socioeconomic development of their societies. In particular, it communicates the capacity of online academic communities in conserving resources, accruing intellectual capital, building social capital, and bridging with the international society. In so doing, we described ASANA as a socio-academic network and articulated on its structure, e-services, and used telecommunication infrastructure to deliver these e-services. We also highlighted the key challenges ASANA faces and demonstrated suggestions for improvement.

We also showed that there are tremendous untapped values embedded in online academic communities. In addition, we suggested that scientists should develop efficient ways to measure and assess these tremendous untapped values. We hope that forward thinking Arab academics would follow the footsteps of ASANA and adopt online resources and electronic facilities to achieve collective actions. We also hope that this chapter brings the attention of practitioners and academics to the capabilities of online communities as a vehicle for social, political, economic, and educational reform. Finally, there is a lot more to be done to improve the performance and sustainability of such knowledge and culture brokers. Their growth and prosperity need the support of governments, developing agencies, foundations, businesses and the society at large.

REFERENCES

Abdelaal, A., Ali, H., & Khazanchi, D. (2009). The Role of Social Capital in the Creation of Community Wireless Networks. *The 42nd Hawai'i International Conference on Systems Sciences*, Waikoloa, HI, January 5-8.

Agarwal, R. (1999). *Individual Acceptance of Information Technologies*. Retrieved March 2010, from http://www.c4ads.seas.gwu.edu/classes/CSci285_Fall_2004/readings/9/required/Agarwal_02.pdf

Allee, V. (2008). Value Network Analysis and Value Conversion of Tangible and Intangible Assets. *Journal of Intellectual Capital*, *9*(1), 5–24. doi:10.1108/14691930810845777

Barras, R. (1984). Towards a theory of innovation in services. *Research Policy*, *15*, 161–173. doi:10.1016/0048-7333(86)90012-0

Brynjolfsson, E., & Hitt, L. (1998). Beyond the Productivity Paradox: Computers are the Catalyst for Bigger Changes, *the Communications of the ACM, 41*(8), 49-56.

Coleman, J. (1988). Social Capital in the Creation of Human Capital. *American Journal of Sociology*, V94.

Dadashzadeh, M. (2002). *Information Technology Management in Developing Countries*, IRM press.

Davis, F. D. (1989). Perceived usefulness, perceived ease of use, and user acceptance of information technology. *Management Information Systems Quarterly*, *13*(3), 319–340. doi:10.2307/249008

Forte, A., & Bruckman, A. (2008). Why Do People Write for Wikipedia? Incentives to Contribute to Open-Content Publishing. In *Proceedings of 41st Annual Hawaii International Conference on Systems Sciences,* Waikoloa, HI.

Hage, J., & Dewar, R. (1973). Elite Values and versus Organizational Structure in Predicting innovation. *Administrative Science Quarterly*, *18*, 279–290. doi:10.2307/2391664

Henten, A. (2001). *Services, E-Services, and Nonservices: Cases on Managing E-Services*, edited by Ada Scupola, (pp. 1-9), 2008.

Hofheinz, A. (2005). The Internet in the Arab World: Playground for Political Liberalization, *International. Politics & Society*, (3): 78–96.

Hofstede, G. (1980). *Culture's Consequences: International Differences in Work-Related Values*. Newbury Park, CA: Sage.

Internet World State. (2009). *Internet World State*. Retrieved on January 28th, 2019, from http://www.internetwd eorldstats.com/stats5.htm

Johnson, W. H. (1999). An Integrative Taxonomy of Intellectual Capital: Measuring the Stock and Flow of Intellectual Capital Components in the Firm. *International Journal of Technology Management, 18*, (5/6/7/8), 1999.

Kluckhohn, C. (1962). *Culture and Behavior*. New York: The Free Press of Glencoe.

Magrassi, P. (2002). A Taxonomy of Intellectual Capital. *Research Note COM*-17-1985.

Mayer, M. D., & Tan, F. B. (2003). Beyond Models of National Culture in Information System Research, *Journal of Global Information Management*, January 01, 2002.

Parker, K., & Chao, J. (2007). Wiki as a Teaching Tool, *Interdisciplinary Journal of Knowledge and Learning Objects* V 3.

Peter, C., Scott, J., & Wasserman, S. (2004). *Models and Methods in Social Network Analysis*. Cambridge, UK: Cambridge University Press.

Pfeil, U., Zaphiris, P., & Ang, C. S. (2006). Cultural differences in collaborative authoring of Wikipedia. *Journal of Computer-Mediated Communication, 12*(1), article 5.

Png, I., Tan, B., & Wee, K. (2001). Dimensions of National Culture and Corporate Adoption of IT Infrastructure. *IEEE Transactions on Engineering Management, 48*(1). doi:10.1109/17.913164

Rheingold, H. (2000). *Virtual Community*. Cambridge, MA: The MIT Press.

Sannwald, W. (2000). Understanding Organizational Culture. *Library Administration & Management, 14*(1), 8–14.

Segars, A. H., & Grover, V. (1993). Re-examining Perceived Ease of Use and Usefulness: A Confirmatory Factor Analysis. *Management Information Systems Quarterly, 17*, 517–525. doi:10.2307/249590

Straub, D. W. (1994). The Effect of Culture on IT Diffusion: E-mail and fax in Japan and the US. *Information Systems Research, 5*, 23–24. doi:10.1287/isre.5.1.23

Teng & Nelson. (1996). The Influence of Organizational Factors on CASE Technology Adoption. *Journal of Information Technology Management, VII, 1*(2).

Tornatzky, L. G., & Klein, R. J. (1982). Innovation Characteristics and Innovation Adoption-Implementation: A meta-analysis of Findings. *IEEE Transactions on Engineering Management, EM-29*, 28–45.

Woolcock, M. (1998). Social Capital and Economic Development: Toward a Theoretical Synthesis and Policy Framework. *Theory and Society, 27*(2), 151–208. doi:10.1023/A:1006884930135

Zakour, A. B. (2004). Cultural Differences and Information Technology Acceptance. In *Proceedings of the 7th Annual Conference of the Southern Association for Information Systems*.

ENDNOTES

This study is based on participatory research conducted by the author in ASANA. The author's role was the president of a local unit for two years and as a manager of the Research Support project for a year.

[1] The real name of the association and its country of origin are concealed for the purpose of anonymity

Chapter 15
Social Software and E-Governments in the Arab World

Salam Abdallah
Abu Dhabi University, UAE

Ashraf Khalil
Abu Dhabi University, UAE

ABSTRACT

The emergence of Web 2.0 social software is changing the social life of people and beginning to infiltrate and impact governments in the Western World. There, the emergence of a new government paradigm is challenging the traditional governments and governance. A shift is at this point required for governments to stay in line with their citizens, who are becoming more computer savvy and being raised and fed on the internet and mobile devises. We identify for specific areas in which Web 2.0 applications are being used innovatively in Western governments. The paper then turns its focus to the Arab World and puts forward thoughts on the potential, opportunities and challenges for both citizens and governments for embracing this new wave of web-based services. We conclude that Web 2.0 applications have great potential to leverage the mission of Arab e-governments through connecting and collaborating with businesses, educational institutions, private and non-government initiatives, and of course individual citizens, many of whom are already beginning to embrace social media.

INTRODUCTION

E-government is the application of Information and Communication Technology (ICT) by government agencies with the intention of enhancing its services to the public and at the same increasing the level of government transparency to encourage people's participation. The emergence of social software tools, is changing the social life of people and beginning to infiltrate and impact governments in the Western World. Businesses have already tapped into social media as a mean for providing new touch points for customer interaction, which is already improving customer satisfaction and giving them leads for new product development and marketing decisions (Aberdeen, 2008). Now discussions are beginning to appear about the emergence of a new government paradigm known

DOI: 10.4018/978-1-60960-048-8.ch015

as "government 2.0". The paradigm challenges the traditional governments and governance. A shift would be required for governments to stay in line with their citizens, who are becoming more computer savvy and being raised and fed on the internet and mobile devises. Social software applications are changing the way we are socializing and learning – it is a cultural change. Government officials need to understand this change, which can hold a great potential for involving citizens to proactively and creatively participate in certain government and community activities. This paper advocates the use of social software tools based on the Web 2.0 philosophy and technologies in the context of e-government websites. These applications can provide opportunities for government to harness the talent and technological skills of citizens who are willing to participate in government and community initiatives. It is an opportunity for government officials and citizens to cooperate and learn from each other.

In this paper we consider the usefulness of Web 2.0 philosophy and technologies as applied to government operations and services with particular focus in the Arab World. We first provide a brief introduction to e-governments, Web 2.0 and then identify major domains of usage by governments in the Western World, including the potential risks and challenges faced when implementing Web 2.0 services. This paper goes on to discuss the potential of Web 2.0 based social software in e-government in the context of Arab World governments and provides a discussion on the readiness to embrace this new wave of web-based services in the Arab World for both citizens and governments.

E-GOVERNMENTS

The concept of e-government is founded on the basis of leveraging information technology in all its forms to foster relations with citizens and businesses and across government agencies (Cloete,

2003). Technologies have been used by governments in a variety of ways to improve services, increase participation and empower citizens by involvement in the policy making process (OCDE, 2003). Governments aim to transform the way they interact with citizens by enhancing transparency, communication and participation (Pascual, 2003; Korac-kakabadse et. al., 2002). Such initiative is sometimes referred to as e-democracy. In turn, according to the World Bank, governments who provide access to information and encourage engagement may lead to "less corruption, increased transparency, greater convenience, revenue growth, and/or cost reductions (World Bank, n.d.)."

In Western countries e-government services had evolved from being only source of information publishing to fully interactive and transactional services. E-government literature usually focuses on goals or values that e-government services produce (Mohamed, 2008) such as efficiency, effectiveness, access, accountability, equity, empowerment/participation, transparency, availability of services, responsiveness and integrity.

Thus far, research has indicated that the adoption level of e-government is still considered low. Some of the common variables affecting the rate of adoption is trust in e-government (Warkentin et al., 2002) and culture, as well as other factors such as perceived ease of use and usefulness. Trust in e-government is seen as an essential enabler of adoption of e-governments (Belanger et al., 2008). The term e-democracy is even more encompassing term that examines trust of government, political party affiliation, and e-citizens who have political agenda and have desired to make changes in government policies.

Despite the fact that traditional e-government services have lately grown significantly, citizens are slow and reluctant to use the services. Only a small percentage of Internet users are currently making use of them. Even in countries where Internet usage is common, the numbers of e-government users are far from satisfying (Wauters et al., 2008). Capgemini reported that only a little over 10% of European citizens use Internet-based public services at a transactional level. Governments in both the western and eastern world have acknowledged that there is ample room for improvement to increase participation and to move more towards e-democracy (Wauters et al., 2008; Al-Fakhri et. al, 2007; Gilbert, 2004).

Governments' attention has been towards promoting awareness programs and to include services that will increase citizens' participation such as the utilization of Web 2.0 applications as illustrated later in this document. Web 2.0 applications, have achieved drastic success in terms of take-up, i.e. the number of Internet users who utilize these applications, even though these services came much later than e-government services (Pascu, 2008). We argue that employing Web 2.0 technology as an integral part of e-government services may provide the right catalyst for generating the level of popularity witnessed by Web 2.0 services. Web 2.0 is an emergent phenomenon that is creating a paradigm shift that is reshaping and reconfiguring society (Zappen et. al., 2008). The next section describes the Web 2.0 environment.

WEB 2.0

Web 2.0 is a term that was coined and promoted by Tim O'Reilly in his Web 2.0 conference in 2004. The name suggests an evolved or refined version of the World Wide Web of the 1990s. Web 2.0 involves a new generation of sites that encourage collaboration, creativity and connectivity. The most common examples of Web 2.0 sites are YouTube, Wikipedia, Flickr, Facebook, LinkedIn, and Myspace. One example of Web 2.0 services is a mash-up service which combines data from one website with data from another site, which is then presented in a third site at which the user can manipulate and control how the data is combined and presented. Web 2.0, with its various sites and services, is a platform for bringing together user-generated content as well as more dynamic and interactive websites on a large scale (O'Reilly, 2005).

Web 2.0 is composed of three major components, namely technologies, applications, and values (O'Reilly, 2005). From the technology point of view, the building blocks of Web 2.0 are Ajax, CSS, XML, Open API, and RSS. Ajax, the most prominent technology, is an acronym for "Asynchronous JavaScript and XML" which enables more dynamic, efficient and natural interaction between users and services (Garrett, 2005). Based on these building blocks many new applications have been developed, such as wikis, blogs, podcasts, tagging (del.icio.us), RSS feeds, massive multiplayer online games (Second Life), and social networks (Facebook, LinkedIn).

Even though technology and applications play a major role without which Web 2.0 would not exist as we know it, the true core building blocks of Web 2.0 are not technological but philosophical and values-based. O'Reilly describes Web 2.0 as having an "architecture of participation, a built-in ethics of cooperation, in which the service acts primarily as an intelligent broker, connecting the edges to each other and harnessing the power of the users themselves" (O'Reilly, 2005). All Web 2.0 applications share the same values. They build on the knowledge of the users and they improve with an increase in users. The total value of users' contributions is greater than the sum of each individual contribution, which is referred to as Collective Intelligence.

As for the users of Web 2.0 applications, Young and Rainie identified three major types of Web 2.0 users with different levels of involvement (Young, 2007; Rainie, 2007). The first type are the core users of Web 2.0 who are credited with producing fundamental Web 2.0 content such as blogs, Wikipedia articles, YouTube videos, and photos on Flickr. This group is generally younger and more tech-savvy. The size of this group is usually small and tends to be made up of users from the United States. An estimated 13 percent of bloggers hail from the U.S., while only 3 percent write from European countries (Young, 2007; Rainie, 2007).

The second identified type of Web 2.0 users represents active users who participate by rating products and leaving feedback, comments, and reviews of existing content. Examples of this kind of contribution include posting a web review about a recent stay in a hotel, or writing a book review at Amazon.com. This group of users is estimated to include 10%-20% of the total users of the Internet (Young, 2007; Rainie, 2007).

The third type is represented by active readers of Web 2.0 content. This group of users read blogs and reviews, view photo-sharing sites and watch YouTube movies produced by the first two groups of users. These users may read articles on Wikipedia but not contribute information or read book reviews at Amazon.com before buying a book. The size of this group of users is estimated to be around 40% of the total Internet users (Young, 2007; Rainie, 2007).

WEB 2.0 DOMAINS FOR E-GOVERNMENT SERVICES IN WESTERN CONTEXT

Pascu and Osimo et al. (2008) have identified two major categories of domains in which Web 2.0 can be utilized in the government context, namely "Back Office Domains" and "Front Office Domains". The front office domains are usually the services that address citizen-government relations whereas the back office domains represent internal government operations or address inter-governmental organizations. In this section, we list the main domains where Web 2.0 services can be utilized in the e-Government. The list of domains is not comprehensive by any means as Web 2.0 philosophy is built on change, creativity and challenge. Thus, we are confident to assert that Web 2.0 services will be used in ways that no one could have imagined.

Regulation

Regulation refers to governmental control and monitoring of services including telecommunications, education and healthcare. Governments worldwide are shifting from a service-providing role toward a more regulatory role. Through this shift, governments can benefit greatly from direct and open engagement with ordinary citizens as well as experts. Accordingly, citizens are expected to assume a more active role in demanding certain regulations by the government.

An example that demonstrates how Web 2.0 can be used in regulation is the initiative led by New Zealand police. The aim of this initiative is to encourage the public to contribute toward drafting the new policing act. The initiative was implemented as a wiki that allows citizens to leave their suggestions, read others suggestions, and leave feedback and comments. This initiative was open for one week as a trial and resulted in thousands of site visits and a large collection of ideas and suggestions (New Zealand Police, n.d).

Collaboration

Government organizations are infamous for their lack of efficiency and effectiveness as many of the responsibilities are delegated among different institutional levels and departments. This is another important domain where Web 2.0 services can play a major role in fostering collaboration and exchange of information within government institutions and among different government branches. Wikis are the most popular Web 2.0 service for use in this domain as many governmental entities have set up internal wikis to enhance collaboration within the organization as well as to streamline internal policymaking across organizations. For example TiWiki is a collaborative website run by the Ministry of Education so that people from various agencies can collaborate and share ideas in the tertiary education sector (Tertiary Education, n.d).

Information Management and Knowledge Transfer

Governmental institutions can be viewed as institutes with a high density of knowledge and information. Information management and knowledge transfer are an integral part of modern organizational management with the goal of increasing efficiency and reducing the risk of duplications. The government of New Zealand is a leader in adapting many Web 2.0 tools in the area of knowledge transfer. Examples of these initiatives are:

- **Principals Electronic Network:** A service run by the New Zealand ministry of education to establish an interactive online community for all New Zealand primary and secondary school principals. The goal of the service is to provide principals with a space for sharing, discussing, reflecting, and facilitating learning from each other's experience and knowledge (The PEN community, n.d).
- **Research e-Labs:** A blog for ICT specialists working for the New Zealand government to explore web trends, open-source software, and technology for the government. The aim of the site is to publish best practices and case studies so that the knowledge of different governmental organizations is shared, learned from, and not duplicated (Research e-labs, n.d.).

Public Engagement

As mentioned earlier, public engagement has long been a strategic goal for various governments. It is also one of the largest domains for Web 2.0 services and can be divided into two subsections as follows:

- **Political participation**: Web 2.0 services have been utilized for political agendas since their beginning and, according to Kohut (2008), is the one of the few areas where Web 2.0 impact is mature and visible. Blogs and YouTube have been playing a crucial role, for example, in the 2008 American presidential election. Senator Barak Obama has relied on social networks as a major pillar of his presidential campaign. His investment in social networks has enabled Senator Obama to reach to young voters in a very effective way that could not be achieved through traditional tools (webtrends, n.d). Web 2.0 tools are also hugely important for politics in France, where presidential candidates opened campaign headquarters in Second Life (3D virtual world). In the UK, E-petition is another example of how citizens are encouraged to participate in political life. E-petition is an initiative launched by the office of the British Prime Minster by which citizens can submit petitions directly to the government agency, or they can see and sign petitions submitted by others (mysociety, n.d).
- **Service Provision:** Web 2.0 services allow citizens to take a more active role in monitoring the delivery and quality of governmental services. The major role of service provision is usually undertaken by a civil society and activists often without government involvement. Below we list some endeavors that use Web 2.0 services in the domain of service provision:
  - Blogs, wikis and mash-up maps were widely used in the aftermath of Hurricane Katrina to organize the efforts of helping survivors and speed up the recovery process (katrinahelp, n.d).
  - Trackitt is a site that enables users to collaboratively track the progress of their immigration applications by sharing information with other applicants. This site is a direct response to

the lack of government transparency in immigration services and the extremely poor customer service usually faced by the applicants. Trackitt aims to offer greater visibility into the application process. The site states that "as more users share their case information, it reduces the uncertainty involved in the processing of their applications, and makes it easier to estimate when their applications will be processed." The site offers wikis, RSS, and analysis tools for users. This example shows how people can collaborate to improve and enhance a poor government service and lack of governmental transparency.

POTENTIAL RISKS AND CHALLENGES OF WEB 2.0

Zimmer (2008) argues that "Web 2.0 represents a blurring of the boundaries between Web users and producers, consumption and participation, authority and amateurism, play and work, data and the network, reality and virtuality." Indeed, Web 2.0 applications empower creativity through collaboration, but certain risks are involved such as exploitation and breach of privacy. Zimmer (2008) further points to "peer surveillance, the exploitation of free labor for commercial gain, and the fear of increased corporatization of online social and collaborative spaces and outputs" as risks of this new paradigm. In addition, social network sites' reliance on search engines in gathering personal information may pose a threat to informational privacy. Osimo (2008, p. 42) take a more focused approach on challenges that are applicable in the e-government context, such as:

- Low participation
- Participation restricted to an elite

- Low quality contributions and additional "noise"
- Loss of control due to excessive transparency
- Destructive behavior by users, e.g. at rate-myteachers.com
- Manipulation of content according to contributors' agendas, e.g. in the case of wikipedia
- Privacy issues, e.g. the publication of sensitive/official information

In addition to the above risks, another major challenge is the extra burden on the IT infrastructure. Web 2.0 services are demanding in this way, especially in the case of streaming video, audio and RSS feeds. Furthermore, users depend on broadband internet access. Government staff should become aware that adopting Web 2.0 services requires extra administrative duties. Finally, we need to determine if government staff and citizens are ready for productive conversation through this new medium and if they understand clearly the implication of privacy.

Factoring in these issues can help governments to manage potential risks when adopting Web 2.0 services and at the same time to enrich users' experience by facilitating greater participation.

E-GOVERNMENTS IN THE ARAB WORLD

In the recent years, the Arab World governments have introduced online presence as a way to increase performance and cut costs, which has led to proliferation of e-government websites. Many of these websites are stagnant at the information level and frequently non-functional and contain out of date information. The status of e-governments in the Arab World may be characterized as confusing, outdated, and overlapping and it seems that e-government is driven by technological capabilities rather than user/citizen demand. This slow development has been blamed on low computer literacy and lack of IT skills. The efforts of the United Arab Emirates, however, are usually hailed at their successes in the Arab region since some of their sites provide transaction services such as paying fines or licenses including e-tendering, but what is not clearly known is the level of citizens' participation.

In order to validate the level of e-government adoption of web 2 we took a cursory look at 50 e-government web sites representing numerous ministries in the Arab World including Saudi Arabia, United Arab Emirates, Lebanon, Tunisia, Morocco, Qatar, Jordan, Yemen, Kuwait, Algeria, Syria, Sudan, Oman, Palestine, and Egypt. The purpose of the examination was to determine if there is currently any integration of Web 2.0 applications such as blogs, RSS, discussion boards, podcasts, wikis or any related sign of collaborative efforts between government officials and the public.

The examination revealed no sign of interaction between the two using social network applications. Many of the visits also returned messages such as "Network error", "Only runs on Internet Explorer", "Under construction", "Coming soon", and "Broken links". The lack of Web 2.0 visibility may be attributed to either lack of awareness of their value by the government or lack of demand from their citizens, or it could be that we are not ready for such engagement, since we have not reached a maturity level yet in terms of providing full transaction services. The latter statement may be refuted by the fact that the United Arab Emirates have reached satisfactory progress at the transactional level, yet they do not have Web 2.0 applications deployed.

To make a contrast, we examined the Oakland County website as one of the highest ranked websites in the United States. The Oakland County had shown interest in information sharing and collaboration through e-government. A glance

at their home page reveals easy-to-spot Web 2.0 components such as blogs, podcasting, RSS feeds, video, forum, mobile, and live virtual meetings. We have also visited other websites from the US, and we were able to note that RSS is common feature on most US websites, and other web 2.0 applications were deployed sporadically.

WEB 2.0 IN THE ARAB WORLD

Next we want to examine the extensive use of Web 2.0 by the public in the Arab World, as a possible contributing factor for the lack of Web 2.0 applications in the e-government of Arab word.

In the Arab blogosphere users have commented that the reason that social media have not gained momentum yet in the Arab World is because the Web 1.0 arena including e-commerce continues to have a low adoption rate. However, there are certain signs of positive trends that users are engaging themselves in social networking applications. For example itoote.net is a blog aggregator of approximately 200 Arab bloggers writing in both Arabic and English discussing range of subjects.

In the last two years we have witnessed Maktoob.com (7 million users in 2007) and Jeeran. com introduce blogs bundled into a host of other services. The Jeeran community has reached over 1.3 million members with 120,000 blogs. Ikbis. com and youtubeislam are newcomers emulating youtube.com but targeting the Arab/Muslim community. Discussion boards have been in existence for many years, many of them having emanated from Saudi Arabia. Discussion boards have high demands in the Middle East, which is used as medium of informal education covering issues such as Islam, technology, economics, literature, medicine, science, and sports.

Another interesting, emerging model in Arab websites is providing free online learning exemplified by the "Open Islamic Academy" (www. islamacademy.net). Learning takes place through live online lectures, and students can join other students in online discussions. The academy has attracted 5000 regular online learners who at the end of the three years will be awarded certificates of recognition from the lecturers, who are considered prominent Islamic scholars.

Mecca.com is another social web application meant to be the Facebook for Muslims. The newly established Questler.com is also inspired by Facebook. Other websites emulate the popular tagging application Digg.com, such as wapher.com and darabet.com. The traffic to these sites is considerably low.

Thousands of Arabs, however, are engaged in Western-established Web 2.0 applications such as Facebook. One Jordanian group has even used Facebook with the aim of taking a proactive role in their country's development. Members of the network established a presence in Facebook to support Amman (Jordan) Municipality's progressive plan to develop the city. Another notable activity was initiated by a blogger in Amman (360east.com) who attempted to engage participants in designing "A world-class bus map for Amman". Participants showed enthusiastic interest and many volunteered to help, though there is no sign of whether the project ever materialized.

FUTURE RESEARCH DIRECTIONS

Multiple other Web 2.0 applications have recently been established in the Arab World, more than can be listed within the scope of this paper. Our aim, however, is to illustrate the positive upward trend by Arab citizens in networking and collaborating online using Web 2.0 services. The quality and the value of the collaboration, as well as how long it will last, are topics for future studies. We envisage that this trend will continue to grow as the concept of globalization driven by internet technology increasingly spreads through the Arab region.

SOLUTIONS AND RECOMMENDATION

E-government is no longer about providing information and services; it is about engaging citizens in new ways to reach the target citizens in a more effective way by involving customers in creation, delivery and dissemination. Governments in the Arab World can only gain from the Web 2.0 interactive environment to build credibility, trust and a long-term relationship with the public. Web 2.0 is being leveraged in e-government in the West and must not be ignored by the public sector in the Arab World.

The Arab World is composed of 22 different countries of diverse social and political backgrounds. However, despite these differences, they share many common values and to a large extent are influenced by similar factors. Thus we expect the opportunities offered by Web 2.0 services for e-government in the Western World and listed in the previous section (regulation, collaboration, information management and knowledge transfer, and public engagement) to be relevant to the Arab World to varying degrees. There are various indications that the domains of regulation, collaboration and information management are gaining ground in the Arab World. However, the domain of public engagement, especially in the realm of political participation, is not likely to gain much momentum in the near future due to the political realities of the Arab World. However, we are witnessing multiple attempts to engage the public in greater participation for the purpose of regulation and collaboration For example, the FIKRA project by Al-Ain municipality, a city in United Arab Emirates, is an electronic suggestion system created to encourage city residents to share ideas for improving quality of life in the city and to take part in decision making. Such initiative should give the government invaluable insights. Even though these attempts such as forums and feedback forms do not fit exactly under the umbrella of Web 2.0 services, they are considered a first step and a precursor of Web 2.0 services.

Many governments in the Arab World are already struggling to succeed with their current e-government initiatives. Adding the Web 2.0 dimension would certainly impose further challenges and obstacles. The caution of governments in the Arab World to engage the public is driven by the risk of implications of transparency and engagement in the political system. Another challenge for governments is having the right IT infrastructure in place. For example there is a need for an identity management system in order to minimize destructive behavior by citizens. An additional factor that must be considered to minimize risks is having effective security and clear privacy policies in place.

For successful interaction between the public sector and citizens there is a need to allow citizens to exercise a certain degree of freedom but with a responsibility for meeting common objectives that serve the well being of the country in order to avoid manipulation of government agendas and to prevent anarchy. This will require training within the public sector to effectively moderate this increase in interaction.

We feel that the public in the Arab World is ready for this wave and the West is exhibiting innovative projects that can be adapted to Arab culture. We do not expect that everyone will be attracted, but we do expect the technology-savvy to be attracted to this new environment. We do not expect to see the adoption of Web 2.0 immediately changing the landscape of e-government. The change, however, will be accelerated when the young generation move into the public sector to drive the change, which also requires flatter organizational structure to allow better collaborative environment, for now such restructuring will face resistance to change.

In the long run, not opening will lose legitimacy and authority; government should seriously consider opening using web 2.0 application to transform passive spectators to contributors to

democracy. For now governments can engage the public online on the domains of regulations, collaboration and, knowledge management while leaving political participation for later stage as the political system allows.

CONCLUSION

Web 2.0 social software has great potential to leverage the mission of Arab e-governments through connecting and collaborating with businesses, educational institutions, private and non-government initiatives, and of course individual citizens, many of whom are already beginning to embrace social media. The aim is to empower citizens in voicing their concerns by allowing for greater participation and to together work towards the well-being of the country. Interacting with citizens may lead to more transparency, better decision-making, laws, policies and calculated actions on the part of government. This opening will also make governments more accountable, and this by itself, is a satisfying reason to consider these tools. Government initiatives to become more trusting and trustworthy support greater citizen participation as well as better governance.

Governments need to be aware that the coming generations will be connected generations cultured by the internet and mobiles. These generations are known to question information and validate it through multiple sources and are also coming to expect a certain level of accessibility to and interaction with their governments. At present, there is no evidence that Arab e-government websites are engaging citizens although it is evident that citizens have begun to harness the power of Web 2.0 tools for multiple purposes.

This paper highlighted the current condition of Arab e-governments and how Web 2.0 can supplement their mission. Risks and challenges of Web 2.0 are real, but so are the opportunities. Before making the paradigm shift, governments must make sure they and their citizens are ready for this new channel and need to determine the best way to proceed towards more dynamic governments that can respond and meet the expectations of their supporting communities. Governments and citizens alike must also consider the extent to which they can be entrusted with the liberty of engaging constructively in a more open and supporting manner.

REFERENCES

Aberdeen Group. (2008). *Customer 2.0 The Business Implication of Social Media*. Retrieved May 15, 2009, from http://www.aberdeen.com/summary/report/benchmark/5195-RA-customer-20-social-media.asp.

Al-Fakhri, O., Cropf, A., Kelly, P., & Higgs, G. (2007). E-Government in Saudi Arabia: Between Promise and Reality. *International Journal of Electronic Government Research*, 4(2), 59–85.

Bélanger, F., & Carter, L. (2008). Trust and risk in e-government adoption. *The Journal of Strategic Information Systems*, 17(2), 165–176. doi:10.1016/j.jsis.2007.12.002

Cloete, F. (2003). Assessing governance with electronic policy management tools. *Public Performance and Management Review*, 26(3), 276–290. doi:10.1177/1530957602250233

Garrett, J. (2005). Ajax: A New Approach to Web Applications. *Adaptive Path*. Retrieved 15 May, 2009, from http://www.adaptivepath.com/ideas/essays/archives/000385.php

Gilbert, D., Balestrini, P., & Littleboy, D. (2004). Barriers and benefits in the adoption of egoverment. *International Journal of Public Sector Management*, 17(4), 286–301. doi:10.1108/09513550410539794

Katrinahelp (n.d). *Katrinahelp*. Retrieved May 15, 2009, from http://katrinahelp.info

Kohut, A. E. (2008). Social Networking and Online Videos Take Off: Internet's broader role in campaign. *T. P. R. Center, The PEW research centre*. Retrieved May 15, 2009, from http://people-press.org/reports/pdf/384.pdf

Korac-Kakabadse, A., & Korac-Kakabadse, N. (2002). Information technology's impact on the quality of democracy: Reinventing the 'democratic vessel.' In Heeks, R. (Ed.), *Reinventing Government in the Information Age: International Practice in IT-Enabled Public Sector Reform (pp. 211-228)*. London: Routledge.

Mohamed, N., AbdelRahman, A., & Fadlalla, A. (2008). A context-based integrative framework for e-government initiatives. *Government Information Quarterly*, 25(3), 448–461. doi:10.1016/j.giq.2007.02.004

Mysociety (n.d). *Mysociety*. Retrieved May 15, 2009, from http://www.mysociety.org/

New Zealand Police (n.d). *New Zealand Police*. Retrieved 15 May, 2009, from http://www.policeact.govt.nz/wiki/

O'Reilly. Tim. (2005). *What Is Web 2.0: Design Patterns and Business Models for the Next Generation of Software: O'Reilly Media*. Retrieved 15 May, 2009, from http://www.oreillynet.com/pub/a/oreilly/tim/news/2005/09/30/what-is-web-20.html.

OECD. (2003). *The e-Government Imperative*. Paris: OECD Publications Service.

Osimo, D. (2008). *Web 2.0 in Government: Why and How?: JRC Scientific and Technical*. Retrieved 15 May, 2009, from http://www.egov.vic.gov.au/index.php?env=-inlink/detail:m2762-1-1-8-s-0:l-10908-1-1--

Pascu, C. (2008). An Empirical Analysis of the Creation, Use and Adoption of Social Computing Applications: IPTS Exploratory Research on Social Computing. *JRC Scientific and Technical Reports*. Retrieved 15 May, 2009, from http://ftp.jrc.es/EURdoc/JRC46431.pdf

Pascu, C., & Osimo, D. (2008). Social computing: implications for the EU social innovation landscape. *Foresight*, *10*(1), 37–52. doi:10.1108/14636680810856017

Pascual, J. (2003). e-Government. *UNDP-APDIP, e-ASEAN Task Force*. Retrieved 4 October, 2010, from http://www.apdip.net/publications/iespprimers/eprimer-egov.pdf

Rainie, L. E. (2007). 'Wikipedia users'. *P. E. W. Internet*. Retrieved 15 May, 2009, from http://www.pewinternet.org/Reports/2007/Wikipedia-users.aspx

Research e-labs. (n.d). *Research e-labs*. Retrieved 15 May, 2009, from http://research.elabs.govt.nz/

Tertiary Education. (n.d). *Tertiary Education*. Retrieved 15 May, 2009, from http://wiki.tertiary.govt.nz/

The PEN community. (n.d). *The PEN community*. Retrieved 15 May, 2009, from http://pen.leadspace.govt.nz/

Warkentin, M., Gefen, D., Pavlou, P., & Rose, G. (2002). Encouraging citizen adoption of e-government by building trust. *Electronic Markets*, *12*(3), 157–162. doi:10.1080/101967802320245929

Wauters, P., & Lörincz, B. (2008). User satisfaction and administrative simplification within the perspective of eGovernment impact: Two faces of the same coin? *European Journal of ePractice, 4*, 1-10.

Webtrends (n.d). *How Barack Obama Is Using Web 2.0 to Run for President*. Retrieved 15 May, 2009, from http://webtrends.about.com/od/web20/a/obama-web.htm

World Bank. (n.d.). *Definition of e-government*. Retrieved 15 May, 2009, from http://web.worldbank.org.ezproxy-m.deakin.edu.au/WBSITE/EXTERNAL/TOPICS/EXTINFORMATION-ANDCOMMUNICATIONANDTECHNOLO-GIES/EXTEGOVERNMENT/0,contentMDK:20507153~menuPK:702592~pagePK:148956~piPK:216618~theSitePK:702586,00.html

Young, O. (2007). *Topic Overview: Web 2.0*. Forrester Research.

Zappen, J., Harrison, T., & Watson, D. (2008). A new paradigm for designing e-government: web 2.0 and experience design. In *Proceedings of the 2008 international conference on Digital government research*.

Zimmer, M. (2008). Preface: Critical perspective on web 2.0. *First Monday*, *13*(3).

Compilation of References

AACTE Committee on Innovation and Technology (Ed.). (2008). *Handbook of Technological Pedagogical Content Knowledge (TPCK) for Educators*. NY: Routledge.

Abdalla, I. A., & Al-Homoud, M. A. (2001). Exploring the implicit leadership theory in the Arabian Gulf States. *Applied Psychology: An International Review*, *50*(4), 506–531. doi:10.1111/1464-0597.00071

Abdallah, S., & Albadri, F. (2009). Strategic information systems planning: A case in the context of United Arab Emirates. *International Conference on Information Resources Management (Conf-IRM2009)*, Dubai, UAE.

Abdelaal, A., Ali, H., & Khazanchi, D. (2009). The Role of Social Capital in the Creation of Community Wireless Networks. *The 42nd Hawai'i International Conference on Systems Sciences*, Waikoloa, HI, January 5-8.

Abdinnour-Helm, S., Lengnick-Hall, M. L., & Lengnick-Hal, A. (2003). Pre-Implementation Attitudes and Organisational Readiness for Implementing Enterprise Resource Planning Systems. *European Journal of Operational Research*, *146*(2), 258–273. doi:10.1016/S0377-2217(02)00548-9

Aberdeen Group. (2008). *Customer 2.0 The Business Implication of Social Media*. Retrieved May 15, 2009, from http://www.aberdeen.com/summary/report/benchmark/5195-RA-customer-20-social-media.asp.

Abouchedid. K., & Eid. G. (2004). E-Learning Challenges in the Arab World. New York: *Emerald Group Publishing Limited*.

Abuhmaid, A. (2009). *ICT Integration across Education Systems: The Experience of Jordan in Educational Reform*. Saarbrücken, Germany: VDM Verlag Dr. Müller.

Agarwal, R. (1999). *Individual Acceptance of Information Technologies*. Retrieved March 2010, from http://www.c4ads.seas.gwu.edu/classes/CSci285_Fall_2004/readings/9/required/Agarwal_02.pdf

Ahmad, N., & Daghfous, A. (2010). Knowledge sharing through inter-organizational knowledge networks: Challenges and opportunities in the United Arab Emirates. *European Business Review*, *22*(2), 153–174. doi:10.1108/09555341011023506

Ainin, S. & Hisham, N. (2008). Applying Importance –performance Analysis to Information Systems: An Exploratory Case Study. *Journal of Information, Information Technology and Organization, 3*.

Akhter, F. (2007). A case study: Adoption of information technology in e-Business of United Arab Emirates. *IEEE international Conference on e-Business Engineering*, Hong Kong, China

Akhter, F., & Kaya, L. (2009). Building secure e-business systems: Technology and culture in the UAE. In *Proceedings of the 2008 ACM symposium on Applied computing*, Ceara, Brazil.

Akkari, A. (2004). Education in the Middle East and North Africa: The Current Situation and Future Challenges. *International Education Journal*, *5*(2), 144–153.

Al Akremi, A., Nasr, M. I., & Sassi, N. (2007, September). *Impact de la culture nationale sur la confiance interpersonnelle en milieu du travail: Analyse comparative entre la France et la Tunisie*. [Impact of the national culture on interpersonal confidence in work context: Comparative analysis between France and Tunisia]. Paper presented at the 18th congrès annuel de l'Association Francophone de Gestion des Ressources Humaine (AGRH) [Paper presented at the 18th annual congress of the Francphoic Association of Human Resource Management], Fribourg, Switzerland.

Aladwani, A. M. (2001). Change Management Strategies for Successful ERP Implementation. *Business Process Management Journal*, *7*(3), 266–275. doi:10.1108/14637150110392764

Al-Ashban & Burney. (2001). Customer adoption of tele-banking technology, the case of Saudi Arabia. *International Journal of Bank Marketing*, *19*(5), 191–201. doi:10.1108/02652320110399683

Albadri, F. (2009). The Integrated Project Risk Model: A Risk Based Model for Managing Information Technology Projects. In Kidd, T. T. (Ed.), *Handbook of Research on Technology Project Management, Planning, and Operations*. Hershey, PA: IGI Global.

Albadri, F., & Abdallah, S. (2009). Characterization and Competency Building in Context of and Oil & Gas Company. In *Proceedings of the 11th. International Business Information Management Conference*, Cairo, Egypt, Jan 4-6.

Albadri, F., & Jordan, E. (2000). The integrated management model: Risk Management driven decision making in IT Projects. In *Proceedings of the 5th International APDSI Conference, Asia-Pacific Region of Decision Institute*, Tokyo, Japan, July 24-27.

Albadri, F., & Jordan, E. (2002). The integrated Project Risk Management Methodology: The effective use of software in IT Projects. In *Proceedings of the 7th International APDSI Conference, Asia-Pacific Region of Decision Institute*, Bangkok, Thailand, July 24-27.

Albadri, F., & Jordan, E. (2003). The role of Controls to Minimize Risks of Information Technology Projects Failure. In *Proceedings of the 14th Australasian Conference on Information Systems*, Perth, Western Australia, November 26-28.

Albadri, F., & Jordan, E. (2004). Project Success: Project Managers Competence. In *Proceedings of the 3rd International Business Information Management Conference*, Amman, Jordan, July 4-6.

Albirini, A. (2006). Cultural perceptions: The missing element in the implementation of ICT in developing countries. *International Journal of Education and Development using ICT, 2*(1).

Al-Fakhri, O., Cropf, A., Kelly, P., & Higgs, G. (2007). E-Government in Saudi Arabia: Between Promise and Reality. *International Journal of Electronic Government Research*, *4*(2), 59–85.

Al-Furaih, I. S. (2002*). Internet regulations: The Saudi Arabia experience*. Retrieved (month,date), from: http://inet2002.org/CD-ROM/lu65rw2n/papers/u05-a.pdf

Al-Gahtani. (2003). Computer Technology adoption Saudi Arabia, Correlates of perceived innovation attributes. *Journal of Information Technology for Development*, *10*(1), 57–69. doi:10.1002/itdj.1590100106

Al-Hinai, H. (2004). *The Sultanate of Oman's E-Government Initiative and its Economic and Social Impact. Unpublished Master disertaion*. Salford, UK: University of Salford.

Ali, A., & Wahabi, R. (1995). Managerial value systems in Morocco. *International Studies of Management and Organization*, *25*(3), 87–102.

Al-Jabri, M. A. (1994). *Introduction à la critique de la raison arabe. Introduction to the criticism of the Arab reason. Translated from Arab by A. Mahfoud & M. Geoffroy*. Paris: La Découverte.

Al-Jaghoub, S., & Westrup, C. (2003). Jordan and ICT-led development: towards a competition state. *Information Technology & People*, *16*(1), 93–110. doi:10.1108/09593840310463032

AlKhawaja, K. (2007, August 15). Seven Thousand Students Move from Private to Governmental Schools. *ALRAI*, P.1.

Allee, V. (2008). Value Network Analysis and Value Conversion of Tangible and Intangible Assets. *Journal of Intellectual Capital*, 9(1), 5–24. doi:10.1108/14691930810845777

Al-Mabrouk, K., & Soar, J. (2009). An analysis of the major issues for successful information technology transfer in Arab countries. *Journal of Enterprise Information Management*, 22(5), 504–522. doi:10.1108/17410390910993518

Al-Maliki, S. G. Y. Al-Khalidi, (2005). *Information Systems Evaluation Process in Private Organisations in Saudi Arabia*, Unpublished doctoral thesis, University of East Anglia, UK.

Al-Mashari, M., & Al-Mudimigh, A. (2003). ERP Implementation: Lessons from a Case Study. *Information Technology & People*, 16(1), 21–33. doi:10.1108/09593840310463005

Al-Mashari, M., Al-Mudimigh, A., & Zairi, M. (2003). Enterprise Resource Planning: A Taxonomy of Critical Success Factors. *European Journal of Operational Research*, 352–364. doi:10.1016/S0377-2217(02)00554-4

Al-Omari, A. (2004). *International Electronic Transactions Measurement Scale System*. Digital publish doctoral thesis, School of Information Technology and Engineering, the George Mason University.

Al-Roubaie, A., & Abdul-Wahab, R. (2009). Data mining and knowledge discovery: An Approach for sustaining development in GCC countries. *International Association of Computer Science and Information Technology* – Spring Conference, Singapore.

Al-Sa'd, A. (2007). *Evaluation of Students' Attitudes Towards Vocational Education in Jordan. Lärarutbildningen*. Malmohogskola.

Al-Solbi, A. & Mayhew, P., (2004, June.). *Development of Information and Communication Technology Indicators* in the Kingdom of Saudi Arabia, ECCO XII'04, Beirut.

Al-Solbi, A. (2007, May), *Evaluating E-Health and Use of ICT: Learning From the Experiences of Saudi Health Organisations*, ECCO 2007 XXth Anniversary Conference European Chapter on Combinatorial Optimization, Limassol, Cyprus, May 24-26, 2007

Al-Solbi, A. & Mayhew Pam, (2006, December). A strategic Framework for Electronic Readiness Assessment. In *Proceeding of the Internet and Information System in the Digital Age Conference*, Brescia, Italy.

Al-Solbi, A., & Mayhew, P. (2004, June.). Development of Information and Communication Technology Indicators in the Kingdom of Saudi Arabia. In Proceedings *of the 17th European conference on combinatorial optimization*, Beirut.

Al-Solbi, A., & Mayhew, P. (2005b, July). Measuring E-Readiness Assessment in Saudi Organisations: Preliminary Results From A Survey Study. In *Proceeding of the First European Conference on Mobile Government*, 10-12 July 2005, University of Sussex, Brighton, UK

Al-Solbi, A., & Pam Mayhew, P. (2006, December*), A strategic Framework for Electronic Readiness Assessment*, Proceeding of the Internet and Information System in the Digital Age Conference, Dec. 14-17, 2006 in Brescia, Italy. ISBN: 0-9753393

Al-Solbi, A., Mayhew Pam, P., & Al-Badi, A. (2005a, July). An Alternative Model For Measuring E-Readiness for Developing Countries: Applied to Saudi Organisations. In *Proceedings of the International Business Information Management Conference*, Lisbon, Portugal.

Alterman, J. (2000). Counting nodes and counting noses: Understanding new media in the Middle East. *The Middle East Journal*, 54(3), 355–363.

Amit, R., & Shoemaker, P. J. H. (1993). Strategic assets and organizational rent. *Strategic Management Journal*, 14(1), 33–46. doi:10.1002/smj.4250140105

Amoako-Gyampah, K., & Salam, A. F. (2004). An extension of the technology acceptance model in an ERP implementation environment. *Information & Management*, 41(6), 731–745. doi:10.1016/j.im.2003.08.010

AMR. (n.d.) *AMR Research (currently part of Gartner Inc.).* Retrieved from http://www.amrresearch.com/Content/View.asp

Anol, B., & Rudy, H. (1997). IT and organizational change: Lessons from client/server technology implementation. *Journal of General Management, 2*(winter), 31., Retrieved August 28, 2004, from http://www.aucegypt.edu/library/libdata/subject.cfm?classid=2

Appleby, R. (2005). *The Spatiality of English Language Teaching, Gender and Context.* Unpublished doctoral dissertation, University of Technology, Sydney.

Appuswamy, R. (2000). *Implementations Issues in ERP.* Paper presented at the 1st International Conference on Systems Thinking in Management (ICSTM2000).

Arab Advisor Group. (2005). *Only 56% of Arab cellular operators provide the MMS service, while Morocco and Lebanon have the highest SMS rates in the Arab world.* Retrieved October 19, 2009, from http://www.arabadvisors.com/Pressers/presser-170105.htm

Arab Advisor Group. (2007). *Lebanon cellular users survey.* Retrieved October 19, 2009, from http://www.arabadvisors.com/Pressers/presser-161207.htm

Arab Open University. (2007). *Arab Open University.* Retrieved, April 2007, from: http://www.arabou.org/

Attiyah, H. S. (1993). *Roots of organization and management problems in Arab countries: Cultural or otherwise?* Paper presented at the Arab Management Conference, Bradford.

Aziz, S. (2009). Dimensions of IT literacy in an Arab region: A study in Barkha (Oman). *Information and Communication Technologies Development International Conference.* Doha, Qatar.

Bagozzi, R. (2007). The legacy of the technology acceptance model and a proposal for a paradigm shift. *Journal of the Association for Information Systems, 8*(4), 244–254.

Baile, S., & Lefièvre, V. (2003, June). *Le succès de l'utilisation de la messagerie électronique: Etude de ses déterminants au sein d'une unité de production aéronautique.* The success of the use of the electronic message: Study of its determinants within a manufacturing unit. Paper presented at the 8th colloque de l'Association Information et Management (AIM). Paper presented at the 8th seminar of the Association of Information and Management, Grenoble.

Baker, B. (1995). The role of feedback in assessing information system planning effectiveness. *The Journal of Strategic Information Systems, 4*(1), 61–80. doi:10.1016/0963-8687(95)80015-I

Bakos, J. Y., & Treacy, M. E. (1986). Information Technology and Corporate Strategy, A Research Perspective. *Management Information Systems Quarterly, 10*(2), 107–119. doi:10.2307/249029

Ballon, D. (1998). Modeling Information Manufacturing Systems to Determine Information Product Quality. *Management Science, 44*(4), 462–484. doi:10.1287/mnsc.44.4.462

Balmer, J. M. T. (1998). Corporate identity and the advent of corporate marketing. *Journal of Marketing Management, 14*, 963–996. doi:10.1362/026725798784867536

Balmer, J. M. T., & Soenen, G. (1999). The acid test TM of corporate identity management. *Journal of Marketing Management, 15*, 69–92. doi:10.1362/026725799784870441

Banker, R. D., & Kauffman, Robert J. (2004). The Evolution of Research on Information Systems, A Fiftieth-Year Survey of the Literature in Management Science. *Management Science, 50*(3), 281–298. doi:10.1287/mnsc.1040.0206

Barakat, H. (1993). *The Arab world: Society, culture and state.* Berkeley, CA: University of CA Press.

Barnes, D., Mieczkowska, S., & Hinton, M. (2003). Integrating operations and information strategy in e-business. *European Management Journal, 21*(5), 626. doi:10.1016/S0263-2373(03)00111-7

Barras, R. (1984). Towards a theory of innovation in services. *Research Policy*, *15*, 161–173. doi:10.1016/0048-7333(86)90012-0

Bates, A. (2005). *Technology, e-learning and distance education*. Boca Raton, FL: Routledge Taylor & Francis Group.

Bawden, D. (2008). Smoother pebbles and the shoulders of giants, the developing foundations of information science. *Journal of Information Science*, *34*(4), 415–426. doi:10.1177/0165551508089717

BECTA. (2004) A review of the research literature on barriers to the uptake of ICT by teachers. British Educational Communications and Technology Agency. Retrieved February 10, 2010, from http://partners.becta.org.uk/page_documents/research/barriers.pdf

Bélanger, F., & Carter, L. (2008). Trust and risk in e-government adoption. *The Journal of Strategic Information Systems*, *17*(2), 165–176. doi:10.1016/j.jsis.2007.12.002

Ben Fadhel, A. (1992). *La dynamique séquentielle culture gestion: Fondements théoriques et analyse empirique du cas tunisien*. Sequential dynamics of management/culture: Theoretical bases and empirical analysis of the Tunisian case. (Unpublished doctoral dissertation), University of Nice, France.

Ben Zina, S. (2002). *Le profil culturel de l'innovateur tunisien: Approche psychologique et comportementale. Cultural profile of the Tunisian innovator: Psychological and behavioral approaches. (Unpublished master's theses), University of Tunis*. Tunisia: ISG.

Benbasat, I., Goldstein, D. K., & Mead, M. (1987). The Case Research Strategy in Studies of Information Systems. *Management Information Systems Quarterly*, *11*(3), 369–386. doi:10.2307/248684

Beyerbach, B., Walsh, C., & Vannatta, R. (2001). From teaching technology to using technology to enhance student learning: pre-service teachers' changing perceptions of technology infusion. *Journal of Technology and Teacher Education*, *9*(1), 105–127.

Binbasioglu, M., & Winston, E. (2002). *Are you solving the right Problem? A Study of Packaged Software Implementation, International Association for Computer Information Systems*. Fort Lauderdale, FL: IACIS.

Bitner, N., & Bitner, J. (2002). Integrating technology into the classroom: eight keys to success. *Journal of Technology and Teacher Education*, *10*(1), 95–100.

Blili, S., & Raymond, L. (1993). Information technology, threats and opportunities for small and medium-sized enterprises. *International Journal of Information Management*, *13*(6), 439–448. doi:10.1016/0268-4012(93)90060-H

Bouattour, S., & El Louadi, M. (2004, October). *Les aspects culturels de l'adoption des technologies de l'information et de la communication dans le monde arabe*. Cultural aspects of the adoption of information and communication technology in the Arab world. Paper presented at the 8th Conférence Internationale de Management des Réseaux (CIMRE), Paper presented at the 8th International Conference of Network Management, Hammamet, Tunisia.

Bouraoui, K., & Chastrette, M. (1999). Conceptions d'élèves et d'étudiants français et tunisiens sur la conduction dans les piles électrochimiques. Designs of French and Tunisian pupils and students on conduction in the electrochemical batteries. *Didaskalia, 14*, 39-60. Retrieved May 12, 2008, from http://documents.irevues.inist.fr/bitstream/handle/2042/23868/DIDASKALIA_1999_14_39.pdf;jsessionid=5E280612129F8D3A4B33ADF7711A4F65?sequence=1

Boynton, A. C., & Zmud, R. W. (1987). 'Information technology planning in the 1990: Direction for practice and research'. *Management Information Systems Quarterly*, *1*(1), 59–71. doi:10.2307/248826

Brand, G. (1997). What research says: training teachers for using technology. *Journal of Staff Development*, *19*(1), 10–13.

Brand, H., & Duke, J. (1982, December). Productivity in commercial banking: computers spur the advance. *Monthly Labor Review*, 19–27.

Brazel J. F., & Dang, Li. (2008). The Effect of ERP System Implementations on the Management of Earnings and Earnings Release Dates. *Journal of Information Systems 22*(2) fall 2008, 1–21.

Brown, I. (2008). Investigating the impact of the external environment on strategic information systems planning: A qualitative inquiry. *Proceedings of the 2008 annual research conference of the South African Institute of Computer Scientists and Information Technologists, South Africa.*

Brynjolfsson, E., & Hitt, L. (1998). Beyond the Productivity Paradox: Computers are the Catalyst for Bigger Changes, *the Communications of the ACM, 41*(8), 49-56.

Bueno, S., & Salmeron, J. (2008). TAM-based success modeling in ERP. *Interacting with Computers, 20*(6), 515–523. doi:10.1016/j.intcom.2008.08.003

Burger, M. J. (2009). Replicating Milgram: Would people still obey today? *The American Psychologist, 64*(1), 1–11. doi:10.1037/a0010932

Caldeira, M., & Ward, J. (2002). Understanding the successful adoption and use of IS/IT in SMEs, an explanation from Portuguese manufacturing industries. *Information Systems Journal, 12*(2), 121–152. doi:10.1046/j.1365-2575.2002.00119.x

Calisir, F., Gumussory, C. A., & Bayram, A. (2009). Predicting the behavioral intention to use enterprise resource planning systems: An exploratory extension of the technology acceptance model. *Management Research News, 32*(7).

Campbell, D. T., & Stanley, J. C. (1963). *Experimental and Quasi-Experimental Designs for Research*. Chicago: Rand McNally.

Carr, N. G. (2003). IT Doesn't Matter. *Harvard Business Review, 81*(5), 41–48.

Cassidy, T., Matthew, M. (2002). Higher Education in the ARAB states: Responding to the Challenges of Globalization. *AMIDEAST conference on Higher Education In ARAB Countries*

Castells, M. (1999). Flows, Networks, and Identities: A Critical Theory of the Informational Society. In Castells, M., Flecha, R., Freire, P., Giroux, H. A., Macedo, D., & Willis, P. (Eds.), *Critical Education in the New Information Age* (pp. 37–64). New York: Rowman & Littlefield Publishers, Inc.

Chatelard, G. (2004). *Jordan: A Refugee Haven, Migration Policy Institute.* Retrieved February 19, 2007, from http://www.migrationinformation.org/Profiles/display.cfm?ID=236

Chatfield, T., & Althujran, O. (2009). A cross-country comparative Analysis of E-government service delivery among Arab Countries. *Information Technology for Development, 15*(3), 151–170. doi:10.1002/itdj.20124

Chatterjee, D. (2002). Shaping up for E-commerce: Institution Enablers of the organizational assimilation of Web technologies. *Management Information Systems Quarterly, 26*(2), 65. doi:10.2307/4132321

Checkfacebook (2009). http://www.checkfacebook.com. Retreieved February 11, 2009 from http://www.checkfacebook.com.

Chen, I. J. (2001). Planning for ERP systems: analysis and future trend. *Business Process Management Journal, 7*(5), 374–386. doi:10.1108/14637150110406768

Chossudovsky, M. (2003). *The Globalization of Poverty and the New World Order* (2nd ed.). Pincourt, Canada: Global Research.

Choucri, N., Maugis, V., Madnick, S., & Siegel, M. (2003). *Global E-Readiness For What? Centre for E-Business at MIT*. Sloan School of Management.

Ciborra, C. (2000). *From control to Drift*. Oxford, UK: Oxford University Press.

Ciborra, C., & Hanseth, O. (1998). From tool to Gestell. Agendas for managing information infrastructures. *Information Technology & People, 11*(4), 305–327. doi:10.1108/09593849810246129

Cisco Learning Institute. (2004). *Case Study: Jordan Education Initiative*. Retrieved July 28, 2008, from http://www.cisco.com/web/about/ac48/pdf/EU_Case_Study_JEI.pdf

Clarke, A. (2008). e-Learning Skills, 2nd edition. New York: *Palgrave Macmillan*.

Cloete, F. (2003). Assessing governance with electronic policy management tools. *Public Performance and Management Review, 26*(3), 276–290. doi:10.1177/1530957602250233

Cohen, A. (2007). One nation, many cultures: A cross-cultural study of the relationship between personal cultural values and commitment in the workplace to in-role performance and organizational citizenship behavior. *Cross-Cultural Research, 41,* 273–300.

Coleman, J. (1988). Social Capital in the Creation of Human Capital. *American Journal of Sociology,* V94.

Conlon, T., & Simpson, M. (2003). Silicon Valley verses Silicon Glen: the impact of computers upon teaching and learning: a comparative study. *British Journal of Educational Technology, 34*(2), 137–150. doi:10.1111/1467-8535.00316

Connect, U. J. (2007). *Information and Communication Technology in Education Diploma*. Retrieved November 10, 2009, from http://www.ju.edu.jo/alumni/UJConnect12/topics/ICT.htm

Cooper & Zmud. (1990). Information Technology Implementation Research, A Technological Diffusion Approach. *Journal of Management Science, 36*(2), 123–139. doi:10.1287/mnsc.36.2.123

Corm, G. (2007). *Histoire du Moyen-Orient: De l'antiquité à nos jours* [History of the Middle-East: From antiquity to our days]. Paris: La Découverte.

Credif. (2005) Centre de recherche, d'études, de documentation et d'informations sur la femme. *La femme tunisienne: Acteur de développement regional- approche Empowerment.* [The Tunisian woman: Actor of regional development- Empowerment approach. Desjeux, D. (2003). Les échelles d'observation de la culture. Scales of the culture's observation. *Communication, 22.* Retrieved April 17, 2009, from http://www.argonautes.fr/sections.php?op=viewarticle&artid=383

CSPP. Org, (2000). *The CSPP readiness guide for living in the networked world.* CSPP.Org, Retrieved May, 2010, from: http://www.cspp.org

Dadashzadeh, M. (2002). *Information Technology Management in Developing Countries*, IRM press.

Davenport, T. (1998). Putting the enterprise into the Enterprise System. *Harvard Business Review, 76*(4), 121–131.

Davenport, T. H. (2000). *Mission Critical: Realizing the Promise of Enterprise Systems*. Boston, USA: Harvard Business School Press.

Davis, F. D. (1989). Perceived usefulness, perceived ease of use, and user acceptance of information technology. *Management Information Systems Quarterly, 13*(3), 319–340. doi:10.2307/249008

Delone, W., H. (1998). Determinants of success for computer usage in small business. *Management Information Systems Quarterly, 12*(1), 51–67. doi:10.2307/248803

DeLone, W., & McLean, E. (1992). Information systems success: the quest for the dependent variable. *Information Systems Research, 3*(1), 60–95. doi:10.1287/isre.3.1.60

DeLone, W. and McLean, E. (2003). The DeLone and McLean Model of Information Systems Success: A Ten-Year Update. *Journal of Management Information Systems, spring 2003, 19*(4), 9-30.

Desjeux, D. (2005). Usages et enjeux du SMS en Chine, en France et en Pologne. Uses and stakes of SMS in China, France and Poland. *Consommations & sociétés.* Retrieved October 13, 2009, from http://www.argonautes.fr/sections.php?op=viewarticle&artid=353

Dewachi, A, (2001). Overview of Internet in Arab States. *International Journal of Education and Development using ICT, 2*(1).

Dhaoui, H. (2000). *Pour une psychanalyse maghrébine: La personnalité*. For a Maghrebin psychoanalysis: The personality. Paris: L'Harmattan.

Díez, E., & McIntosh, B. S. (2009). A review of the factors which influence the use and usefulness of information systems. *Environmental Modelling & Software, 24*(5), 588–602. doi:10.1016/j.envsoft.2008.10.009

Donner, J. (2007). The rules of beeping: Exchanging messages via intentional missed calls on mobile phones. *Journal of Computer-Mediated Communication, 13*(1). Retrieved January 20, 2009, from http://jcmc.indiana.edu/vol13/issue1/donner.html

Dutta, S., Lanvin, B., & Paua, F. (2003, January). *The Global Information Technology Report 2002-2003: Readiness for the Networked World*, Oxford University Press Published.

Dwairy, M., Achoui, M., Abouserie, R., Farah, A., Sakhleh, A., Fayad, M., & Khan, H. K. (2006). Parenting styles in Arab societies: A first cross-regional research study. *Journal of Cross-Cultural Psychology, 37*(3), 230–247. doi:10.1177/0022022106286922

Earl, M. J. (1993). Experiences in strategic information systems planning. *Management Information Systems Quarterly, 17*(1), 1–24. doi:10.2307/249507

Edwards, J. E. (2003). *The relationship of e-commerce readiness to technology acceptance: The case of Barbados*. Digital publish doctoral thesis. NOVA South-eastern University.

Ehie, I. C., & Madsen, M. (2005). Identifying Critical Issues in Enterprise Resource Planning (ERP) Implementation. *Computers in Industry, 50*(6), 545–557. doi:10.1016/j.compind.2005.02.006

Ein-Dor, P., Myers, M., & Raman, K. S. (1997). Information technology in three small developed countries. *Journal of Management Information Systems, 13*(4), 61–89.

El Louadi, M. (2004). Cultures et communication électronique dans le monde arabe. Cultures and electronic communication in the Arab world. *Revue Systèmes d'Information et Management, 9*(3), 117–143.

El Sawah, S., Tharwat, A. A. E., & Rasmy, M. H. (2008). A quantitative model to predict the Egyptian ERP implementation success index. *Business Process Management Journal, 14*(3), 288–306. doi:10.1108/14637150810876643

Elango, R, and Selvam, M. (2008). Quality of e-Learning: An Analysis Based on e-Learners' Perception of e-Learning. *The Electronic Journal of e-Learning, 6*(1), 31-44.

Elmandjra, M. (1990, May). *Future of the Islamic world. Future studies: Needs, facts and prospects.* Paper presented at the Future of the Islamic World Symposium, Algiers, Algeria.

Escobar, A. (2004). Development, Violence and the New Imperial Order. *Development, 47*(1), 15–21. doi:10.1057/palgrave.development.1100014

ESCWA. (2009) *Economic and Social Commission for Western Asia, United Nations Economic and Social Commission for Western Asia publication.* Retrieved April 2010, from http://www.escwa.un.org/information/publications/edit/upload/ictd-09-12.pdf

Esteves, J., & Pastor, J. (2000). *Towards the Unification of Critical Success Factors of ERP Implementation.* Paper presented at the 10 th Annual Business Information Technology (BIT) 2000.

European SchoolNet. (2005). *Assessment Schemes for Teachers' Ict Competence- a Policy Analysis.* Retrieved August 20, 2006, from http://www.eun.org/insight-pdf/special_reports/PIC_Report_Assessment%20schemes_insightn.pdf

Fabry, D., & Higgs, J. (1997). Barriers to the effective use of technology in education. *Journal of Educational Computing, 17*, 385–395. doi:10.2190/C770-AWA1-CMQR-YTYV

Fan, M. (2000). The adoption and design methodologies of component-based enterprise systems. *European Journal of Information Systems, 9*, 25–35. doi:10.1057/palgrave.ejis.3000343

Finney, S., & Corbett, M. (2007). ERP Implementation: A Compilation and Analysis of Critical Success Factors. *Business Process Management Journal, 13*(3), 329–347. doi:10.1108/14637150710752272

Fischer, H. (2006). *Digital Shock: Confronting the New Reality* (Mullins, R., Trans.). Montreal: McGill-Queen's University Press.

Fitouri, C. (1991). L'acculturation, premier facteur du sous-développement. [Acculturation, the first factor of the underdevelopment]. *Série Sociologie, 17*, 47–57.

Fleck, J. (1994). Learning by trying: the implementation of configurational technology. *Research Policy, 23*, 637–652. doi:10.1016/0048-7333(94)90014-0

Flower, J. (1996a). *The five fundamentals of dealing with change*. Retrieved July 21, 2004, from www.well.com/user/bbear/change2.html

Flower, J. (1996b). *The root ideas in dealing with change*. Retrieved July 22, 2004, from www.well.com/user/bbear/change2.html

Forman, R. (1994). Australia in Laos: In-Country Delivery of University Programs. In Thomas, P. (Ed), *Teaching for Development: An international Review of Australian Formal and Non-formal Education for Asia and the Pacific*. (pp 98-104). Canberra, ACT.

Forrester, J. W. (1968). *Principles of Systems*. Cambridge, MA: Wright Allen.

Forte, A., & Bruckman, A. (2008). Why Do People Write for Wikipedia? Incentives to Contribute to Open-Content Publishing. In *Proceedings of 41st Annual Hawaii International Conference on Systems Sciences*, Waikoloa, HI.

Francalanci, C. (2001). Predicting the implementation effort of ERP projects: empirical evidences on SAP R/3. *Journal of Information Technology, 16*(1), 33–48. doi:10.1080/02683960010035943

Franklin, T. (2007). *A Cultural Bridge: Instructional Technology Program Goes Global*. Retrieved February 10, 2010, from http://www.coe.ohiou.edu/news-events/jordanicte.htm

Franklin, T., & Franklin, D. (2007). *ICTE Final Report.* (Unpublished report), University of Jordan, Jordan.

Franklin, T., & Franklin, D. (2008). *ICTE Year 2 Evaluation Report.* (Unpublished report), University of Jordan, Jordan

Franklin, T., & Franklin, D. (2009). *ICTE Year 3 Evaluation Report.* (Unpublished report), University of Jordan, Jordan

French, W. L., & Bell, C. H. (1978). *Organization development* (2nd ed.). Englewood Cliffs, N.J.: Prentice Hall.

French, W. L., Bell, C. H., & Zawacki, R. A. (Eds.). (2000). *Organization development and transformation.* New York: Irwin McGraw-Hill.

Fryling, M. (2005). ERP Implementation Dynamics. *Information Science and Policy University at Albany. State University of New York, Oct.2005*

Fulk, J., Schmitz, J., & Steinfield, C. W. (1990). A social influence model of technology use. In Fulk, J., & Steinfield, C. (Eds.), *Organizations and Communication Technology* (pp. 117–142). Sage publications.

Fullan, M. (2000a). The Return of Large-Scale Reform. *Journal of Educational Change, 1*(1), 5–28. doi:10.1023/A:1010068703786

Fullan, M. (2000b). The Three Stories of Education Reform. *Phi Delta Kappa International, 81*(8), 581–584.

Fullan, M. (2001). *The New Meaning of Educational Change* (3rd ed.). New York: RoutledgeFalmer.

Gable, G. (2001). Large packaged application software maintenance: a research framework. *Journal of Software Maintenance and Evolution: Research and Practice, 13*(6), 351–371. doi:10.1002/smr.237

Gable, G., et al. (2002,). Enterprise Resources Planning Systems Impacts: A Delphi study of Australian public sector. In *Proceedings of the sixth Australasian Conference of Information Systems (PACIS 2002), 2-4 September 2002, Tokyo, Japan.*

Gable, G., Sedera, D., & Chan, T. (2003). *Enterprise Systems Success: A Meaurement Model*. Paper presented at the 24th International Conference in Information Systems.

Gaines, S. O., Henderson, M. C., Kim, M., Gilstrap, S., Yi, J., & Rusbult, C. E. (2005). Cultural value orientation, internalized homophobia, and accommodation in romantic relationships. *Journal of Homosexuality, 50*(1), 97–117. doi:10.1300/J082v50n01_05

Garrett, J. (2005). Ajax: A New Approach to Web Applications. *Adaptive Path*. Retrieved 15 May, 2009, from http://www.adaptivepath.com/ideas/essays/archives/000385.php

Gavlak, D. (2007, July 27). Jordan Appeals for Help in Dealing with Iraqi Refugees. *Washington Post (Washington, D.C.)*, A16.

Geertz, C. (2000). *Interpretation of cultures: Selected essays*. New York: Basic Books.

George, A., & Davoodi, H. (2003). *Challenges of growth and globalization in the Middle East and North Africa*. International Monetary Fund, Publication Services.

Gerstein, M., & Reisman, H. (1982). Creating Competitive Advantage with Computer Technology. *The Journal of Business Strategy, 3*(1), 53–60. doi:10.1108/eb038956

Gilbert, D., Balestrini, P., & Littleboy, D. (2004). Barriers and benefits in the adoption of egovernment. *International Journal of Public Sector Management, 17*(4), 286–301. doi:10.1108/09513550410539794

Guessoum, N. (2006). Science et société dans le contexte arabe. *Science and society in the Arab context*. Retrieved November 13, 2009, from http://science-islam.net/breve.php3?id_breve=546&var_recherche=Guessoum&lang=fr

Guha, S. (2003). Are we all technically prepared? Teachers' perspective on the causes of comfort or discomfort in using computers at elementary grade teaching. *Information Technology in Childhood Education Annual*, 317–349.

Hage, J., & Dewar, R. (1973). Elite Values and versus Organizational Structure in Predicting innovation. *Administrative Science Quarterly, 18*, 279–290. doi:10.2307/2391664

Hamade, S. (2009). Information and communication technology in Arab countries". *Problems and Solutions. Sixth International Conference on Information Technology:* New Generations. Las Vegas, NV.

Hamilton, S., & Chervany, N. L. (1981a). Evaluating Information System Effectiveness - Part I, Comparing Evaluation Ap-proaches. *Management Information Systems Quarterly, 5*(3), 55–69. doi:10.2307/249291

Hamilton, S., & Chervany, N. L. (1981b). Evaluating Information System Effectiveness - Part II, Comparing Evaluation Viewpoints. *Management Information Systems Quarterly, 5*(4), 79–86. doi:10.2307/249329

Hammo, B., Hudaib, A., Huneiti, A., & Al-Zoubi, M. B. (2007). *Preparing Jordanian Teachers to Face the Challenges in Applying Information and Communication Technology in Education*. In the Fourth Annual Conference of Learning International Networks Consortium, MIT-LINC 2007, October 28-30, Amman, Jordan.

Haraldsson, H., Belyazid, S., & Sverdrup, H. (2006, June). *Causal Loop Diagrams–promoting deep learning of complex systems in engineering education*. Paper presented at the 4th Pedagogical Inspiration Conference 1. Lund University, Sweden.

Hartono, E., Lederer, A. L., Sethi, V., & Zhuang, Y. (2003). Key predictors of the implementation of strategic information systems plans. *The Data Base for Advances in Information Systems, 34*(3), 41–53.

Hasan, O. (2000). Improving the Quality of Learning: Global Education as a Vehicle for School Reform. *Theory into Practice, 39*(2), 97–103. doi:10.1207/s15430421tip3902_6

Hasan, H., & Ditsa, G. (1999). The impact of culture on the adoption of IT: an interpretive study. *Journal of Global Information Management, 7*(1), 5–15.

Hatem, T. (2003). *Understanding cultural differences between Americans and Arabs: The Egyptian case.* In the Proceedings of the tenth American University in Cairo Research Conference, Globalization Revisited: Challenges and Opportunities, April 6-7: 249-271. The American University in Cairo, Egypt.

Hear, N. V. (1995). The Impact of the Involuntary Mass 'Return' to Jordan in the Wake of the Gulf Crisis. *The International Migration Review, 29*(2), 352. doi:10.2307/2546785

Heeks, R. (2002). Information systems and developing countries: Failure success, and local improvisations. *The Information Society, 18,* 101–112. doi:10.1080/01972240290075039

Heeks, R., & Bhatnagar, S. (1999). Understanding Success and Failure in Information Age Reform. In Heeks, R. (Ed.), *Reinventing Government in the Information Age: International Practice in IT-Enabled Public Sector Reform* (*Vol. 1*, pp. 49–74). London: Routledge.

Henten, A. (2001). *Services, E-Services, and Nonservices: Cases on Managing E-Services*, edited by Ada Scupola, (pp. 1-9), 2008.

Heresh, J., & Twal, N. (2002, August 30). IT Awareness and e-Education Making Progress. *Jordan Times*, p. 18. Amman.

Hill, C. E., Loch, K. D., Straub, D. W., & El-Sheshai, K. (1998). A qualitative assessment of Arab culture and information technology transfer. *Journal of Global Information Management, 6*(3), 29–38.

Hill, F. M., & Collins, L. K. (1999). The quality management and business process re-engineering: A study of incremental and radical approaches to change management at BTNI. *Total Quality Management, 10*(1) (January): 37. Retrieved August 28, 2004, from http://www.aucegypt.edu/library/libdata/subject.cfm?classid=2

Hitt, M. A., Ireland, R. D., & Hoskisson, R. E. (2001). *Strategic Management: Competitiveness and Globalization.* United States: South-Western College Publishing.

Hofheinz, A. (2005). The Internet in the Arab World: Playground for Political Liberalization, *International. Politics & Society,* (3): 78–96.

Hofstede, G. (1980). *Culture's consequences: International differences in work-related values.* Beverly Hills, CA: Sage.

Hofstede, G., & Hofstede, G. J. (2005). *Cultures and organizations, software of the mind: Intercultural cooperation and its importance for survival.* New York: McGraw-Hill.

Hofstede, G. (1991). *Cultures and organizations: software of mind.* London: McGraw Hill.

Hofstede, G. (1980). *Culture's Consequences: International Differences in Work-Related Values.* Newbury Park, CA: Sage.

Holsapple, C. W., Wang, Y., & Wu, J. (2006). Empirically Testing User Characteristics and Fitness Factors in Enterprise Resource Planning Success. *International Journal of Human-Computer Interaction, 9*(3), 323–342.

Hong, K., & Kim, Y. (2002). The critical success factors for ERP implementation: an organizational fit perspective. *Information & Management, 40*(1), 25–40. doi:10.1016/S0378-7206(01)00134-3

Hruskocy, C., Cennamo, K., Ertmer, P., & Johnson, T. (2000). Creating a community of technology users: students become technology experts for teachers and peers. *Journal of Technology and Teacher Education, 8*(1), 69–84.

Hunaiti, Z., Masa'deh, R., & Mansour, M. (2009). Electronic commerce adoption barrier in small and medium-sized enterprises (SMEs) in developing countries: The case of Libya, *IBIMA. Business Review (Federal Reserve Bank of Philadelphia), 2,* 37–45.

Ibn Khaldoun, A. (1863). *Les prolégomènes.* An introduction to history. Translated in French by W. Mac Guckin & B. De Slane. Paris: Librairie orientaliste Paul Geuthner, 1936.

Ibrahim, R. (1985). Computer usage in developing countries: Case Study Kuwait. *Information & Management, 8,* 103–112. doi:10.1016/0378-7206(85)90038-2

ICT, (2002), Information and Communication Technology and Development in the Arab Countries Report.

IDC. (n.d.). *IDC*. Retrieved from http://www.idc.com/getdoc.jsp?containerId=IDC_P336

Ifinedo, P. (2006a), Extending the Gable et al. enterprise systems success measurement model: a preliminary study. *Journal of Information Technology Management, 17*(1), 14-33.

Ifinedo, P. (2007). Investigating the relationships among ERP systems success dimensions: a structural equation model. *Issues in Information Systems, 8*(2).

Ifinedo, P., & Nahar, N. (2007). ERP system success: an empirical analysis of how two organizational stakeholder groups prioritize and evaluate relevant measures. *Enterprise Information Systems, 1*(February), 25–48. doi:10.1080/17517570601088539

Internet World State. (2009). *Internet World State*. Retrieved on January 28th, 2019, from http://www.internetwdeorldstats.com/stats5.htm

Internet World Statistics. (2009), Retrieved November, 2009, from http://www.internetworldstats.com/

ITU. (2009). *Measuring the information society - The ICT development Index*. New York: International Telecommunication Union.

Ives, B., & Jarvenpaa, S. L. (1991). Applications of Global Information Technology: Key Issues for Management. *Management Information Systems Quarterly, 15*(1), 33–49. doi:10.2307/249433

Jaber, W. (1997). A survey of factors which influence teachers' use of computer-based technology. Dissertation Virginia Polytechnic Institute and State University.

Jang, W., Lin, C., & Pan, M. (2009). Business strategies and the adoption of ERP: Evidence from Taiwan's communications industry. *Journal of Manufacturing Technology Management, 20*(8), 1084–1098. doi:10.1108/17410380910997227

Jaradat, E. (1992). The Philosophy of Educational Development in Jordan. *Risalat Al Muallim, 33*(3), 5–62.

Jeffrey, A., Arunabha, G., & Rias, M. (2007). *Fundamentals of WiMAX: Understanding Broadband Wireless Networking*. Upper Saddle River, NJ: Prentice Hall Communications Engineering and Emerging Technologies Series.

Jennett, P. (2002, September). *A readiness Model For Telehealth*. In Proceedings of International Society for Telemedicine (ISFT) in the seventh International Conference on Telemedicine, Regensberg, Germany Sep. 22-25th, 2002.

Johnson, B. L. (2000). Great Expectations but Politics as Usual: The Rise and Fall of a State-Level Evaluation Initiative. *Journal of Personnel Evaluation in Education, 13*(4), 361–381. doi:10.1023/A:1008157300684

Johnson, P., Fidler, C. S., & Rogerson, S. (1998). Management communication: A technological revolution? *Management Decision, 36*(3), 160. doi:10.1108/00251749810208940

Johnson, W. H. (1999). An Integrative Taxonomy of Intellectual Capital: Measuring the Stock and Flow of Intellectual Capital Components in the Firm. *International Journal of Technology Management, 18*, (5/6/7/8), 1999.

Katrinahelp (n.d). *Katrinahelp*. Retrieved May 15, 2009, from http://katrinahelp.info

Keen, P. G. W. (1981). Information Systems and Organizational Change. *Communications of the ACM, 24*(1), 24–33. doi:10.1145/358527.358543

Khawaja, M. (2003). Migration and the Reproduction of Poverty: The Refugee Camps in Jordan. *International Migration (Geneva, Switzerland), 41*(2), 29–57. doi:10.1111/1468-2435.00234

Kim, M. & Weiss, J. (1989). Total productivity growth in banking: the Israeli banking sector 1979-1982. *Journal of Productivity Analysis*, (1, 2), 239-153.

Kiridis, A., Drossos, V., & Tsakiridou, H. (2006). Teachers facing information and communication technology (ICT): the case study of Greece. *Journal of Technology and Teacher Education, 14*(1), 75–96.

Kirkwood, C. W. (1998). *System Dynamics Methods: A Quick Introduction*. Ventana Systems, Inc.

Kluckhohn, C. (1962). *Culture and Behavior*. New York: The Free Press of Glencoe.

Kohut, A. E. (2008). Social Networking and Online Videos Take Off: Internet's broader role in campaign. *T. P. R. Center, The PEW research centre*. Retrieved May 15, 2009, from http://people-press.org/reports/pdf/384.pdf

Korac-Kakabadse, A., & Korac-Kakabadse, N. (2002). Information technology's impact on the quality of democracy: Reinventing the 'democratic vessel.' In Heeks, R. (Ed.), *Reinventing Government in the Information Age: International Practice in IT-Enabled Public Sector Reform (pp.* 211-228*)*. London: Routledge.

Kriem, M. S. (2009). Mobile telephony in Morocco: A changing sociality. *Media Culture & Society, 31*(4), 617–632. doi:10.1177/0163443709335729

Krippendorff, K. (2004). *Content analysis: an introduction to its methodology*. New York: Sage Publications Ltd.

Krumbholz, M. (2000). Implementing enterprise resource planning packages in different corporate and national cultures. *Journal of Information Technology, 15*(4), 267–280. doi:10.1080/02683960010008962

Kuchinsky, A. (1996). Transfer means more than just technology. *Communications of the ACM archive, 39*(9), 28-29.

Kumar, K., & Van Hillegersberg. (2000). ERP: Experience and evolution. *Communications of the ACM, 43*(April), 23–26.

Kumar, V. (2001). *An investigation of critical management issues in ERP implementation: empirical evidences from Canadian organizations*. Technovation.

La Porte, T. M., Demchak, C., & Friis, C. (2001). Webbing governance: Global trends across national-level public agencies. *Association for Computing Machinery, Communications of the ACM, 44*(1) (January 2001), 63. Retrieved August 28, 2004, from http://www.aucegypt.edu/library/libdata/subject.cfm?classid=2

Lapp, C., & Ossimitz, G. (2008, June). Proposing a classification of feedback loops in four types. *Scientific Inquiry, 9*(1), 29–36.

Lassoued, K. (2005). *Technologies de l'information et culture nationale: Cas de la Tunisie*. Information technology and national culture: The Tunisian case. Retrieved December 12, 2008, from http://medforist.ensias.ma/Contenus/Conference%20Tunisia%20IEBC%202005/papers/June25/43.pdf

Learning Object Network with CAPA project. (2008). *Learning Object Network with CAPA project*. Retrieved April, 2008, from http://www.lon-capa.org/

Leavitt, H. J. (1965). Applying Organizational Change in Industry: Structural, Technological, and Humanistic Approaches. In March, J. G. (Ed.), *Handbook of Organizations*. Chicago: Rand McNally.

Lederer, A. L., & Sethi, V. (1988). The implementation of strategic information systems planning methodologies'. *Management Information Systems Quarterly, 12*(3), 444–461. doi:10.2307/249212

Lee, Z., & Lee, J. (2000). An ERP implementation study for a knowledge transfer perspective. *Journal of Information Technology, 15*(4), 281–288. doi:10.1080/02683960010009060

Light, B. (2001). The maintenance implications of the customization of ERP software. *Journal of software maintenance and Evolution. Research and Practice, 13*(6), 415–429.

Loudon, K. (1997). *Compiler Construction Principles and Practice*. New York: PWS publication.

Loveless, T. (1996). Why aren't computers used more in schools? *Educational Policy, 10*(4), 448–467. doi:10.1177/0895904896010004002

LTAC Project Group. (2009). Student Induction to E-Learning (SIEL), *IMS Global Learning Consortium*, Retrieved November, 2009, from http://www.imsglobal.org/siel.cfm

Lucas, H. C. Jr. (1975). Performance and the Use of an Information System. *Management Science, 21*(8), 908–919. doi:10.1287/mnsc.21.8.908

Luqman, Y., & Quraesh. (1992). Facilitating the adoption of information technology in a developing country. *Journal of International Management, 23*(2), 75–82.

Lynn, M., & Gelb, B. D. (1996). Identifying innovative national markets for technical consumer goods. *International Marketing Review, 13*, 43–57. doi:10.1108/02651339610151917

Madapusi, A., & Kuo, C. (2007). Assessing Data and Information Quality in ERP Systems. In *Proceedings of the Decision Sciences Institute Annual Meeting, Arizona.*

Madar Research Group. (2008). 2007 Arab ICT use Index Growth. *Madar Research Journal, 5*(6). Retrieved April 14, 2009, from http://www.madarresearch.com/

Magnusson, J., Nilsson, A., & Carlsson, F. (2004). *Forecasting ERP Implementation Success – Towards A Grounded Framework.* Paper presented at the 13th European Conference on Information Systems (ECIS 2004), Turku, Finland.

Magrassi, P. (2002). A Taxonomy of Intellectual Capital. *Research Note COM*-17-1985.

Mahmood, M. A., Hall, L., & Daniel, L. S. (2001). Factors Affecting Information Technology Usage, A Meta-Analysis of the Empirical Literature. *Journal of Organizational Computing and Electronic Commerce, 11*(2), 107–130. doi:10.1207/S15327744JOCE1102_02

Manternach-Wigans, L., & Bender, C. L., & Maushak, N. J. (1999) Technology integration in Iowa high schools: perceptions of teachers and students. A paper presented at *the American Education Communications & Technology Conference,* San Antonio, TX.

Markus, M. L., Tanis, T., & van Fenema, P. C. (2000). Enterprise Resource Planning: Multisite ERP Implementations. *Communications of the ACM, 43*(4), 42–46. doi:10.1145/332051.332068

Markus, M., & Tanis, C. (2001b). The Enterprise System Experience – From adoption to success. In Zmud, R. (Ed.), *Framing the Domains of IT management: Projecting the Future through the Past* (pp. 173–207). OH: Pinnflex Educational Resources Cincinnati.

Markus, M. L., & Tanis, C. (2000). the enterprise system experience - from adoption to success. In R. W. Zmud (Ed.), *Framing the Domains of IT Management: Projecting the Future Through the Past* (pp. 173-207). Cincinnatti, OH, USA: Pinnaflex Educational Resources, Inc.

Massell, D. (1998). *State Strategies for Building Capacity in Education: Progress and Continuing Challenges.* Paper presented at the Consortium for Policy Research in Education, Philadelphia, PA.

Matar, N., Hunaiti, Z., Huneiti, Z., & Al-Naafa, M. (2007). E-Learning Status in Arab Counties. In *Proceeding of the International conference on Information Society (i-Society 2007),* Indiana.

Matar, N., Hunaiti, Z., & Matar, S. (2009). Promoting Flexible Unified E-Learning Structure. *The 4th International Conference on Mobile and Computer Aided Learning,* IMCL2009

Matsui, Y. (2002). Contribution of manufacturing departments to technology development: An empirical analysis for machinery, electrical and electronics, and automobile plants in Japan. *International Journal of Production Economics, 80*(2), 185. doi:10.1016/S0925-5273(02)00317-1

Matsumato, D. (2006). Culture and cultural worldviews: Do verbal descriptions about culture reflect anything other than verbal descriptions of culture? *Culture and Psychology, 12*(1), 33–62. doi:10.1177/1354067X06061592

Mayer, M. D., & Tan, F. B. (2003). Beyond Models of National Culture in Information System Research, *Journal of Global Information Management,* January 01, 2002.

McFarlan, F. W., & McKenney, J. (1983). *Corporate Information Systems Management.* Homewood, IL: Richard D. Irwin Inc.

McFarlan, F. W., & McKenney, J. L. (1981). *Information Systems Planning, A Contingent Focus. Harvard Business School Case* (pp. 181–128). Boston, Massachusetts: HBS Case Services.

McKinsey & Company. (2005). *Building Effective Public-Private Partnerships: Lessons Learned from the Jordan Education Initiative.* Retrieved August 22, 2007, from http://www.jei.org.jo/KnowledgeCenterfiles/McKinsey%20Final%20Report_May%2005.pdf

Melewar, T. C., & Navalekar, A. (2002). Leveraging corporate identity in the digital age. *Marketing Intelligence & Planning, 20*(2). doi:10.1108/02634500210418518

Mensching, J., & Corbitt, G. (2004). EPR data archiving- a critical analysis. *Journal of Enterprise Information Management, 17*(2), 131–141. doi:10.1108/17410390410518772

Mentzas, G. N. (1997). Implementing an IS Strategy - A Team Approach. *Long Range Planning, 30*(1), 84–95. doi:10.1016/S0024-6301(96)00099-4

Mezher, T. (2007), Information and Communication Technology in the Arab Region in Nazli Choucri, D. Mistree, F. Haghseta, T. Mezher, W.R. Baker, C.I. Ortiz (Eds), on *Mapping Sustainability Alliance for Global Sustainability,* Springer-Verlag New York Inc.

Milgram, S. (1974). *Obedience to Authority: An Experimental View.* New York: Harper & Row.

Ministry of Education. (2001). *E-Learning: A Strategic Framework.* Amman.

Ministry of Education. (2004, September). *The Development of Education: National Report of the Hashemite Kingdom of Jordan.* Paper presented at the International Conference on Education, Geneva.

Ministry of Education. (2006). *The National Strategy for Education.* Amman: The Department of Educational Research and Development.

Ministry of Education. (2007). *A Brief About the Ministry of Education.* Retrieved January 9, 2008 from http://www.moe.gov.jo/glm.pdf.

Miniwatts Marketing Group. (2010). *Middle East Internet Usage and Population Statistics.* Retrieved April from http://www.internetworldstats.com/stats5.htm

Minsky, B. D., & Marin, D. B. (1999). Why faculty members use E-mail: The role of individual differences in channel choice. *Journal of Business Communication, 3*(2), 194–217. doi:10.1177/002194369903600204

MOE. (2003). Incorporating information and communication technology (ICT) in technology learning process. *Teacher's Journal, 42*(1).

MOE. (2009). Jordan Ministry of Education's e-Learning Initiative. Retrieved October, 15, 2009, from http://www.moe.gov.jo

Mofleh, S., Wanous, M., & Strachan, P. (2008). Developing Countries & ICT Initiatives: Lessons Learnt for Jordan's Experience. [EJISDC]. *Electronic Journal on Information Systems in Developing Countries, 34*(5), 1–17.

Mohamed, N., AbdelRahman, A., & Fadlalla, A. (2008). A context-based integrative framework for e-government initiatives. *Government Information Quarterly, 25*(3), 448–461. doi:10.1016/j.giq.2007.02.004

Moon, Y. B. (2007). Enterprise Resource Planning (ERP): a review of the literature. *International Journal of Management and Enterprise Development, 4*(3), 235–264. doi:10.1504/IJMED.2007.012679

Moore, P. (2004). *Doing Business in the Middle East: Politics and Economic Crisis in Jordan and Kuwait.* Cambridge, New York: Cambridge University Press. doi:10.1017/CBO9780511492242

Morisi, T. L. (1996). Commercial banking transformed by computer technology. *Monthly Labor Review,* (August): 30–36.

Morrisson, C., & Friedrich, S. (2004). La condition des femmes en Inde, Kenya, Soudan et Tunisie. The condition of the women in India, Kenya, Sudan and Tunisia. *Centre de développement de l'OCDE,* document 235. Retrieved September 19, 2005, from http://www.oecd.org/dataoecd/41/48/33645983.pdf

Mouza, C. (2002). Learning to teach with new technology: implications for professional development. *Journal of Research on Technology in Education, 35*(2), 272–289.

Murphy, E. (2006). The political impact of information technologies in the gulf Arab states. *Third World Quarterly, 37*(6), 1059–1083. doi:10.1080/01436590600850376

Mu'tamen, M. (2007). The Role of the Jordanian Educational System in Development Towards the Knowledge Economy. *Risalat Al Muallim, 43*(1), 12–21.

Mysociety (n.d). *Mysociety.* Retrieved May 15, 2009, from http://www.mysociety.org/

Nah, F. H. (2001a). Characteristics of ERP software maintenance: a multi-cause study. *Journal of software maintenance and Evolution. Research and Practice, 13*(6), 339–414.

Nah, F. F., Lau, J. L., & Kaung, J. (2001). Critical factors for successful implementation of enterprise systems. *Business Process Management Journal, 7*(21), 285–296. doi:10.1108/14637150110392782

National Centre for Human Resources Development. (2005). *Terms of References for the Formative Evaluation Study of School Social & National Education Curriculum and Its Implementation.* Retrieved May 25, 2007, from http://www.nchrd.gov.jo/TORSNE.pdf.

New Zealand Police (n.d). *New Zealand Police.* Retrieved 15 May, 2009, from http://www.policeact.govt.nz/wiki/

Newkirk, H. E. 1; Lederer A.L., & Srinivasan C. (2003). Strategic information systems planning: too little or too much? *The Journal of Strategic Information Systems, 12*(3), 201–228. doi:10.1016/j.jsis.2003.09.001

News, U. J. (2009). *About UJ.* Retrieved November 10, 2009, from http://www.ju.edu.jo/Pages/AboutUJ.aspx

Ngai, E., Law, C., & Wat, F. (2008). Examining the Critical Success Factors in the Adoption of Enterprise Resource Planning. *Computers in Industry*, 548–564. doi:10.1016/j.compind.2007.12.001

Nickols, F. (2004). A Primer. In *Themanager.org.* Retrieved July 21, 2004, from www.themanager.org/knowledgebase/management/change.htm

Nicolaou, I., Bhattacharya, S. (2007). Organizational performance effects of ERP systems usage: The impact of post-implementation changes. *International Journal of Accounting Information Systems. On-line edition, 7*(1), 18-35.

Norris, P. (2001). *Digital Divide: Civic engagement, information poverty, and the Internet worldwide.* Cambridge: Cambridge University Press.

Northumbria University. (2006). *The CAMEL Project.* Higher Education Funding Council For England & JISC InfoNet.

OECD. (2003). *The e-Government Imperative.* Paris: OECD Publications Service.

OECD. (2006). *Programme for International Student Assessment.* Paris: OECD.

O'Reilly. Tim. (2005). *What Is Web 2.0: Design Patterns and Business Models for the Next Generation of Software*: *O'Reilly Media.* Retrieved 15 May, 2009, from http://www.oreillynet.com/pub/a/oreilly/tim/news/2005/09/30/what-is-web-20.html.

Osimo, D. (2008). *Web 2.0 in Government: Why and How?: JRC Scientific and Technical.* Retrieved 15 May, 2009, from http://www.egov.vic.gov.au/index.php?env=-inlink/detail:m2762-1-1-8-s-0:l-10908-1-1--

Oster, A., & Antioch, L. (1995). Measuring productivity in the Australian banking sector. In P. Andersen, J. Dwyer, & D. Gruen (Eds), *Productivity and Growth:* 201-12. Conference Proceedings, 10-11 July, Reserve Bank of Australia, Sydney.

Oyserman, D. (1993). The lens of personhood: Viewing the self and others in a multicultural society. *Journal of Personality and Social Psychology, 65*, 993–1009. doi:10.1037/0022-3514.65.5.993

Palvia, P., Shailendra, C., Palvia, J., & Whitworth, J. (2002). *Global information technology management environment: representative world issues, in Palvia and Roche (eds.)* Global information technology and electronic commerce: Ivy League publishing.

Parker, K., & Chao, J. (2007). Wiki as a Teaching Tool, *Interdisciplinary Journal of Knowledge and Learning Objects* V 3.

Parr, A., & Schanks, G. (2000). A model of ERP project implementation. *Journal of Information Technology, 15*(4), 289–304. doi:10.1080/02683960010009051

Pascu, C., & Osimo, D. (2008). Social computing: implications for the EU social innovation landscape. *Foresight, 10*(1), 37–52. doi:10.1108/14636680810856017

Pascu, C. (2008). An Empirical Analysis of the Creation, Use and Adoption of Social Computing Applications: IPTS Exploratory Research on Social Computing. *JRC Scientific and Technical Reports*. Retrieved 15 May, 2009, from http://ftp.jrc.es/EURdoc/JRC46431.pdf

Pascual, J. (2003). e-Government. *UNDP-APDIP, e-ASEAN Task Force*. Retrieved 4 October, 2010, from http://www.apdip.net/publications/iespprimers/eprimer-egov.pdf

Pelgrum, W. J. (2001). Obstacles to the integration of ICT in education: results from a worldwide educational assessment. *Computers & Education, 37*, 163–178. doi:10.1016/S0360-1315(01)00045-8

Penrose, E. T. (1959). *The theory of growth of the firm.* London: Basil Blackwell.

Peter, C., Scott, J., & Wasserman, S. (2004). *Models and Methods in Social Network Analysis.* Cambridge, UK: Cambridge University Press.

Pfeil, U., Zaphiris, P., & Ang, C. S. (2006). Cultural differences in collaborative authoring of Wikipedia. *Journal of Computer-Mediated Communication, 12*(1), article 5.

Philip, G. (2007). IS strategic planning for operational efficiency. *Information Systems Management, 24*(3), 247–264. doi:10.1080/10580530701404504

Piccoli, G. (2008). *Information Systems for Managers. Text & Cases.* New York: John Wiley & Sons, Inc.

Pickett, L. (2004). Focus on technology misses the mark. *Industrial and Commercial Training, 36*(6/7), 247. Retrieved August 28, 2005, from http://www.aucegypt.edu/library/libdata/subject.cfm? Classid=2

Png, I., Tan, B., & Wee, K. (2001). Dimensions of National Culture and Corporate Adoption of IT Infrastructure. *IEEE Transactions on Engineering Management, 48*(1). doi:10.1109/17.913164

Porter, M. E. (1985). *Competitive Advantage* (pp. 11–15). New York: The Free Press.

Porter, M. E. (1990). The competitive advantage of nations. *Harvard Business Review, 90211*, 73–91.

Porter, M. E. (1990). *The competitive advantage of nations.* New York: Free Press.

Porter, M. E. (2001). Strategy and the internet. *Harvard Business Review, 75*(3), 63–78.

Prahalad, C. K., & Krishnan, M. S. (2002). The dynamic synchronization of strategy and information technology. *MIT Sloan Management Review 083, 43*(4) (summer), 23-34.

Premkumar, G., & King, R. W. (1991). Assessing strategic information systems planning. *Long Range Planning, 24*(5), 41–58. doi:10.1016/0024-6301(91)90251-I

Preston, C., Cox, M., & Cox, K. (2000). *Teachers as innovators: an evaluation of the motivation of teachers to use Information and Communications Technology.* London: MirandaNet.

Pricewaterhouse Coopers. (2008). *Why isn't IT spending creating more value?* Retrieved March 2010, from http://www.pwc.com/en_US/us/increasing-it-effectiveness/assets/it_spending_creating_value.pdf

Proctor, T., & Doukakis, I. (2003). Change management: The role of internal communication and employee development. *Corporate Communication, 8*(4) (2003), 268. Retrieved on 28 August 2003, from http://www.aucegypt.edu/library/libdata/subject.cfm?classid=2

Qadamani, E. (2003, September 29). King Launches 1st E-Learning Forum. *Alrai*, P.3. Amman.

Radijan, J. (1996). Bring in danet, bring in da bucks. *US Banker, New York, 106*(9), 18.

Ragowsky, A., Ahituv, N., & Neumann, S. (2000). *The Benefits of Using Information Systems. Communications of the ACM, 43*(11), 303–311. doi:10.1145/352515.352532

Rainie, L. E. (2007). 'Wikipedia users'. *P. E. W. Internet*. Retrieved 15 May, 2009, from http://www.pewinternet.org/Reports/2007/Wikipedia-users.aspx

Ramani, A. (2006, March). Hospital information system: PULSE [Implementing IT in health-care]. In *Proceeding of 18th National Computer Conference* (NCC18), March, 2006, Riyadh, Saudi Arabia.

Ramizer, P., & Garcia, R. (2005). *Success of ERP Systems in Chile: An Imperical Study*. Paper presented at the European, Mediterranean and Middle Eastern Conference in Information Systems (EMCIS2005).

Realo, A., Allik, J., & Vadi, M. (1997). The hierarchical structure of collectivism. *Journal of Research in Personality, 31*(1), 93–116. doi:10.1006/jrpe.1997.2170

Rengersen, L. (2006). Sustainable Software Implementation: Using Emergent Change to Increase Enterprise Recourse Planning (ERP) Systems' Implementation, management and Governance. Retrieved 15 March, 2009, from http://www.os.utwente.nl/tsr/files/42-47.pdf

Report, A. K. (2009). Towards Productive Intercommunication for Knowledge. *Mohammed Bin Rashid Al-Maktoum Foundation & United Nations Development Program,* Retrieved April, 2007 From: http://www.arabstrategyforum.org/asf2009en/attachments/144_programme-english.pdf

Research e-labs. (n.d). *Research e-labs*. Retrieved 15 May, 2009, from http://research.elabs.govt.nz/

Rhee, E., Uleman, J. S., & Lee, H. K. (1996). Variations in collectivism and individualism by in-groups and culture: Confirmatory factor analysis. *Journal of Personality and Social Psychology, 71*(5), 1037–1054. doi:10.1037/0022-3514.71.5.1037

Rheingold, H. (2000). *Virtual Community*. Cambridge, MA: The MIT Press.

Rizk, N. (2004, September). E-Readiness Assessment of Small and Medium Enterprises in Egypt: A Micro Study. *Topics in Middle Eastern and North African Economies, electronic journal,* (6), Middle East Economic Association and Loyola University Chicago.

Rockart, J. F., & Scott Morton, M. S. (1984). Implications of Changes in Information Technology for Corporate Strategy. *Interfaces, 14*(1), 84–95. doi:10.1287/inte.14.1.84

Rogers, E. M. (1983). *Diffusion of Innovations*. New York: The Free Press.

Rondinelli, D. A., Middleton, J., & Verspoor, A. (1990). *Planning Education Reforms in Developing Countries: The Contingency Approach*. Durham: Duke University Press.

Ross, J. W., & Weill, P. (2002). Six IT Decisions Your IT People Shouldn't Make. *Harvard Business Review*, 84–91.

Russell, G., & Bradley, G. (1997). Teachers' computer anxiety: implications for professional development. *Education and Information Technologies, 2*, 17–30. doi:10.1023/A:1018680322904

Sabherwal, R., et al (2006). Information system success: individual and organizational determinants. *Management Science* DEC-06.

Saleh, H. (1991). Water Resources and Food Production in Jordan. In Wilson, R. (Ed.), *Politics and the Economy in Jordan* (pp. 14–54). London: Routledge.

Sandholtz, J. H. (2001). Learning to teach with technology: A comparison of teacher development programs. *Journal of Technology and Teacher Education, 9*(3), 349–374.

Sannwald, W. (2000). Understanding Organizational Culture. *Library Administration & Management, 14*(1), 8–14.

Sarker, S., & Lee, A. S. (2003). Using a case study to test the role of three key social enablers in ERP implementation. *Information & Management, 40*(8), 813–829. doi:10.1016/S0378-7206(02)00103-9

Sawyer, S., & Southwick, R. (2002). Temporal issues in information and communication technology enabled organizational change. *The Information Society, 18*, 263–280. doi:10.1080/01972240290075110

Schmid, R., Fesmire, M., & Lisner, M. (2001). Riding up the learning curve- elementary technology lessons. *Learning and Leading with Technology, 28*(7), 36–39.

Schneider, P. et al. (1999). ERPeople skills. *CIO magazine, 12*(10), 30-37.

Schultz, B. (2005). Out with the old, in with the new data center. *Network World.* Framingham, 22, no. 22 (June 6),38. Retrieved August 28, 2005, from http://www.aucegypt.edu/library/libdata/subject.cfm?classid=2

Schumpeter, J. A. (1934). *The theory of economic development: an inquiry into profits, capital, credit, interest, and business cycle.* Cambridge, MA: Harvard University Press.

Scouarnec, A., & Silva, F. (2008). Les pratiques de management en Euro Méditerranée: Proposition d'un cadre d'analyse. Practices of management in Euro Mediterranean: Proposal of an analysis framework. Retrieved April 12, 2009, from www.reims-ms.fr/agrh/docs/actes-agrh/pdf-des-actes/2008scouarnec-silva.pdf

Seddon, P. B. (1997). A re-specification and extension of the DeLone and McLean model of IS success. *Information Systems Research, 18*(3), 240–253. doi:10.1287/isre.8.3.240

Sedera, D., et al. (2002). Enterprise resource planning systems impacts: a Delphi study of Australian public sector organizations. In *Proceedings of the Pacific Asia Conference on Information Systems (PACIS)*, 584-600, Tokyo, Japan.

Sedera, D., et al. (2004). Measuring enterprise systems success: the importance of a multiple stakeholder perspective. In *Proceedings of the 12th European Conference on Information Systems* (pp. 1-13). Turku, Finland.

Segars, A. H., & Grover, V. (1993). Re-examining Perceived Ease of Use and Usefulness: A Confirmatory Factor Analysis. *Management Information Systems Quarterly, 17*, 517–525. doi:10.2307/249590

Seijaparova, D., & Pellekaan, J. W. H. (2004). *Jordan: An Evaluation of World Bank Assistance for Poverty Reduction, Health and Education.* Washington, DC: The World Bank Operations Evaluation Department, World Bank.

Senge, P. (1994). *The fifth discipline field book: Strategies and tools for building a learning organization.* New York: Doubleday.

Shafi, I. M. (2002). *Assessment of the impact of Internet technology use among Saudi business organisation,* Unpublished doctoral thesis, faculty of Mississippi State University, Mississippi State, Mississippi.

Sharabi, H. (1988). *Neo patriarchy: A theory of distorted change in Arab society.* Oxford, UK: Oxford University Press.

Shariati, A. (1971). *Fatemeh Fatemeh Ast. Fatimah is Fatimah.* Tehran: Husseinieh Ershad.

Shiau, W., Hsu, P., & Wang, J. (2009). Development of measures to assess the ERP adoption of small and medium enterprises. *Journal of Enterprise Information Management, 22*(1), 99–118. doi:10.1108/17410390910922859

Shirazi, F., Gholami, R., & Higon, D. (2009). The impact of information and communication technology (ICT), education and regulation on economic freedom in Islamic middle eastern countries. *Information & Management, 46*, 426–433. doi:10.1016/j.im.2009.08.003

Shore, B., & Venkatachalam, V. (1994). Prototyping: a metaphor for cross-cultural transfer and implementation of IS applications. *Information & Management, 3*(27), 175–184. doi:10.1016/0378-7206(94)90045-0

Skok, W. (2001). *Potential Impact of Cultural Differences on Enterprise Resource Planning (ERP) Projects.* The Electronic Journal on Information Systems in Developing Countries.

Slater, M., Antley, A., Davison, A., Swapp, D., Guger, C., Barker, C., et al. (2006). A virtual reprise of the Stanley Milgram obedience experiments. *PLoS ONE, 1*(1): e39. Retrieved July 17, 2009, from http://www.plosone.org/article/info:doi%2F10.1371%2Fjournal.pone.0000039

Smida, A., & Latiri, R. (2004, October). *L'attitude du manager tunisien face à l'avenir.* The attitude of the Tunisian manager toward future. Paper presented at CIDEGEF colloque, University Saint-Joseph, Beirut.

Smith, D., & Hardaker, G. (2000). e-Learning Innovation through the Implementation of an Internet Supported Learning Environment. *Journal of Educational Technology & Society, 3*(3), 2000.

Snoeyink, R., & Ertmer, P. (2001). Thrust into technology: how veteran teachers respond. *Journal of Educational Technology Systems, 30,* 85–111. doi:10.2190/YDL7-XH09-RLJ6-MTP1

Soh, C. (2000). Cultural Fits and Misfits: Is ERP a Universal Solution? *Communications of the ACM, 43*(4), 47–51. doi:10.1145/332051.332070

Soumitra, D., Augusto, L., & Irene, M. (2006). The Global Information Technology Report 2005-2006: Leveraging ICT for Development, 5th Edition. New York: *Palgrave Macmillan*

Soumitra, D., Shalhoub, Z., & Samuels, G. (2007). Promoting Technology and Innovation: Recommendations to Improve Arab ICT Competitiveness. *World Economic Forum: The Arab World Competitiveness Report,* 2007

Soyah, T., & Magroun, W. (2004, October). *Influence du contexte culturel tunisien sur l'orientation des systèmes d'information des banques tunisiennes.* Influence of the Tunisian cultural's context on the orientation of the information systems of the Tunisian banks. Paper presented at CIDEGEF colloque, University Saint-Joseph, Beirut.

Spott, D. (2000). Componentizing the enterprise applications packages. *Communications of the ACM, 43*(4), 63–90. doi:10.1145/332051.332074

Stapleton, G., & Rezak, C. J. (2004). Change Management Underpins a Successful ERP Implementation at Marathon Oil. *Journal of Organizational Excellence, 23*(4), 15–22. doi:10.1002/npr.20022

Stewart, G., Milford, M., Jewels, T., Hunter, T., & Hunter, B. (2000). *Organisational Readiness for ERP Implementation.* Paper presented at the Americas Conference on Information Systems (AMCIS2000), USA.

Straub, D. W., Loch, K. D., & Hill, C. E. (2001). Transfer of information technology to the Arab world: A test of cultural influence modelling. *Journal of Global Information Management, 9,* 6–28.

Straub, D. W. (1994). The Effect of Culture on IT Diffusion: E-mail and fax in Japan and the US. *Information Systems Research, 5,* 23–24. doi:10.1287/isre.5.1.23

Swierczek, F. W., Pritam, K. S., & Bechter, C. (2005). Information technology, productivity and pofitability in Asia-Pacific banks. *Journal of Global Information Technology Management, 8*(1).

Symantec EMEA Internet Security Threat Report. (2009). *Symantec EMEA Internet Security Threat Report.* Retrieved April 2010, from http://eval.symantec.com/mktginfo/enterprise/white_papers/b-whitepaper_emea_internet_security_threat_report_04-2009.en-us.pdf

Synnott, W. R., & Gruber, W. H. 1981. *Information resource management,* New York.

Talal, A. B. (1998, September). *Teacher Education and School Reform in a Changing World.* Paper presented at The 43rd World Assembly, International Council of Education for Teaching, Amman.

Tassin, E. (1994). Pouvoir, autorité et violence: La critique arendtienne de la domination. Power, authority and violence: Criticism of the Arendt's domination. In Goddard, J. C., & Mabille, B. (Eds.), *Le Pouvoir* (pp. 266–301). Paris: Vrin.

Tawalbeh, M. (2001). The Policy and Management of Information Technology in Jordanian Schools. *British Journal of Educational Technology, 32*(2), 133–140. doi:10.1111/1467-8535.00184

Techwarelabs. (2009). *Techwarelabs.* Retrieved November, 2009, from http://www.techwarelabs.com/articles/other/wimax_wifi

Tella, A., Tella, A., Toyobo, O. M., Adika, L. O., & Adeyinka, A. A. (2007). An Assessment of Secondary School Teachers Uses of ICTs: Implications for Further Development of ICT's Use in Nigerian Secondary Schools. *The Turkish Online Journal of Educational Technology, 6*(3), 12.

Teng & Nelson. (1996). The Influence of Organizational Factors on CASE Technology Adoption. *Journal of Information Technology Management, VII, 1*(2).

Teo, T. S. H., & Ang, J. S. K. (2001). An examination of major IS problems. *International Journal of Information Management, 21*, 457–470. doi:10.1016/S0268-4012(01)00036-6

Tertiary Education. (n.d). *Tertiary Education.* Retrieved 15 May, 2009, from http://wiki.tertiary.govt.nz/

The PEN community. (n.d). *The PEN community.* Retrieved 15 May, 2009, from http://pen.leadspace.govt.nz/

Todd, E. (1983). *La troisième planète: Structures familiales et systèmes idéologiques. The third planet: Family structures and ideological systems.* Paris: Le Seuil.

Toole, T. (2005). *A Project Management Causal Loop Diagram.* Paper presented at ARCOM Conference, London UK., Sept. 5-7.

Tornatzky, L. G., & Klein, R. J. (1982). Innovation Characteristics and Innovation Adoption-Implementation: A meta-analysis of Findings. *IEEE Transactions on Engineering Management, EM-29*, 28–45.

Totter, A., Stutz, D., & Grote, G. (2006). ICT and Schools: Identification of Factors Influencing the use of new Media in Vocational Training Schools. *The Electronic Journal of e-Learning, 1*(4), 95-102.

Triandis, H. C. (1989). The self and social behavior in different cultural context. *Psychological Review, 96*, 506–520. doi:10.1037/0033-295X.96.3.506

Triandis, H. C. (2001). Individualism-Collectivism and personality. *Journal of Personality, 69*, 907–924. doi:10.1111/1467-6494.696169

Tricker, R. I. (1988). Information resource management - a cross-cultural perspective. *Information & Management, 15*(1), 37–46. doi:10.1016/0378-7206(88)90028-6

Umble, E. J., Haft, R. R., & Umble, M. M. (2003). Enterprise Resource Planning: Implementation Procedures and Critical Success Factors. *European Journal of Operational Research, 146*(2), 241–257. doi:10.1016/S0377-2217(02)00547-7

UNCDF. (2006). *UNCDF.* Retrieved January 10, 2010, from http://www.uncdf.org/english/local_development/documents_and_reports/thematic_papers/devcom/200611_bridge/

UNDP. (1999). *Human Development Report 1999.* New York, Oxford: Oxford University Press.

UNDP. (2003). *Arab Human Development Report 2003. Building a Knowledge Society.* United Nations Development Programme.

UNDP. (2004). *Arab Human Development Report 2004. Towards Freedom in the Arab World.* United Nations Development Programme.

UNDP. (2005). *Arab Human Development Report 2005. Towards the Rise of Women in the Arab World.* United Nations Development Programme.

UNESCO. (2005). *Unesco science report 2005.* Paris: UNESCO Publishing.

UNESCO Regional Bureau. (2009). A Decade of Higher Education in the Arab States: Achievements & Challenges. *UNESCO Regional Bureau for Education in the Arab states,* Beirut.

UNISCO International Science. (2003). Science and Technology Education in the Arab world in the 21st Century. Technology & Environmental education Newsletter, XXVIII (3-4).

United Nations. (2005). *UN Millennium Development Goals.* Retrieved May 25, 2007, from http://www.un.org/millenniumgoals

Vannatta, R. (2000). Evaluation to planning: technology integration in a school of education. *Journal of Technology and Teacher Education, 8*(3), 231–246.

Venkatesh, V., Morris, M. G., Davis, G. B., & Davis, F. D. (2003). User Acceptance of Information Technology: Toward a Unified View. *Management Information Systems Quarterly*, *27*(3), 425–478.

Wah, S. S. (2002). *Behavioral attributes of the transformational Chinese leader.* Unpublished doctoral dissertation. The Netherlands: Maastricht School of Management.

Wainwright, D., & Waring, T. (2004). Three domains for implementing integrated information systems: redressing the balance between technology, strategic and organisational analysis. *International Journal of Information Management*, *24*(4), 329–346. doi:10.1016/j.ijinfomgt.2004.04.001

Wang, E. T. G., & Chen, J. H. F. (2006). Effects of internal support and consultant quality on the consulting process and ERP system quality. *Decision Support Systems*, *42*(2), 1029–1041. doi:10.1016/j.dss.2005.08.005

Wang, E. T. G., Shih, S.-P., Jiang, J. J., & Klein, G. (2008). The consistency among facilitating factors and ERP implementation success: A holistic view of fit. *Journal of Systems and Software*, *81*(9), 1609–1621. doi:10.1016/j.jss.2007.11.722

Ward, J., Hemingway, C., & Daniel, E. (2005). A framework for addressing the organisational issues of enterprise systems implementation. *The Journal of Strategic Information Systems*, *14*(2), 97–119. doi:10.1016/j.jsis.2005.04.005

Warkentin, M., Gefen, D., Pavlou, P., & Rose, G. (2002). Encouraging citizen adoption of e-government by building trust. *Electronic Markets*, *12*(3), 157–162. doi:10.1080/101967802320245929

Wauters, P., & Lőrincz, B. (2008). User satisfaction and administrative simplification within the perspective of eGovernment impact: Two faces of the same coin? *European Journal of ePractice*, *4*, 1-10.

Webtrends (n.d). *How Barack Obama Is Using Web 2.0 to Run for President*. Retrieved 15 May, 2009, from http://webtrends.about.com/od/web20/a/obama-web.htm

Weiss, J. W., & Anderson, D., Jr. (2004). CIOs and IT professionals as change agents, risk and stakeholder managers: A field study. *Engineering Management Journal*, *16*(2) (June), 13. Retrieved August 24, 2004, from http://www.aucegypt.edu/library/libdata/subject.cfm?classid=2

William, D., Pauline, C., & Namkee, P. (2004). The Social Shaping of a Virtual Learning Environment, *Electronic Journal of e-Learning. 2*(1), 69-80.

Wilson, P. R. (1968). Perceptual distortion of height as a function of ascribed academic status. *The Journal of Social Psychology*, *74*, 97–102. doi:10.1080/00224545.1968.9919806

Wilson, E. J. (2004). *The Information Revolution and Developing Countries*. Cambridge, Massachusetts: MIT Press.

Woolcock, M. (1998). Social Capital and Economic Development: Toward a Theoretical Synthesis and Policy Framework. *Theory and Society*, *27*(2), 151–208. doi:10.1023/A:1006884930135

World Bank. (2002). *Lifelong Learning in the Global Knowledge Economy: Challenges for Developing Countries*. Washington, DC: World Bank.

World Bank. (2005). *Monitoring and Evaluation Toolkit for E-Strategies Results*. Washington, DC: Global Information and Communication Technologies Department, World Bank.

World Bank. (1999). *Education in the Middle East & North Africa: A Strategy Towards Learning for Development.* Retrieved February 22, 2005, from http://www.worldbank.org/education/strategy/MENA-E.pdf.

World Bank. (2000). *Partnership for Education in Jordan.* Retrieved January 8, 2005, from http://lnweb90.worldbank.org/oed/oeddoclib.nsf/24cc3bb1f94ae11c85256808006a0046/ca2a4e7ab293853c852568a8004fc84b/$FILE/193precis.pdf.

World Bank. (2003). *Project Appraisal Document on a Proposed Loan to the Hashemite Kingdom of Jordan for an Education Reform for Knowledge Economy I Program.* Retrieved September 22, 2005, from http://www-wds. worldbank.org/servlet/WDSContentServer/WDSP/IB /2003/05/10/000094946_03043004015982/Rendered/ PDF/multi0page.pdf

World Bank. (n.d.). *Definition of e-government.* Retrieved 15 May, 2009, from http://web.worldbank.org.ezproxy-m. deakin.edu.au/WBSITE/EXTERNAL/TOPICS/EXTIN-FORMATIONANDCOMMUNICATIONANDTECH-NOLOGIES/EXTEGOVERNMENT/0,contentMDK:2 0507153~menuPK:702592~pagePK:148956~piPK:216 618~theSitePK:702586,00.html

World Health Organization, (2004), *Strategy 2004-2007 eHealth for Health – care Delivery, Report 2004.*

World Summit on the information society. (2003). *Declaration of Principles - Building the Information Society: a global challenge in the new Millennium.* Retrieved April 2010, from http://www.itu.int/wsis/docs/geneva/ official/dop.html

Xu, H. (2002). Data Quality Issues in Implementing an ERP. *Industrial Management & Data Systems, 102*(1), 47–58. doi:10.1108/02635570210414668

Yamaguchi, S. (1994). Collectivism among the Japanese: A perspective from the self. In U. Kim, H. C. Triandis, C. Kagitcibasi, S. C., Choi, G. & Yoon (Eds.), *Individualism and Collectivism: Theory, method, and applications* (pp.175-188). Thousand Oaks, CA: Sage.

Yasin, M., & Yava, U. (2007). An Analysis of e-business practices in the Arab culture: Current inhibitors and future strategies. *Cross Cultural Management: an International Journal, 14*(1), 68–73. doi:10.1108/13527600710718840

Young, O. (2007). *Topic Overview: Web 2.0.* Forrester Research.

Zakour, A. B. (2004). Cultural Differences and Information Technology Acceptance. In *Proceedings of the 7th Annual Conference of the Southern Association for Information Systems.*

Zappen, J., Harrison, T., & Watson, D. (2008). A new paradigm for designing e-government: web 2.0 and experience design. In *Proceedings of the 2008 international conference on Digital government research.*

Zghal, R. (2002). Acquis sociaux de la femme tunisienne et inerties culturelles et institutionnelles: Education, emploi et attitudes envers le statut social de la femme. Social rights of the Tunisian woman and cultural and institutional inertias: Education, employment and attitudes towards the social status of the woman. In Nations Unies (Ed.), *Disparités entre femmes et hommes et culture en Afrique du Nord* (pp. 41-55). Centre de Développement Sous-régional pour l'Afrique du Nord.

Zhao, Y., Pugh, K., Sheldon, S., & Byers, J. (2002). Conditions for classroom technology innovations. *Teachers College Record, 104*(3), 482–515. doi:10.1111/1467-9620.00170

Zimmer, M. (2008). Preface: Critical perspective on web 2.0. *First Monday, 13*(3).

Zlitni, S. (2003). Utilisation des NTIC dans les entreprises: Au-delà des usages professionnels. [Use of the new ICT in the companies: Beyond the professional uses]. *Revue Internationale de sociologie et de sciences sociales, Esprit critique, 5*(4). Retrieved May 20, 2006, from http://www.vcampus.univ-perp.fr/espritcritique/0504/ esp0504article02.html

Zogby, J. (2003). Why do they hate us? *The Link, 36*(4), 2–13.

Zoltán, D. (2003). *Questionnaires in second language research: construction, administration, and processing.* New York: Routledge.

About the Contributors

Salam Abdallah is an IS&T academic and practitioner. Dr. Abdallah has a PhD in Information Systems from Curtin University of Technology, Australia. He has worked for several years with UN Relief and Works Agency (UNRWA) overseeing ICT facilities and curriculum development at schools and vocational training centers in UNRWA's entire field of operations. He is an active participant on the Global Text Project, and a founder member of Special Interest Group of the Association of Information Systems: ICT and Global Development. Dr. Abdallah is also an active researcher in the field of Information Systems and has published articles in local and international conferences and journals. Currently he is the Coordinator for the College of Business and Administration at Abu Dhabi University, Al Ain, UAE.

Fayez Albadri is a well established consultant, manager and educator for over two decades. He holds a Doctorate in Management from MGSM Macquarie University in Sydney Australia, Masters in Intelligent Information Processing Systems from University of Western Australia in Perth, Graduate Certificate in Computer Instructional Design from Edith Cowan University in Perth, and Bachelor degree in Engineering from University of Westminster in London, UK. He is recognized as IS&T Specialist and Management Expert for his record in managing IT projects, implementing ERP systems and e-business solutions. Dr. Albadri is a pioneer researcher and academic with important contributions in the areas of entrepreneurship, e-business, IT strategic planning, IT project management and risk management. He is renowned for his development of (IPRM) the Integrated Project-Risk Model and the introduction of (IELCM) the Integrated ERP Life-Cycle Management approach. He has also delivered numerous seminars and training workshops to hundreds of academics and professionals in Australia and the Middle East.

* * *

Abdelnasser Abdelaal is an assistant professor of IT at Sur College of Applied Sciences in Oman. He earned his Ph.D. in the area of socioeconomic implications of wireless networks from University of Nebraska, USA. Dr. Abdelaal has published more than a dozen of publications in international journals and conferences on social capital, and in service pricing and business models in the domain of community wireless networks. Dr. Abdelaal also writes for Alyahat newspaper on the socioeconomic transformations of technology. He also participated in a number of projects that support knowledge transfer to the Arab World and provide IT solutions to small business and underserved communities.

Atef Abuhumaid is an Assistant Professor at the Faculty of Education and Psychology in Amman Arab University for Graduate Studies. His research interests include ICT in education reform, e-learning, ICT teacher professional development. He is the author of ICT Integration across Education Systems. Before obtaining his PhD from the University of Technology in Sydney, he taught 1-12 teachers in both Jordan and Saudi Arabia, giving him valuable opportunities to practice and compare educational practices from different countries.

Norita Ahmad is an Assistant Professor of MIS at the American University of Sharjah in UAE. She earned her PhD in Decision Science Engineering Systems from Rensselaer Polytechnic Institute (RPI), and received her BS in Computer Sciences and MS in Telecommunications and Network Management both from Syracuse University. Her research interest includes decision analysis, database management systems, and knowledge management. She has also presented her research internationally and published in several scholarly journals such as International Journal of Information Technology & Decision Making, Journal of Intelligent Manufacturing, Journal of Business Ethics, International Journal of Enterprise Information Systems, and International Journal of Services Operations and Informatics.

Sami Akabawi is a Professor of Information Systems in the Management Department at the School of Business, at the American University in Cairo, Egypt. He was the director of the Business Administration program from 1986 to 1992. He was early adopter of the concept diffusion of IT components in the teaching and course curricula. Dr. Akabawi is actively pursuing consultations and advisory services to local and regional business institutions and organizations in ICT area. His research interests are in the architecting, design, development and implementation of ICT systems. Adoption and institutionalization of Enterprise systems and their impacts on human agency's behavior, particularly in developing countries environment, are also part of his research themes. He earned his PhD in Computer Science from the City University in London, in1977, Master of Science in Computer Science from London University in 1973, and Bachelor of Science in Industrial Engineering from Cairo University, in 1967.

Alanoud Alhaj earned her Bachelor's Degree in Biological Sciences and Masters Degree in Genomics and Bioinformatics both from the George Washington University in USA. Ms. Alhaj is currently pursuing a second Masters Degree in Business Administration at the American University of Sharjah. She has a strong interest in medical research and has already published two research papers in Oncogene, one of the world's leading cancer journals. Alanoud is also interested in the managerial aspects of business.

Hamed Al-Hinai is a PhD student at the School of Computing and Technology at the University of Sunderland in the UK. He obtained his BSc Computer Science from King Fahd of Petroleum and Minerals University in Saudi Arabia in 1990 and his MSc in Managing IT from University of Salford in UK in 2004. In addition, he is a certified PRINCE2 practitioner and Oracle certified professional (development track). Mr. Al-Hinai has led and supervised the development and integration of numerous software solutions to successful implementation, especially supply chain management systems. He has also worked as technical team member in two ERP implementations. His research focuses on enterprise systems implementation and their impact in organizations and stakeholders.

Ali Al Kinani is a director of the Enterprise Resources Planning (ERP) of the National Guard Health Affairs at King Abdulaziz Medical City in KSA. He works as an assistant Professor in Health Public and Health Informatics College at King Saud bin Abdulaziz University for Health Sciences in Riyadh, Saudi Arabia. He has over 15 years of professional experience in Computer Science and Information Technology. He earned his undergraduate degree in Computer Sciences from King Abdul Aziz University in KSA in 1996, and PhD degree in computer science from School of Computer Sciences at the University of East Anglia in UK in 2006. He has carried out many successful consultations in ICT and e-government projects for several organizations. He published 33 papers in the International conferences across 17 countries over the world, and participated in 76 conferences and workshops in ICT.

Mohamed Baka is the ICT Advisor to the Abu-Dhabi Water and Electricity Authority (ADWEA). He is also the Chairman of the Emirates Center for Innovation and Entrepreneurship at the UAE University. He earned his PhD in Computing from Bristol University in the UK and received executive education at MIT Sloan and Harvard Business School. Dr. Baka has more than 20 years experience in the field of Information Technology Management. He has led several global ICT Programs for blue-chip firms as part of the European Strategic Project in Information Technology (ESPRIT). His academic experience includes a research tenure at Imperial College – London and is currently a visiting Scholar at Waterloo University in Canada. Dr Baka is on the editorial board of the "Springer" book series on Technology and Knowledge Management.

Helen Edwards is a professor of software engineering at the University of Sunderland in the UK, and a Principal Consultant for Hazel Insight. Her original background is in mathematics (she has a BSc in Applied Mathematics from UCW Aberystwyth, and an MSc in Engineering Mathematics from the University of Newcastle). She focused, for her PhD (awarded by the CNAA) in the area of software engineering methods. She is a Chartered Engineer (CEng), a Chartered IT Professional (CITP), and a Fellow of the BCS. Dr. Edwards research focuses on sociotechnical systems – and she is particularly interested in qualitative field studies evaluating the impact systems and technologies in organizations and social situations. She has supervised a number of doctoral candidates who have based their research in the Arabian Gulf context. She has published widely in the area of software engineering and information systems.

Sonda Fakhfakh is a doctoral candidate at the University of Tunis. Her current research interests concern the Arab culture and the ICT use in the Arab world. Ms Fakhfak has earned her undergraduate degree (maître) at the University of Carthage and her Masters degree (DEA) from the University of Sfax. Her master's thesis examined the determinants of IT outsourcing in the case of Tunisian Banks. She taught courses such as Introduction to Management and Business Systems at the University of Sfax.

Shahid Haling is a Senior Consultant with extensive experience working for many consultancy organizations both in the UK and Middle East. His interests cover adaptive e-learning systems and unified e-learning structures, as well as information security management and IT auditing frameworks.

Bassam Hammo is an Associate Professor at the University of Jordan since 2003, and currently holding the post of chairman of the Computer Information Systems Department. Dr. Hammo holds a PhD. in Natural Language Processing from DePaul University (USA). He is an active researcher in Arabic Natural Language Processing (ANLP). Dr. Hammo is leading a research team working on developing tools for Arabic NLP including: Arabic corpora, morphological analyzers, parsers, thesaurus and ontology. He has supervised over 12 Master theses. He has long been a supporter of free software tools and resources for the Arabic language.

Tarek Hatem is a Professor of Strategic Management and Entrepreneurship at the American University in Cairo (Egypt), with nearly 20 years of experience in teaching, consultancy and training in management in Egypt. Dr. Hatem earned his PhD in Strategic Management from the University of Colorado in the United States in 1986. Earlier, in 1981, he obtained his masters degree in Public Administration from the same university. He is a Certified Management Consultant from the Institute of Management Consultancy in the United Kingdom, and he was the first chairman of the Management Consultants Association in Egypt. Dr. Hatem was the Associate Dean for Executive Education and a Senior Research Fellow at Dubai School of Government (2007-2009). Dr Hatem has several publications and has also been an active participant in both local and international conferences and seminars.

Ziad Hunaiti earned his PhD in 2005 from the School of Engineering and Design at Brunel University, United Kingdom. He is currently with the Department of Computing/Design and Technology, Faculty of Science and Technology at Anglia Ruskin University, UK

Ashraf Khalil is an Assistant Professor at Abu Dhabi University with a PhD from Indiana University in Bloomington, USA. During the last 10 years, he has conducted extensive research on all the main components of the research proposal, namely, social and mobile computing, persuasive computing, human computer interaction and privacy issues. The results of his research studies are published in the top conferences in the field such as CHI, CSCW, and INTERACT. His work has also been featured in USA public media such as National Public Radio and Slashdot. Since he moved to the United Arab Emirates in 2006, he is an active member of the Gulf community of academicians. During his research career, Dr. Khalil has accumulated vast knowledge in applying and developing usability testing techniques to mobile applications. He is interested in applying all the technological innovations in addressing pertinent problems in the Middle East in general and in the Gulf countries in particular.

Sonia Kawas holds the position of Manager of Knowledge and Information Services at the British Council in Jordan, managing information, education, science and outreach projects with governmental, NGO and higher education institutions. During her years of experience at the British Council, Ms. Kawas worked on developing information products targeted to specific audiences. As part of her work on one of the British Council's major regional education projects, she was directly responsible for developing and managing the project's online community of practice and training Jordan's Ministry of Education on activating the community. Ms Sonia has also trained staff of the Ministry of Interior of Iraq, on establishing and managing information resources centers, in addition to extensive training courses delivered to different audiences on various electronic resources and British Council information online products. Ms. Sonia holds a postgraduate certificate in Management from Leicester University in UK and MSc in Information and Knowledge Management from North London University (London Metropolitan University) in UK.

Nasim Matar earned his PhD in 2010 from the school of Computing and Information Technology at Anglia Ruskin University, United Kingdom. Dr. Matar published 7 different books in Arabic language covering different aspects of computer science. His research interests are in the field of Adaptive

E-learning Technology, Unified e-learning and Learning Object Networks (LON). He is worked in different Jordanian universities as a full time lecturer for the colleges of Computer Science, Computer Information Systems, Management Information Systems and Education College. His working and research experience has been specifically oriented towards the Middle East region.

Sadi Matar is currently working at the Ministry of Communication and Transport of Bosnia and Herzegovina as an Expert Advisor for Information Society and as a delegate of Bosnia and Herzegovina at the Broadband South Eastern Europe - (bSEE) initiative. Mr. Matar is also an MCP, MCTS, and MCITP.

Elham Metwally is an Assistant Professor of Strategic Management at Misr International University. She is also an adjunct faculty of Administration of Public Personnel at the American University in Cairo, Egypt. Dr. Metwally earned her Doctorate of Business Administration (2009) and Master of Philosophy (2006) with distinction, from Maastricht School of Management, in the Netherlands. She earned her Master of Business Administration, in 2002, and Bachelor of Arts in Economics, in 1982, with honors, from the American University in Cairo. She is a member of the Academy of Management, and the National Association of Student Financial Aid Administrators in USA. She is also a member of the Middle East Council for Small Business & Entrepreneurship (MCSBE). Dr. Metwally is also the Associate Director for Scholarships and Budgeting at the American University in Cairo. Earlier, she gained more than twelve years of experience at the Hong Kong and Shanghai Banking Corporation (HSBC).

Basem Saraireh is currently holding the post of Regional Director of the Arabian and Education Training Group of Al-Faisal International Academy, which is located at the University of Jordan. For the past three years, he was actively involved as a member in three joint programs in the fields of ICT Education, and Information Technology and E-Government..Dr. Saraireh holds a PhD degree in Reading and Language Arts with a cognate area, focusing on the use of technology in Education. He was a faculty member at Yarmouk University since 2003 and has taught various courses in the use of Technology in education and supervised more than 6 post-graduate theses.

Breinigsville, PA USA
23 November 2010
249730BV00006B/1-72/P